Studies in Systems, Decision and Control

Volume 347

Series Editor

Janusz Kacprzyk, Systems Research Institute, Polish Academy of Sciences,
Warsaw, Poland

The series "Studies in Systems, Decision and Control" (SSDC) covers both new developments and advances, as well as the state of the art, in the various areas of broadly perceived systems, decision making and control–quickly, up to date and with a high quality. The intent is to cover the theory, applications, and perspectives on the state of the art and future developments relevant to systems, decision making, control, complex processes and related areas, as embedded in the fields of engineering, computer science, physics, economics, social and life sciences, as well as the paradigms and methodologies behind them. The series contains monographs, textbooks, lecture notes and edited volumes in systems, decision making and control spanning the areas of Cyber-Physical Systems, Autonomous Systems, Sensor Networks, Control Systems, Energy Systems, Automotive Systems, Biological Systems, Vehicular Networking and Connected Vehicles, Aerospace Systems, Automation, Manufacturing, Smart Grids, Nonlinear Systems, Power Systems, Robotics, Social Systems, Economic Systems and other. Of particular value to both the contributors and the readership are the short publication timeframe and the world-wide distribution and exposure which enable both a wide and rapid dissemination of research output.

Indexed by SCOPUS, DBLP, WTI Frankfurt eG, zbMATH, SCImago.

All books published in the series are submitted for consideration in Web of Science.

More information about this series at http://www.springer.com/series/13304

Alberto Ochoa-Zezzatti · Diego Oliva ·
Aboul Ella Hassanien
Editors

Technological and Industrial Applications Associated With Industry 4.0

Springer

Editors
Alberto Ochoa-Zezzatti
Universidad Autonoma de Ciudad Juarez
Ciudad Juarez, Mexico

Aboul Ella Hassanien
Faculty of Computers and Artificial
Intelligence Information Technology
Department
Cairo University
Giza, Egypt

Diego Oliva
Departamento de Ciencias
Computacionales
Universidad de Guadajalara
Guadajalara, Jalisco, Mexico

ISSN 2198-4182 ISSN 2198-4190 (electronic)
Studies in Systems, Decision and Control
ISBN 978-3-030-68662-8 ISBN 978-3-030-68663-5 (eBook)
https://doi.org/10.1007/978-3-030-68663-5

This Springer imprint is published by the registered company Springer Nature Switzerland AG
The registered company address is: Gewerbestrasse 11, 6330 Cham, Switzerland

Introduction

Technological and Industrial Applications Associated With Industry 4.0 is a novel, innovative, and adequate source of information that compiles interdisciplinary perspectives about diverse issues related with Intelligent Logistics on different ways about Intelligent Optimization, Industrial Applications in real world, Social applications, and Technology applications, each one with a different perspective about the correct solution of this kind of methodologies. This book is a collective effort to introduce new ideas and paradigms from a variety of perspectives using innovative techniques related to Bioinspired Algorithms and methodologies associated with Artificial Intelligence. An innovative and detail book specialized in optimization that considers diverse aspects to realize a more relevant "Intelligent Optimization to improve Logistics" tries to improve with innovative techniques and methodologies in different daily aspects of our lives, in each one of them is possible understand the necessity of improve distances, time, costs, spaces and a plethora of features associated with the modern life (labor associated with delivery of goods, materials or products). We received manuscripts from renowned researchers from all around the world associated with Theoretical foundations of Logistics to understand many paradigms on different Optimization implementation kinds. In addition, we received many manuscripts with expertise on improving optimization related to Logistics of products and services, Optimization of different elements in time and location, Social Applications to enjoy our life in a better way, and finally, Technology Applications of diverse ways to increase our Life Quality. The book starts with a section entitled Mobile Applications and Web Applications to Improve Competitiveness in Industry 4.0, featuring seven chapters on the theoretical ideas related to the correct implementation of a diverse range of Logistics applications in real world Industries. The first chapter of this section is "Implementation of an Intelligent Model Based on Convolutional Neural Network for the Detection of Diseases in Citrus Crops Caused by Bird Pests Using an Intelligent Drone" which is aboard the theoretical fundamentals of an adequate decision support system under uncertainty. In the chapter "Intelligent Application to Detection of Arachnid Bites in Children Implementing Deep Learning Techniques, an AmI-Based Solution" is shown a model based on Bioinspired

Algorithms to improve a specific stochastic model demand. The chapter "Evacuation Route Optimization in the Plaza de la Mexicanidad, Using Humanitarian Logistics" is presented with a mathematical resolution to a business company. The chapter "Automatic Fall Detection for the Care of Older Adults in Smart Cities" determines actions to improvement an innovative model of humanitarian logistics considering uncertainty. The chapter entitled "Automatic Tumor Segmentation in Mammogram Images for Healthcare Systems in Smart Cities" explains different models to prevent problems with geopolitics associated with their decision support system. The chapter "Impact of Industry 4.0: Improving Hybrid Laser-Arc Welding with Big Data for Subsequent Functionality in Underwater Welding" details different strategies to models of uncertainty scenarios. Finally, the last chapter of this section entitled "Metaheuristics for Order Picking Optimisation: A Comparison Among Three Swarm-Intelligence Algorithms" explains different models to organize transportation with extra dimensions to reduce spaces.

The next section of this book is named Modern Technology Applications Including Metaheuristics and Artificial Intelligence Based Applications for Industry 4.0, featuring seven chapters related to different comparatives of Logistics in the search to improve resources in diverse aspects of our transportations or improve process to this. The chapter entitled "Brainwaves Behavior During the Learning Curve Associated with the Manufacturing of a Product with Legos" explains a model related to its routes of delay using Mathematical models and specialized software. The chapter "Audio Features Extraction to Develop a Child Activity Recognition Model Using Support Vector Machine to Monitoring Security in a Smart City" proposes an innovative Visualization Tool to analyze heuristics related to Intelligent Logistics. The chapter entitled "Sentiment Analysis Using Natural Language Processing Through a Speech Recognition System Using a Hybrid Mobile App" describes different models associated with the reactive Logistics under uncertainty situations. The chapter "Logistics of Hospitalization Patients with COVID and Ambulances Required" explains a real problem about the logistics to organize times in a novel model of humanitarian logistics in a Smart City. The chapter "A Heuristic Method for Oil Distribution Networks Applied to the Switching Behavior in the Oil Industry" details novel specifications to reduce time to distribution in a Mathematical Model. The chapter entitled "Metaheuristics for Order Picking Optimisation: A Comparison Among Three Swarm-Intelligence Algorithms" explains a model hybrid to solve problems with time and space. Finally, the last chapter in this section is "Implementation of an Intelligent Framework for the Analysis of Body Movements Through an Avatar Adapted to the Context of Industry 4.0 for the Recruitment of Personnel" which analyzes a model to control the different exogenous aspects in the selection of personnel using an intelligent avatar.

The third and final section is named Industry 4.0 Optimization and Its Future Effects on Z Generation Focused on the Paradigm Shift of an Innovation Ecosystem, featuring seven chapters related to different comparatives of Logistics Models in the search to improve resources in diverse aspects of companies and to improve our lives. The first chapter of this section is "Selection of Factors Influencing for Reliable Electrical Power Transmission Design in Industry 4.0" which explains different models to solve general problems in Logistics. The chapter entitled "Analysis of Transport Logistics Operations at a Link in a Reverse Supply Chain that Values Used Cooking Oil" details a novel model of design of experiments to reduce costs in Logistics of services. The chapter "The Transformation of Supply Chains in the Circular Economy from International Experiences to the Mexican Cases" explains a real problem with the support to an isolated society and the best options to organize this service on the time takes as an important factor the locations and ubiquities in the Container Ship and problems of each point to send the correct goods. The chapter "Nanostores' Density and Geographical Location: An Empirical Study Under Urban Logistics Approach" describes a model to improve times and costs associated with the management in an Aerospatiale project. The chapter "Implementation of a Blockchain Model Implementation to Select the Best Bid in an Industrial Supply Chain" explains a specific topic related to problems on the exogenous costs in a Vehicle routing problem in a Logistics System and its distribution of reordering industrial materials. The chapter "Sociodemographic Analysis of the Location of MSW Collection Centers in Mexico City" details a model to improve Logistics in a Smart City. Finally, in this section is presented the chapter "Classification System to Detect Diseases in Apples by Using a Convolutional Neural Network," which proposes new ideas related to the delivery of products and details a novel technique to analyze the restriction of Time Windows.

The research community must be alert to investigate all these issues in a timely fashion, opening avenues for subsequent edition of this interesting book. The chapters were selected following a rigorous analysis done by the book editors, and each chapter was double- or triple-blind peer-reviewed by at least two experts in the area.

For the content of the research associated with each of the chapters included in this book, the authors are solely responsible. The views, opinions, or positions expressed by the chapter authors are solely those of the authors, and do not necessarily reflect the views, opinions, or positions of the editors. All trademarks, trade names, service marks, and logos referenced in the chapters of this book belong to their respective companies.

Dr. Diego Oliva
diego.oliva@cucei.udg.mx

Dr. Aboul Ella Hassanien
Alberto Ochoa-Zezzatti

Contents

Modern Technology Applications Including Metaheuristics and Artificial Intelligence Based Applications for Industry 4.0

Industry 4.0 Optimization and Its Future Effects on Z Generation Focused on the Paradigm Shift of an Innovation Ecosystem

Mobile Applications and Web Applications to Improve Competitiveness in Industry 4.0

Implementation of an Intelligent Model Based on Convolutional Neural Network for the Detection of Diseases in Citrus Crops Caused by Bird Pests Using an Intelligent Drone

Antonio Romero, Eddy Sánchez-DelaCruz, and Alberto Ochoa

Abstract The incorporation of an intelligent drone that replaces one of the main tasks of farmers, direct visual inspection in order to reduce future damage to citrus production generated by bird pests, where the use of a convolutional neuronal network will serve to extract characteristics and generate a classification model capable of distinguishing between healthy tree foliage from that which is not and which consequently presents some type of pest produced by birds taking into account the coloration of the leaves, as well as, extracting the coloration of the leaves that present diseases, with the aim of optimizing and improving the activities immersed in the cultivation of citrus fruits. In addition, to reduce the damage of bird pests, which will significantly increase productivity in the agricultural sector annually, therefore, through the implementation of unmanned aerial vehicles, provide visual recognition to reduce the indirect effects generated by the type of pests and with this, optimize the activities and techniques involved in citrus cultivation.

Keywords Smart drone · Citrus cultivation · Bird pests · Convolutional neural network

1 Introduction

One of the main problems that farmers have to face is pests in their crops that generate great losses and threaten food security, as well as economic problems for the farmer due to losses. Currently, the technique used to detect bird attacks on crops is a traditional method of control in the fields, direct visual inspection, which consists of personally supervising the planting, that is, the farmer has to go to the area of interest in order to watch and monitor the crops, this method is slow and is not applicable to

A. Romero (✉) · E. Sánchez-DelaCruz
Tecnológico Nacional de México, Campus Misantla, Misantla, Veracruz, Mexico
e-mail: romeroa_cv123@hotmail.com

A. Ochoa
Universidad Autónoma de Ciudad Juárez, Ciudad Juárez, Mexico

A. Ochoa-Zezzatti et al. (eds.), *Technological and Industrial Applications Associated With Industry 4.0*, Studies in Systems, Decision and Control 347,
https://doi.org/10.1007/978-3-030-68663-5_1

large areas of land [1]. However, there are other methods that experience efficiency, but at a higher cost, therefore, the detection of attacks on crops is suitable for viewing from the altitude with the help of an application that allows easy use and detection of the damaged crop.

On the other hand, drone technology provides a solution to the conflicts or difficulties that currently exist for a farmer because they allow for a more extensive and complete expansion of inspections, thus helping to work in the agricultural sector, in addition, manual work is reduced with the implementation of this collaboration, since it benefits them in the field activities due to the fact that the unmanned aerial vehicle becomes the eyes of the citrus grower in order to make this task easier, since with the help of the cameras it captures images showing if the crop has pests and showing the state of the plantation and improving the crop control techniques [13].

On the other hand, there are a variety of artificial intelligence techniques that can be implemented and that can help improve the agricultural sector, so we worked with a Neuronal Convolutional Network, which was trained with images of citrus crops to achieve greater accuracy in pest detection and that after this, this classification model was intended for a mobile application for greater control and also to allow better management [3].

In contrast, once the detection of the disease caused by bird pests has been made, this analysis serves as a repository for a new classification, which is stored directly in the cloud. The different diseases that this pest causes to citrus plantations were analyzed and recognized by expert growers in the area of Misantla, obtaining a coincidence of recognition that exceeds 98% of identification in which the coloring of the foliage is taken into account, as well as the citrus fruit.

2 Project Development

2.1 Object Classifier and Object Detector

Firstly, it is important to know how to differentiate between an object classifier and an object detector. An image classifier is an algorithm that is dedicated to classifying images within a specific category. For example, a group of images of a crop with pests were assigned and this only mentions that the predominant object in the photo is the particular damage, while an object detector is an algorithm that is responsible for identifying various elements within the image and thus classifying them, for example, they receive an image containing a citrus crop and identify that there is a pest, bird, tree, etc. within the image [5].

2.2 *Artificial Neural Network*

Artificial Neural Networks are part of Artificial Intelligence, they are also networks trained through the inputs obtained from external or internal scenarios in the system and these inputs are multiplied by randomly assigned weights, a neural network is an integration of various learning systems, that is why they have the ability to learn through previous training, the ANNs are programming objects that mimic the functioning of biological neurons [10].

On the other hand, NNAs have many advantages such as adaptive learning in which you learn to perform tasks from a set of data being that in the process of learning these data are represented as inputs and weights, In addition, self-organization as you can create your own organization or representation of information received, the neural networks self-organize their information they receive during the learning of the operation using the mathematical methods Adeline, Madeline and Perceptron among others. Additionally, tolerance to partial failures knowing that the partial destruction of the network damages its operation, but does not destroy it completely. This is due to the redundancy of the information contained, that is, this leads to the information not being lost since it works like the human body. Similarly, their operation is in real time, as they can be carried out by computers or special hardware devices to take advantage of the capacity of the NNAs [6].

2.3 *Convolutional Neural Network*

CNNs are similar to ordinary neural networks such as the multi-layer perceptron, they are composed of neurons that have weights and biases that they can learn. Each neuron receives some input, performs a scalar product and then applies a trigger function. What distinguishes CNNs is that they explicitly assume that the inputs are images, which allows us to encode certain properties in the architecture; favoring efficiency gains and reducing the amount of parameters in the network. CNNs even solved the problem that ordinary neural networks do not scale well for high definition images. CNNs work by consecutively modeling small pieces of information, and then mixing the information in the deeper layers of the network, so they are able to model complex variations and behaviors giving fairly accurate predictions [2].

The CNN is a type of Artificial Neural Network with supervised learning that processes its layers imitating the visual cortex of the human eye to identify different characteristics in the inputs that ultimately make it able to identify objects and "see". For this, CNN contains several specialized hidden layers with a hierarchy: this means that the first layers can detect lines, curves and specialize until they reach deeper layers that recognize complex shapes such as a face or the silhouette of an animal [8].

Table 1 Comparison between ANN and CNN

Characteristic	Artificial Neural Network	Convolutional Neural Network (CNN)
Input data in initial layer	The characteristics that were analyzed for example: width, height, thickness, etc.	The pixels in the image. If it is in color, it will be 3 layers for RGB (red, green, blue)
Hidden layers	A number of neurons are chosen for the hidden layers	Available in type: Convolution (with a kernel size and a number of filters) Subsampling
Output layers	The number of neurons to be classified, for "healthy" or "pests" will be 2 neurons	Must flatten the last convolution with one or more layers of hidden "traditional" neurons and make an exit through SoftMax to the exit layer that classifies "healthy" and "plagues" will be 2 neurons
Learning	Supervised	Supervised
Interconnections	Between layers, all the neurons in one layer with the next	There are far fewer connections needed, as the weights we adjust will actually be those of the filters/kernels being used
Meaning of the number of hidden layers	It really is something unknown and does not represent something in itself	Hidden layers are image feature detection maps and have a hierarchy: first layers detect lines, then curves and increasingly elaborate shapes
Backpropagation	Used to adjust the weights of all layer interconnections	It is used to adjust the weights of the kernels

2.4 Comparison Between an Artificial Neural Network and a CNN

The following Table 1 shows some of the characteristics that encompass both artificial intelligence techniques, on the right side the common Artificial Neuronal Network and on the left the Convolutional Neural Network (see Table 1).

2.5 Pixels and Neurons

To start, the network takes the pixels in the image as input. Here you have an image with 4000 × 3000 pixels high and wide, that's equivalent to 12,000,000 neurons. And that's in case you only have 1 color (grayscale). If you had a color image, you

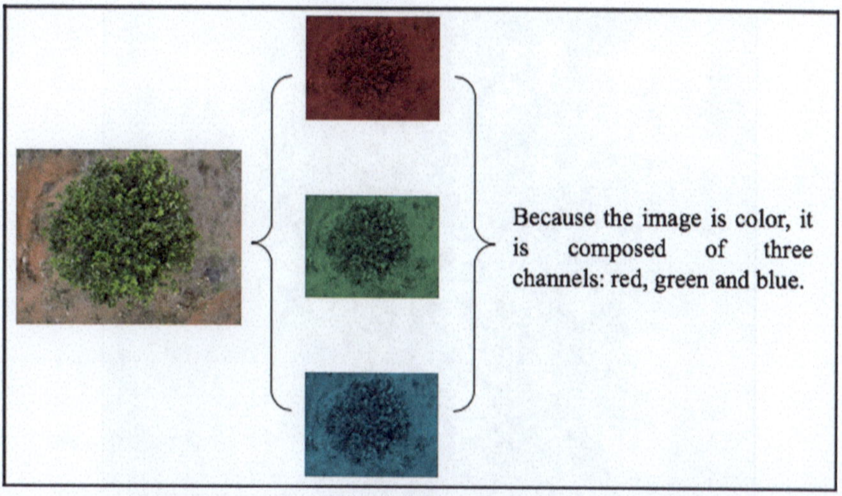

Fig. 1 RGB example

would need 3 RGB channels (network, green, blue) and then you would use 4000 × 3000 × 3 = 36,000,000 neurons as input (see Fig. 1).

Before feeding the grid, it is convenient to normalize the values as input. The colors of the pixels have values ranging from 0 to 255, a transformation of each pixel will be carried out: "value/255" and there will always be a value between 0 and 1. This range, due to the operation of convolutional networks, makes it possible to identify the characteristics for each vector incorrectly. To correct this, it is convenient to previously normalize the image, in a process known as zero mean and unit variance normalization. To summarize, it refers to a 3-dimensional matrix where each dimension represents a layer of color, and each of these layers has values ranging from 0 to 255, and represent the intensity of that color, where the computer to display takes a value and draws a pixel of the resulting color of the mixture (see Fig. 2).

2.6 *Choice of the Crop to Be Analyzed*

Below, you will find statistical information on agricultural production in Misantla, Ver. on an annual basis in the products it generates, the information offered is on area sown, area harvested and the value of production, for seasonal crops in the area [11]. The study includes crops with better production in the municipality, which are ordered in a descending manner, that is, from more to less production, which will help define the crop to be covered by the project (see Fig. 3).

It is possible to identify in the previous graph, four different types of products, which are: the lemon, the orange and the tangerine, being these first two elements those that produce more the region of Misantla, which gives to understand that as

Fig. 2 Persian lemon tree

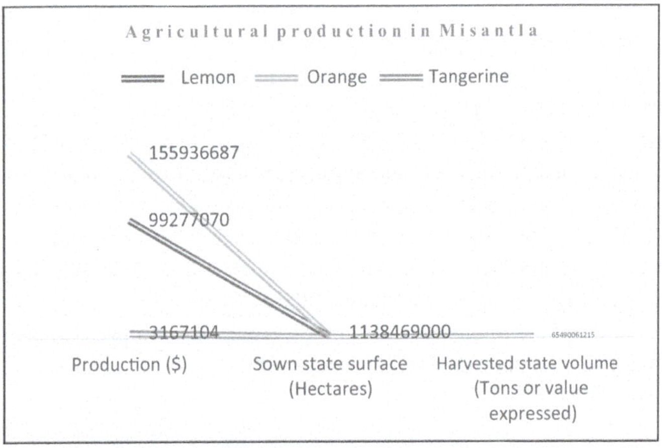

Fig. 3 Agriculture production in Misantla

much the orange and the lemon are two of the fruits that but are cultivated, for that reason, they are taken like reference to be compared with the other products, in where the planted surface takes part, the production in weights, which is taken to determine the type of culture that will be to which the intelligent drone is implemented (see Fig. 4).

Based on the data provided by the Secretariat of Agricultural, Rural and Fishing Development (SEDARPA) and taking as reference the analysis of the graph, it is concluded that the crop to work on the project and which will be implemented the use of drone, will be the citrus, since it presents a greater production in the city of

Fig. 4 Lemon planted area

Misantla, as well as a larger area of cultivation than the other fruits. On the other hand, it is proposed that in this type of citrus fruit is also affected by bird pests.

2.7 Process to Be Automated

According to citrus growers in the Misantla area, Ver., the greening of citrus fruits can cause the shoots and foliage to die, giving them a scarce and untidy appearance, as well as not growing as much as they should or blooming at the wrong time of year, for example in summer or autumn instead of spring. Citrus greening, also known as Huanglongbing, is a bacterial infection transmitted by an insect called the Asian citrus psyllid [7].

In summary, the damage caused by HLB is as follows:

1. Economic death of the plant.
2. Severe fruit and leaf drop.
3. Decrease in fruit weight.
4. Decrease in the sugar level.
5. Increase of the acidity level.
6. Decrease in size and percentage of juice
7. A young plant does not produce fruits.

The symptoms can be found throughout the year, however, are easier to see from September to March because in this season the color of the leaves changes and the pest becomes visible, the disease affects all parts of the tree's crown: leaves, twigs and fruits, as the disease progresses, it will cause the whole tree to decrease. In addition, leaf symptoms include mottled spots, yellow veins, clogged veins or green islands, yellow veins, clogged veins or green islands are not diagnostic alone, on the other hand mottled spotting is the best diagnostic symptom of green leaf. And spotting is a random pattern of yellowing that occurs on leaves that is not the same on the right and left sides of the leaf [4] (see Fig. 5).

A simple procedure to determine if the symptoms are the same in both halves of a leaf is the following, draw two circles in opposite halves of the leaf in order to appreciate if the pattern is the same in both circles as in the following figure (see Fig. 6).

Figure 6 shows two types of problems, on the left side a green problem and on the right-side problems with nutrient deficiency with a line as suggested above.

In addition, the pest problem does not only affect the appearance of the leaf. The external appearance of the fruit can also be affected, whether it is deformed or looks small and green. Taste problems, a variation of salty and bitter. And the internal appearance may have aborted seeds, a yellow spot under the calyx button and/or a curved central core (see Fig. 7).

Finally, the identification of the problems that include all the tree, beginning to present/display yellow buds, death in its branches, delays in the growth, present/display flowering out of season and a remarkable general reduction of the tree (see Fig. 8).

Fig. 5 Leaf simptoms

Fig. 6 Greening and nutrient deficiency problems

Fig. 7 Fruit problems

Fig. 8 Foliage problems

3 Materials and Methods

The following Table 2 describes the characteristics of the laptop, as well as the drone and the system requirements of the two devices mentioned (see Table 2).

The following process was determined from the literature review that was done taking into account work related to digital image processing for the detection and analysis of images in precision agriculture, where the stress that plants present is taken into account to analyze the images.

First, the dataset was built up from images obtained by the DJI Mavic Mini drone. Once the dataset was built, a pre-processing of images was applied to filter and delimit regions of interest, followed by the application of segmentation to the images to identify the regions affected by the pests, in order to extract the characteristics and, here, a label was made to obtain the most relevant characteristics and identify them.

Table 2 Material specifications

Personal computer	
Model	MacBook Pro (13-inch, Early 2011)
System	macOS High Sierra
Type of system	64 bits
Procesador	2.3 GHz Intel Core i5
Graphics card	Intel HD Graphics 3000 512 MB
RAM	8 GB 1600 MHz DDR3
Hard disk drive	240 GB SSD
Drone	
Model	DJI Mavic Mini
Take-off weight	Folded: 140 × 82 × 57 mm (length × width × height) Unfolded: 160 × 202 × 55 mm (length × width × height) Unfolded (with propellers): 245 × 290 × 55 mm (length × width × height) Camera
Sensor	1/2.3″ CMOS Effective Pixels: 12PM/2.3″ CMOS
ISO range	Video: 100–3200 (automatic) 100–3200 (manual) Photo: 100–1600 (automatic) 100–3200 (manual)
Photo size	4:3: 4000 × 3000 16:9: 4000 × 2250
Video resolution	FHD: 1920 × 1080 25/30/50/60 p 2.7 K: 2720 × 1530 25/30 p
Maximum bit rate	40 Mbps
Photo formats	JPEG

Fig. 9 Persian lemon harvest

Finally, to achieve the classification, the crops with pest in order to later determine the damage they present.

3.1 *Acquisition of the Images*

Were carried out 8 flights with a DJI mavic mini, the images taken by the drone are stored in jpeg format with a size of 5,438,160 bytes and dimensions of 4000 × 3000 pixels each, have a focal length of 4.49, are sectioned by folders according to the type of citrus, including 2 main, the Persian lemon and late orange valencia and a sub-folder within each type of citrus which refer to whether the crop has or not pest (see Fig. 9).

3.2 *Development of the Application*

The development of the application is the process of classifying images or recognizing them. In this process, an image is shown to the device's camera sensor and it will tell you which class the image belongs to. By means of image classification and training a deep learning model of convolutional neural network and the help of a google online tool, the model is exported to the TensorFlow lite version that is compatible with the Android device in order to implement it in the application (see Fig. 10).

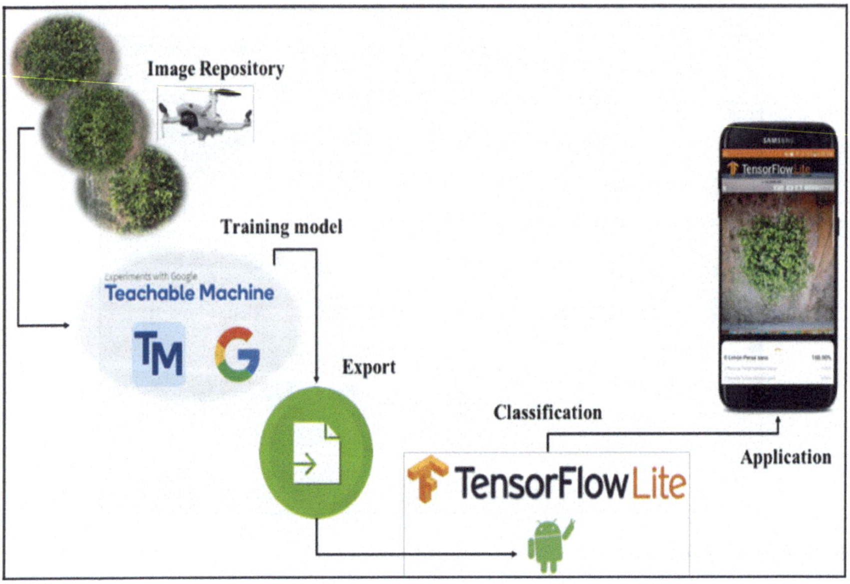

Fig. 10 General description

To perform the training of the model the platform to perform the image classifier is teachable machine from google. At the end of the training, you can export the model of your choice. The model was exported to the Lite version of TensorFlow. Google's Teachable Machine was used, which is an Artificial Intelligence platform that allows the user to train independently to a neural network, while the system can react to known images with sounds or GIF animations. At least 30 frames are required for the Teachable Machine to learn movement recognition. To start the training, the user must hold down the "train" button for a few seconds and make a certain gesture or show something to the computer [12].

The tool allows you to immerse yourself in the definition of the model, algorithm and data processing as well as, focus only on the deployment of the model that is generated, the tool works in the browser, with a webcam or with files that are hosted locally or in Google Drive, and in a few minutes you can quickly understand how a model "learns" through a simple classification demonstration.

To start training the classification model, first different categories or classes were created to carry out the training and have it learn. Four classes were created as shown in Fig. 3, which are two types of citrus fruits with their respective healthy and pest variables. With the classes or labels defined, from the local storage the samples of each class are uploaded to the platform so that the training can begin. More than 200 images were uploaded with 4000 × 3000 resolution per class (see Fig. 11).

After having the classes ready, it is already possible to train the classification model. The model takes only a few minutes to train. Not only is the speed of training

Fig. 11 Citrus harvest

excellent, but you are given valuable dynamic metrics such as accuracy, training loss and dataset test with the observed class accuracy.

The network "input" block comes first. Here, the digital media enters the system (in Fig. 12 on the top left). Then there is the "learning" block (top, middle), where the model learns and interprets the input. The "output" block (top right), is where the interpretation of the model does something like recognizing the inputs and categorizing the results, so it is important to prepare and load the data set in the teachable machine platform with the Google site as well as define the number of classes, for the case of the project were defined 4 classes. The image classification model was trained there and finally, it was exported.

During the training, you can change the number of hyper-parameters as:

1. Number of periods
2. Lot size
3. Learning rate.

Then, it was necessary to export the model in TensorFlow lite format for implementation in Android devices. The ".tflite" files must be placed in the assets folder of the Android project directory and renamed in the java file that is reading it. In addition, you can download the tflite quantified and FLOAT file format.

4 Results

The identification of the type of culture has been obtained by means of the model as well as the classification of the images by means of the convolutional neural network as shown in Fig. 13. Summarizing: it is possible to say that the important

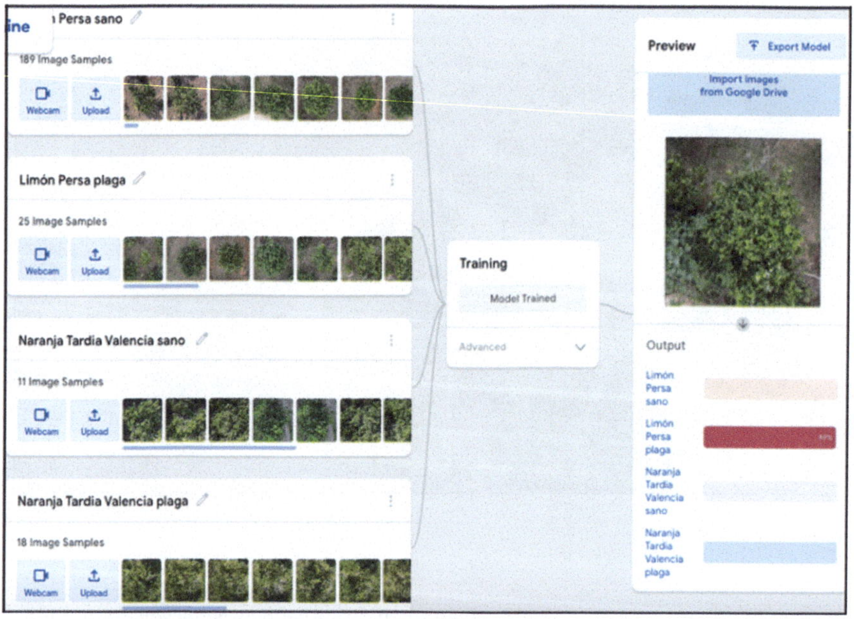

Fig. 12 Training the model

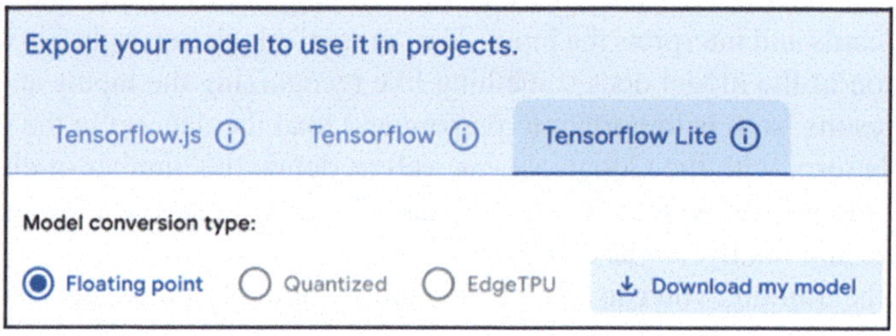

Fig. 13 Exporting the model

elements that were used to create the CNN were the input layer, this is the pixels of the image, providing height, width and depth working with 3 for Red, Green, Blue. The Convolution Layer then performs the process of the output of neurons that are connected in "local regions" of input (i.e. nearby pixels), calculating the product scale between their weights (pixel value) and a small region to which they are connected in the input volume. On the other hand, the Relu layer will apply the activation function on the elements of the matrix and the sampling or subsampling will be in charge of making a reduction in the height and width dimensions, but the depth is maintained, finally the traditional layer of feedforward neuron network that will connect with the last layer of subsampling and will end with the amount of neurons that we want to classify (see Fig. 14).

Fig. 14 Image processing

Due to the incorporation of the images and the training of the model it was possible to obtain an identification of the type of culture of 98%, while for the classification of healthy culture and culture with plague the precision in minor, obtaining 88% since at the moment the set of data is unbalanced and this affects the accuracy of the model (see Fig. 15).

Previously in the Fig. 10, the operation of the system is appreciated in a general way, starting from the top left, the detection of the type of citrus tree, the sensors of the drone are activated, later it makes the specification of the type of fruit to be analyzed within a range determined by the user, later, it carries out the identification of the damage of the foliage from taking characteristics of color, as following step a multivariate analysis with the purpose of determining the contribution of the factors.

5 Experimentation

The factors that intervene in the operation of the system and from which they could affect or improve the performance of the same one is analyzed, for it, the following table is taken into account, in which the amount of space and what a photo in JPEG will occupy is deduced. In this table, it is possible to estimate an average occupation value under normal conditions, and therefore a card capacity.

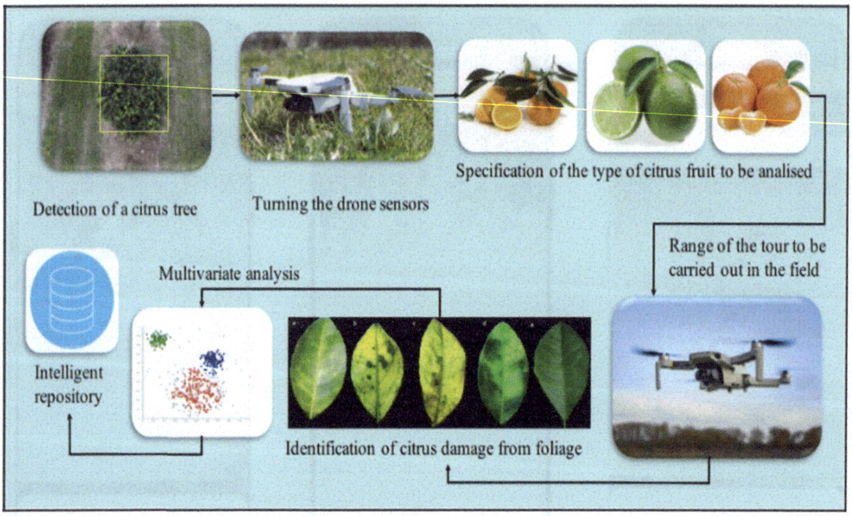

Fig. 15 System layout

A general use table could be the following (note: a compression ratio of 1:8 has been taken for such calculations) (see Table 3).

Nowadays it is of vital importance to develop computer applications with different functionalities and high complexity in a fast and efficient way in all aspects. Performing tests to the software during the development process is a guarantee for the operation of the systems; besides corroborating the degree of reliability before being delivered to the final users; which reduces the defects by using appropriate techniques that enable effective software development processes, minimizing time and costs (see Table 4).

Table 3 Estimate number of photos

Number of estimated photos					
	Camera resolution				
SD Card	12 MP	14 MP	16 MP	20 MP	24 MP
1	227	195	171	137	113
2	455	390	341	273	227
4	910	780	683	546	455
8	1.820	1.560	1.365	1.092	910
16	3.641	3.121	2.731	2.185	1.820
32	7.282	6.483	5.461	4.369	3.641
64	14.564	12.483	10.923	8.738	7.282

Table 4 System factors

Factors affecting system performance				
Factor	Description	Level 1	Level 2	Level 3
A	Battery life	30 min	20 min	18 min
B	Flight time	90 min	70 min	68 min
C	Processing	10 MB/s	6 MB/s	4 MB/s
D	Storage	7.282	6.242	5.461
E	Camera resolution	1080 mp	720 mp	720 mp
F	Flying distance	2 km	1.5 km	1 km
G	Speed without wind	14 km/h	10 km/h	8 km/h
H	Height	120 m	100 m	80 m

On the other hand, the orthogonal table will be applied as an alternative for the continuous improvement of the testing strategies and the decrease of the time required in their execution [9].

When evaluating the literature on robust testing based on orthogonal arrays, emphasis was placed on the techniques employed by Dr. G. Taguchi for the improvement of product and process quality; applying them to engineering processes for software development. Planning and executing the test cases with the previous proposal guarantees the detection of the highest number of errors in the double and triple interactions of the system's functionalities (see Table 5).

For the table above, the type of design is selected, which corresponds to 3 levels and the number of factors which is 8 in order to obtain an experimental table or orthogonal arrangement. In order to enter the parameters to obtain the best factor for the operation of the system (see Table 6).

From the previous data, the following figure is obtained, which contains a graph of points in which one experiments with the parameters exposed in Table 6, with the purpose of obtaining the best results (see Fig. 16).

6 Conclusions and Future Research

The original approach of the project has been the study in the fields of deep learning as well as convolutional neural networks. The different methods and techniques used during the development of systems for image recognition have been explored. During the development several of the original decisions were alternated such as taking into account the recognition of objects within the image. It has been possible to verify the importance of the performance of a neural network since it is not only based on the architecture of the network but the data set available alters the final result, so a balanced data set with a large amount of samples is critical.

Table 5 Experiments

Battery life	Flight time	Processing	Storage	Camera resolution	Flying distance	Speed without wind	Height
30	90	10	7.282	1080	2	14	120
30	90	10	7.282	720	1.5	10	100
30	90	10	7.282	720	1	8	80
30	70	6	6.242	1080	2	14	100
30	70	6	6.242	720	1.5	10	80
30	70	6	6.242	720	1	8	120
30	68	4	5.461	1080	2	14	80
30	68	4	5.461	720	1.5	10	120
30	68	4	5.461	720	1	8	100
20	90	6	5.461	1080	1.5	8	120
20	90	6	5.461	720	1	14	100
20	90	6	5.461	720	2	10	80
20	70	4	7.282	1080	1.5	8	100
20	70	4	7.282	720	1	14	80
20	70	4	7.282	720	2	10	120
20	68	10	6.242	1080	1.5	8	80
20	68	10	6.242	720	1	14	120
20	68	10	6.242	720	2	10	100
10	90	4	6.242	1080	1	10	120
10	90	4	6.242	720	2	8	100
10	90	4	6.242	720	1.5	14	80
10	70	10	5.461	1080	1	10	100
10	70	10	5.461	720	2	8	80
10	70	10	5.461	720	1.5	14	120
10	68	6	7.282	1080	1	10	80
10	68	6	7.282	720	2	8	120
10	68	6	7.282	720	1.5	14	100

Table 6 Optimal parameters

Optimal system parameters							
B. life	F. time	Proc.	Storage	Camera R.	Flying D.	Speed no W.	Height
30 min	90 min	10 MB/s	7.282	1080 mp	2 km	14 km/h	120 m

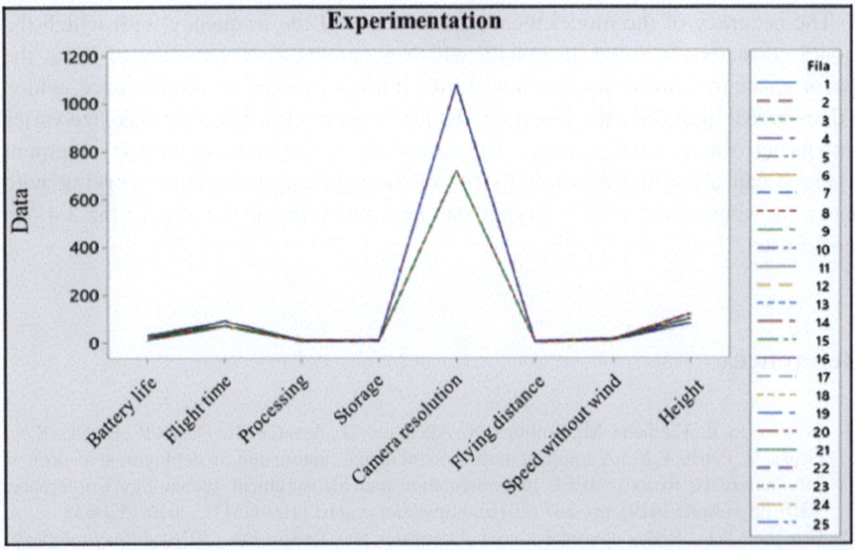

Fig. 16 Experimentation results

The main complications presented are divided into two more relevant ones, image processing and the model. The constructed data set has largely overwhelmed the final accuracy of the application, requiring a much larger data increase than what has been done currently. However, despite providing non-optimal results in some cases, it offers a clear insight into the importance of the available data, as well as the functioning of the neural networks in terms of image perception and how they are treated.

The task of predicting what an image symbolizes is called image classification. An image classification model is trained to recognize various kinds of images. An example of this is the model that has been worked with previously which was trained to recognize images of citrus crops. Thus, when a new image is provided as input to the model, it will generate the probabilities that the image represents each of the crop types it was trained with. In addition, during training, an image classification model receives images and their associated labels, where each label is the name of a different concept, or class, that the model will learn to recognize with enough training data often hundreds or thousands of images per label, an image classification model can learn to predict whether the new images belong to one of the classes in which it has trained to what this prediction process is called inference.

To perform the inference, an image is passed as input to a model. Then, the model will generate a series of probabilities between 0 and 1. Each number in the output corresponds to a label in the training data. By associating the output with the labels with which the model was trained, it was possible to see that the model predicted a high probability that the image represents a pest crop.

The accuracy of the model measured in terms of the frequency with which the model correctly classified an image, which generates a set accuracy of 98%, the performance in terms of the amount of time it takes a model to run inference, which it does in real time since the less time, the faster the model. Also, the disk size varies with its performance and accuracy. The size can be important for mobile development where it can affect the download sizes of the application or when working with hardware where the available storage can be limited, this is the case of the 3.4 Mb model used.

References

1. Bokolonga, E., Hauhana, M., Rollings, N., Aitchison, D., Assaf, M.H., Das, S.R., Biswas, S.N., Groza, V., Petriu, E.M.: A compact multispectral image capture unit for deployment on drones. In: Conference Record—IEEE Instrumentation and Measurement Technology Conference, 2016-July (April 2019), pp. 2–7 (2016). https://doi.org/10.1109/I2MTC.2016.7520445
2. Briega, R.E.L.: Redes neuronales convolucionales con TensorFlow (2016). https://relopezbriega.github.io/blog/2016/08/02/redes-neuronales-convolucionales-con-tensorflow/
3. De Rango, F., Potrino, G., Tropea, M., Santamaria, A.F., Palmieri, N.: Simulation, modeling and technologies for drones coordination techniques in precision agriculture. In: Advances in Intelligent Systems and Computing, vol. 873. Springer International Publishing (2019). https://doi.org/10.1007/978-3-030-01470-4_5
4. Florida, U.: UF/IFAS Citrus Extension: Plant Pathology (2018). https://crec.ifas.ufl.edu/extension/greening/symptoms.shtml
5. Gonzalez, T.F.: Handbook of Approximation Algorithms and Metaheuristics, pp. 1–1432 (2007). https://doi.org/10.1201/9781420010749
6. Matich, D.J.: Cátedra: Informática Aplicada a la Ingeniería de Procesos-Orientación I Redes Neuronales: Conceptos Básicos y Aplicaciones (2001)
7. Mead, F.W.: Oficinas Regionales SAG-SENASA ENVERDECIMIENTO DE LOS CITRICOS (HUANGLONGBING) (2020)
8. Na8: Convolutional Neural Networks: La Teoría explicada en Español | Aprende Machine Learning (2018). https://www.aprendemachinelearning.com/como-funcionan-las-convolutional-neural-networks-vision-por-ordenador/
9. Pentón Saucedo, Á., del Castillo Serpa, A.: Aplicación de la tabla ortogonal en el diseño de los casos de prueba de software. Avanzada Científica **15**(2), 1–12 (2012)
10. Rivas-Asanza, W., Mazon-Olivo, B., Mejía-Peñafiel, E.: Capítulo 1: Generalidades de las redes neuronales artificiales. Redes Neuronales Artificiales Aplicadas Al Reconocimiento de Patrones, June, 11–35 (2018). http://repositorio.utmachala.edu.ec/handle/48000/12499
11. SEDARPA: Veracruz Estadísticas Agrícolas 2017 | SEDARPA (2017). http://www.veracruz.gob.mx/agropecuario/estadisticas-agricolas/
12. Ulasovich, K.: Teachable Machine, la inteligencia artificial de Google, te responde con un GIF | N + 1: artículos científicos, noticias de ciencia, cosmos, gadgets, tecnología (2016). https://nmas1.org/news/2017/10/09/inteligencia-artificial
13. Vatalaro, F., Mazzenga, F., Giuliano, R.: The sub-band vectoring technique for multi-operator environments. IEEE Access **4**(June), 3310–3321 (2016). https://doi.org/10.1109/ACCESS.2016.2580198

Intelligent Application to Detection of Arachnid Bites in Children Implementing Deep Learning Techniques, an AmI-Based Solution

Ivette Mendoza, Eddy Sánchez-DelaCruz, and Alberto Ochoa

Abstract Misantla, is a city located in the state of Veracruz, Mexico. It is located in the mountainous region of the central zone of the state and is a region of warm-humid-regular climate, it has a great ecosystem with great diversity of species. Here the arthropods play an important role in the ecosystem. The arachnids are chelated arthropods and represent one of the largest classes of invertebrates that exist within the classification of animals. Most spider venoms do not harm people by being too weak, and can cause a sharp or stabbing pain, such as a bee sting. However, in cases of severe bites, the surrounding skin can die within a few hours. The reason for the project is the detection of accidents caused by arachnid bites, especially among children in the Misantla region, who are often severely damaged by the venom of a spider bite. It is worth mentioning that there are very few studies devoted to the analysis of the detection of any arachnid worldwide. And, in the country, it is not known, until today, a work of investigation whose objective has been; to identify the bite of a spider, as well as of its species and to suggest at the moment of the bite, some center of specialties to be treated. The importance of environmental intelligence within this study is related to the fact that it is currently in charge of solving this type of problems related to daily life, and with the correct functioning of the implementation of Deep Learning techniques for the detection process, to obtain a classification with the best precision and with a minimum margin of error. For all the above reasons, it is necessary to implement an intelligent application for the citizens of the city of Misantla. Citizens would then have more information about possible arachnid accidents and which medical facilities to go to before an accident occurs.

Keywords AmI · Spider Bites · Deep Learning and Smart Application

I. Mendoza (✉) · E. Sánchez-DelaCruz
Tecnológico Nacional de México, Campus Misantla, Veracruz, México
e-mail: aries.035@hotmail.com

A. Ochoa
Universidad Autónoma de Ciudad Juárez, Ciudad Juárez, México

A. Ochoa-Zezzatti et al. (eds.), *Technological and Industrial Applications Associated With Industry 4.0*, Studies in Systems, Decision and Control 347,
https://doi.org/10.1007/978-3-030-68663-5_2

1 Introduction

According to the literature, it is feasible to use a mobile application to determine the possibility of a spider bite accident, using the data associated with the most frequented areas. Users will then be aware of the locations within the city of Misantla where such incidents occur, as well as what to do before an accident. The use of technological tools [1], would reduce the chances of infants suffering an accident caused by arachnids. The objective of this project is to create a classification model that uses these tools about accidents by the main poisonous spiders (black widow spider, violinist spider, hobo spider, brown widow spider and golden silk weaver spider), of Misantla [2]. An exhaustive analysis of other similar research was carried out, the only similar context is explained in [3], where the authors created a geographic viewer, a tool for the prevention and care of accidents caused by arachnids, which allows the community, health workers, and other actors in general to be kept informed about aspects related to the origin and dimension of health problems caused by accidents caused by spiders and scorpions; As well as on the effects obtained from the actions oriented to the protection and prevention of these toxicological accidents and their best alternatives of treatment and mobility for their opportune attention; and in this way to guide, support and improve the management of the health services, but this investigation does not consider its development in application if not, simply in a web tool besides that it is not developed in Mexico, if not in Colombia.

This project was developed as a proposal to improve accidents in infants since they are more affected in summer seasons when children are exposed to the environment [4], thus causing an incident regarding an arachnid bite, since, in those seasons, these insects go out to hunt or to reproduce, or if the spider has just moved, eat or mate, then the venom will have a greater effect on the infant [1]. All this, through the classification of personalized images by training a simple model of deep learning with the help of an online tool from Google [5]: teachable machine, and then exporting the model to the TensorFlow Lite version that is compatible with the Android device. Finally, this model was implemented on an Android device. The tool, which consists of three different stages, is the result of an analysis and research carried out as part of a Master's program in Computer Systems at the Tecnológico Nacional de México, Misantla Campus. The objective of this experiment is to identify how the proposal works, using a large number of images and making use of the Neuronal Convolutional Networks (CNN) and Ten-sorFlow Lite [6]. Through this experiment, a beta application was used [7], so that volunteer users could make use of it and recognize with the application from their gardens or in the environment where they encounter a spider, what type of species it is or in case of suffering from an arachnid accident, know who caused it and where to go for its case study. This application is still in the development phase and not in a final version. To qualify the project as successful, the application had to be taken to field practices to provide an accurate identification of some of the species to be identified.

2 Implementation of the Intelligent Application

There are several aspects to consider when designing mobile applications: both limited screen size, different resolution and screen sizes between devices. Therefore, the de-signer has to develop the interface in a uniform way to fit most of the devices to be used. In addition, the number of users using Android has had a clear growth of 87% in recent years [8].

For this work, the Android Studio platform for the Android API was used. The programming interface we worked on is XML with the Java-based language.

With respect to the choice of the five different species of arachnids [9], selected for this paper, these were chosen because they are the most dangerous in the Misantla region and because they concentrate the largest number of bites as they are the most common (see Fig. 1).

On the other hand, a classification of images or image recognition was required since it is a concept in which you show an image to the device's camera sensor and it will tell you what is present in that image or tell us what class it belongs to. Therefore, through the classification of hundreds of images by training a simple model of deep learning you get an intelligent application which identifies the types of arachnid species and their bites, from an Android device (see Fig. 2).

The proposed classification model, as can be seen in Fig. 2, as a first step you must have hundreds of images of what you want to classify, that is, a varied dataset of images, in this case you download images of three types; arachnids, spider webs and their bites, of at least five types of the most poisonous arachnid species of the city of Misantla. After that, the images are loaded from Google's Teachable Machine platform in which the classes to be classified are labeled and after that, the model

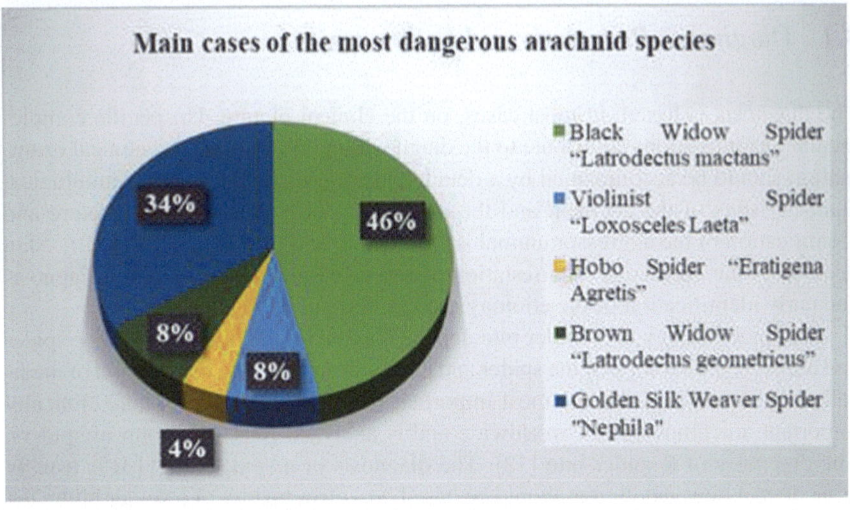

Fig. 1 Most dangerous arachnids in the region of Misantla, Veracruz

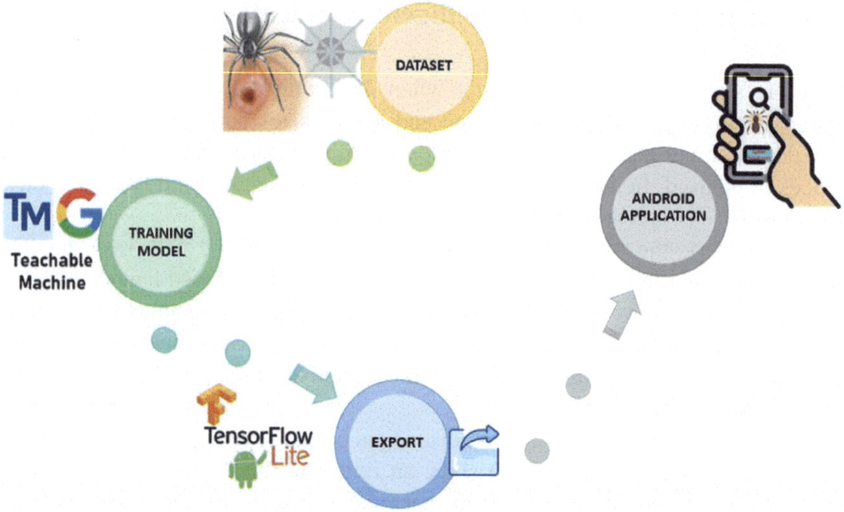

Fig. 2 Proposed classification model

is trained. Thirdly, the trained models are exported as TensorFlow Lite and, finally, they are migrated to the mobile application developed from Android Studio for its correct operation.

3 Spider Bite Recognition

3.1 Diagnosis, Prognosis and Follow-up

The diagnosis is based, in most cases, on the clinical picture. Unspecific complementary examinations contribute to the diagnosis and follow-up. The clinical examination should be accompanied by a detailed history, including the epidemiological characteristics of the accident and the chronology of symptoms. The capture and identification of the aggressor animal is recorded in less than 5% of cases [10]. Skin necrosis is the cutaneous manifestation of several clinical entities. Early diagnosis and early identification of the etiology are key to treatment [11].

The epidemiology of a spider bite depends on the interaction between the spider and humans, the ecology of the spider and the environment. The distribution of medically important spiders is the most important factor in identifying where clinically important arachnids occurs worldwide and is analyzed for each group of spiders. The diagnosis of a spider bite [12]. The diagnosis of a spider bite [13] is usually clinical, and bites indefinites should be based on a clear history of a spider biting the person and then identified by degrees of intoxication. The following Table 1 gives a summary of the above.

Table 1 Degrees of intoxication from spider bites

Degree of intoxication	Clinical picture
Grade I/mild	Pain at site of injury, in lumbosacral region, abdomen. As thenia, adynamia, diaphoresis, sialorrhea, hyperreflexia
Grade II/moderate	Accentuated dyspnea, epiphora, headache, spasm, muscle contracture or stiffness, priapism
Grade III/severe	Myosis, mydriasis, trismus, heart rhythm disorders, bronchospasm

The development these conditions [14] and the release of mediators from the inflammatory process have been associated with a deficiency in the enzyme glucose-6-phos-phate dehydrogenase, although it is not clear why some people only develop skin conditions while others develop systemic poisoning. However, the ELISA for detecting circulating poison and in the lesion presents the greatest advantages, due to its speed and simpler instrumentation [10].

The latency period between the bite and the appearance of clinical signs can vary from minutes to a few hours in the cutaneous form and up to 12 to 48 h in the systemic form. Poisoning can occur in two clinical forms: Cutaneous or cutaneous-necrotic loxoscelism and systemic, visceral cutaneous or viscera-haemolytic loxoscelism [11]. In most cases, the diagnosis should be based on the clinical picture, associated with the compatible laboratory, because the history of the bite usually goes unnoticed. The clinical examination should be accompanied by a detailed history, including the epidemiological characteristics of the accident and the chronology of signs and symptoms [14].

3.2 *Treatment*

There is no unanimous view on the most appropriate treatment for humans. Several factors affect this: time between the accident and the consultation, clinical form (cutaneous or systemic), evolutionary stage (pre-necrotic, necrotic, ulcerative) and the appearance of complications (renal failure, superinfections) [10].

Antivenins are an important therapeutic intervention for poisoning syndromes, and antivenins are available for many groups of spiders [15]. However, antivenins have been less successful in treating arachnidism than those for snake or scorpion poisoning. Antivenin use is based on clinical experience, which has led to discrepancies in the pro-portion of patients treated [14].

The following Table 2 provides an overview of the two types of treatments, according to the spider bite.

It should be noted that the approximate dose of antivenin is always the same for children and adults [14].

Table 2 General and specific treatments for spider bites

Treatment and general measures	Remove all elements that may compress and aggravate the edema in the affected area In injuries located in the members maintain the Trendelenburg position Perform local antisepsis Assess the need for tetanus prophylaxis Use analgesics if necessary The use of steroids is controversial, but may be useful in the early hours specially to reduce the inflammatory component Use of antibiotics in cases of bacterial superinfection
Specific treatments	The rapid establishment of the specific treatment, which is the application of the corresponding antivenom. Its usefulness would be related to the indication of the adequate dose and its administration in due time (precocity) in order to avoid or limit the necrotic picture and/or eliminate or reduce hemolysis in systemic cases

3.3 Prevention

Regular cleaning of homes and removal of intradomestic arthropods is essential. Likewise, if spiders are known to exist in neighboring gardens or farms, it is important to meticulously examine objects (hanging clothes, chairs or other furniture, toys, etc.) that are brought into the home from those sites. The custom of shaking clothes and extended tablecloths to dry in the sun before folding them is a good measure. In addition, it is advisable to keep beds away from walls and windows, as well as the presence of mosquito nets on windows and doors, especially if there is a history of these arthropods in the area. As recommended to prevent other accidents by poisonous animals, when moving objects that have been stacked or accumulated for a while (boxes, firewood, dis-carded objects, clothing, among others), this should be done with gloves and appropriate footwear.

The use of chemical products for combat must be done very carefully because arachnids are resistant to substances commonly used for the elimination of intradomestic arthropods. It is advisable to consult before using biocides. Sometimes no active ingredients, formulations or appropriate forms of application are used, since unless the hard-to-reach fabric where Loxosceles resides is sprayed intensively, its contact with any residual insecticide is minimal. Given the sensitivity of these spiders to environmental moisture, infested peridomestic sites can be treated by watering or flooding [10].

3.4 Methodological Proposal for the Recognition of an Arachnid Bite

According to the evidence, it is feasible to use a mobile application to identify the bite of a spider and know which has been the aggressor, using photographs in real time for the cases. This investigation tries to recommend the patient with this problem, to the place where it is necessary to resort to this type of emergency and with it, to avoid amputations or even the death by the bite of a poisonous spider in some children. Since, it is impossible to know what type of spider has been the aggressor, to the moment in which this one bites and the idea of not knowing which species has been is one of the main problems for this work. That is why with AmI, a series of tools arise to achieve this problem, which in this case is technology, information, image processing, among others (see Fig. 3).

The case study presented of the model proposed to this research, leads to children under 12 years, suffer from the bite of a spider in one of the extremities of the body, which requires then the identification of the species as the bite of this and through the proposal to develop a mobile application of environmental intelligence to determine the behavior of the spider bite in children. Focused on a diagnosis and prognosis for medical follow-up through a geolocalized repository of other patients with similar spider bites, it will be feasible.

From the analysis of the different conditions that cause skin necrosis, it can be concluded that it is very important that a doctor, when faced with a skin necrosis syndrome, thinks about and identifies the causes that can cause this similar clinical manifestation, since the beginning of an adequate and timely therapy will avoid complications and favor the patient's prognosis. Most of these conditions require

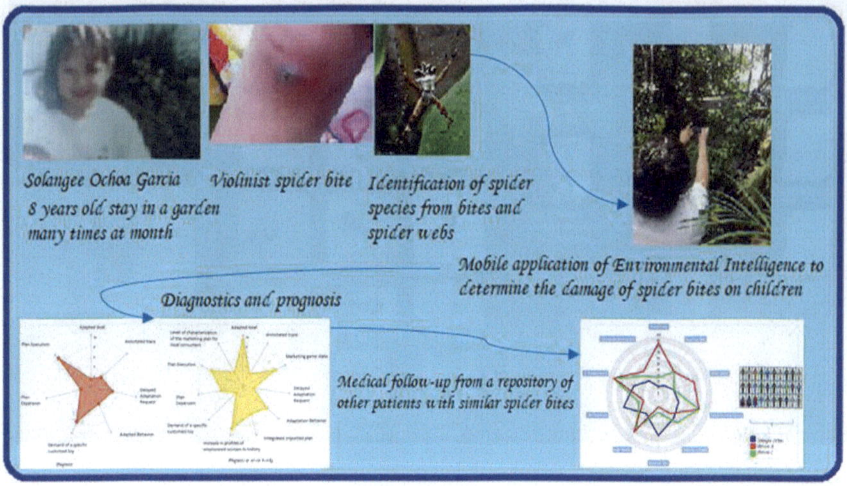

Fig. 3 Proposal of the model associated with research

hospitalization in more complex units, since they can evolve rapidly with multi-organ failure. The role of the dermatologist and dermatopathology is of special importance, since the skin biopsy can be fundamental to determine the early diagnosis.

4 Test Development

The data with which the model was trained was prepared. For this purpose, about 1500 images were downloaded from the web (see Fig. 4) of five different types of spiders, among them; the black widow spider, violin spider, hobo spider, brown widow spider and the golden silk spider (100 images for each species mentioned), as well as their bites of each of these species and their webs, for each of the objects that are required to be detected.

On the other hand, in terms of image processing and resizing, these are left at a fixed size of 400 wide by 350 high. Three folders were created, one of the arachnids and, within this one, five more folders of the five types of spiders, another folder of the spider webs and its five more folders of the five different species of webs and finally the third folder of the bites and, within these, five more folders with their bites of the different species to work with.

Next, the characteristics to be taken from each type of spider, its bite and the web it weaves are described, which was taken into account for its pre-processing in the classification model.

The following Table 3 gives a summary of the characteristics of each species that were taken into account.

Fig. 4 Dataset built with images from the web; arachnids, bites and spider webs

Table 3 Characteristics that the CNN model takes into account in the classification

Species	Characteristics to be identified
Black widow spider (*Latrodectus Mactans*)	It is distinguished by its black color and a red hourglass on the abdomen. Its web is irregularly shaped, tangled and sometimes funnel-shaped. When it bites, it leaves a reddening of the skin, a swelling. A central blister forms at the site of the bite, extravasated blood under the skin (like a bruise). The lesion may enlarge into deep, pus-filled ulcerations
Violinist spider (*Loxosceles Laeta*)	It can be distinguished by the violin in the central part of the body. Its web has a disorderly and irregular design, and is located in right-angled profiles (corners), where it weaves a horizontal net in the shape of a short hammock. When it bites, it leaves a red or lilac coloration in a circle that will be in the middle of the bite, or it creates an ulcer in the first hours of the bite
Hobo spider (*Eratigena Agretis*)	It is distinguished by its brown body with yellow spots on its abdomen, as well as having slightly longer legs and a hairy body. Their web is funnel-shaped, they do not climb vertically, they usually build their webs at ground level or under the floor in basements. When they bite, they leave a small red bite that looks like a mosquito bite, and within the first few hours, blisters will appear around the bite, then burst leaving open wounds
Brown widow spider (*Latrodectus Geometricus*)	It is distinguished by its brown or beige color with a spotted pattern and by having grooves on its legs. It also has an hour-glass on its lower abdomen, but it is bright yellow or orange. Its web is three-dimensional (instead of flat) and these create sticky webs. When it bites, it leaves a small red mark in the area of the bite, causing an infection, such as a rash, pus, or sores
Golden silk spider (*Tricho-nephila*)	It is distinguished by the shape of its body, its abdomen is like a rigid, elongated armor with combinations of dots or stripes in black, white and gold. Their web is distinguished by the color of the silk they produce, the threads of their web shine like gold in the sunlight. When chopped, it leaves a slight reddening of the skin that will disappear completely over time

Once the convolutional neural network takes into account the characteristics described in the table above, a model is obtained that can be migrated to the application. There-fore, the CNN was used to carry out the training. From Google's tool, Teachable Ma-chine, for Machine Learning, this allowed to abstract the definition of the model, algorithm and data processing, since it focuses the deployment of the model to be generated. This tool works in the browser, in this case, with files that

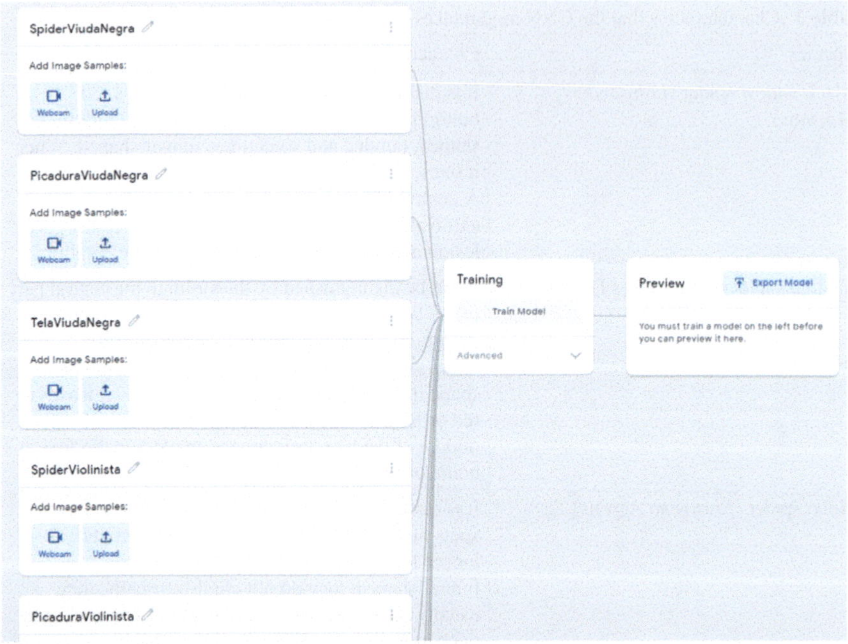

Fig. 5 Starting to train the model for a classification

are hosted on the computer, and in a few minutes, you can quickly understand how a model learns through a simple demonstration of classification.

To start training the classification model, first of all, the categories or classes to teach it must be created (see Fig. 5). For this training, classes were given on the types of arachnids to be identified, as well as the stings that these insects generate.

Once the labels are well defined, the samples for each class are uploaded from the local area so that training can begin. And, creating the training with the classes would look like this (see Fig. 6).

The model takes time to train, sometimes it gets stuck, but the result is quite good. And, after the model training phase is finished, the models are exported as TensorFlow Lite (see Fig. 7), of two different types; the first as *model.tflite* and the second as *model_unquant.tflite*, and the third as *labels.txt.*

On the other hand, making the development of the application on the Android Studio platform, it is as follows (see Fig. 8).

The model *label.txt*, contains all the classes for the training of the CNN so that there is a correct identification of what is intended and, are used to be called from the Android mobile application (see Fig. 9).

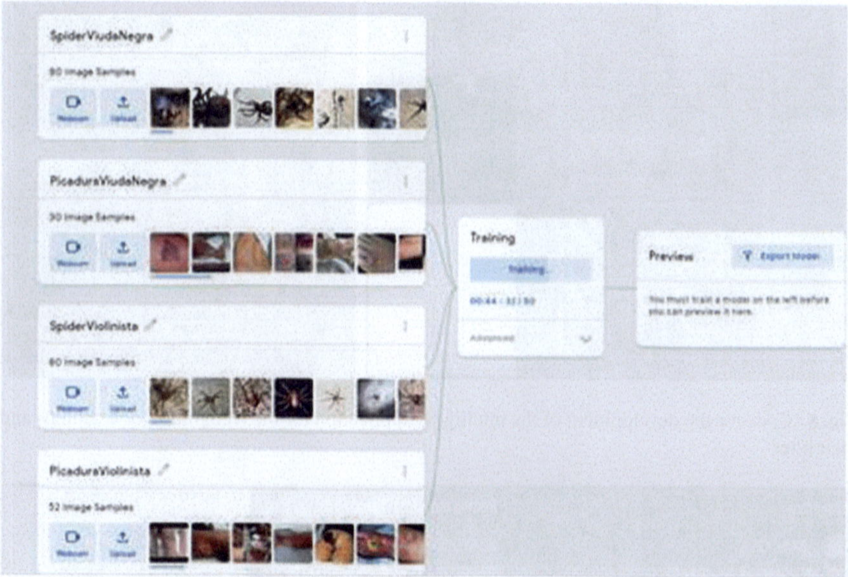

Fig. 6 Classification model training

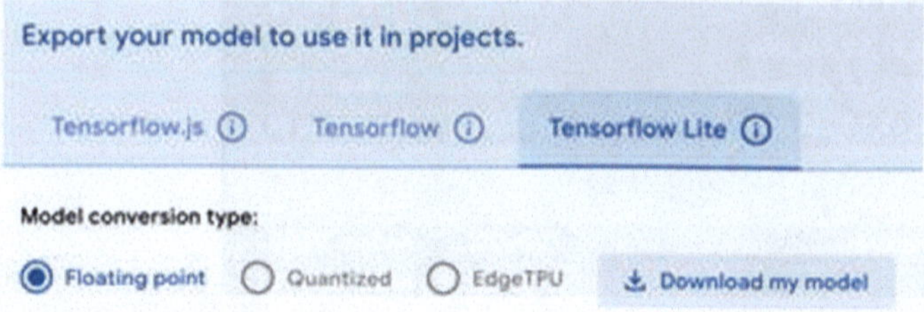

Fig. 7 Exporting the TensorFlow Lite model

4.1 Module for the Recognition of Arachnids and their Bites

Convolutional Neural Networks (CNN), are the algorithm to give the computer the ability to "see". Thanks to this, it is now possible to classify images, detect various types of tumors automatically, teach autonomous cars to drive and a host of other applications. The inputs are images, which allows to code certain properties in the architecture; allowing to gain in efficiency and to reduce the number of parameters in the network. Convolutional neural networks are efficient because they scale well for high definition images [16].

What a convolutional neural network does is too simple, you pass it an image, and it classifies it for you (see Fig. 10).

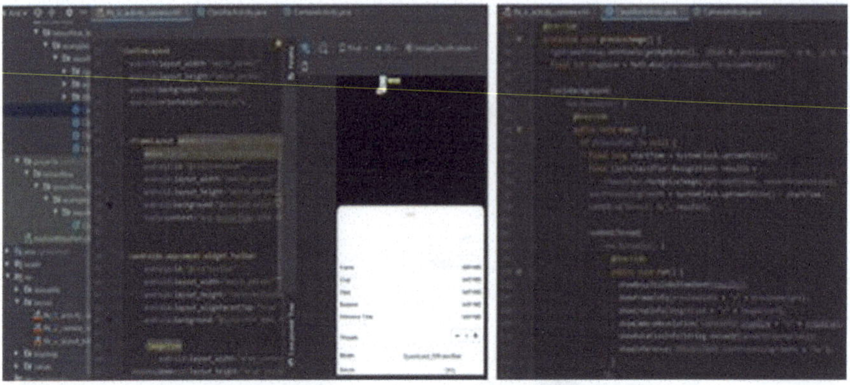

Fig. 8 Code for the development of the intelligent application for the recognition of arachnids and their bites

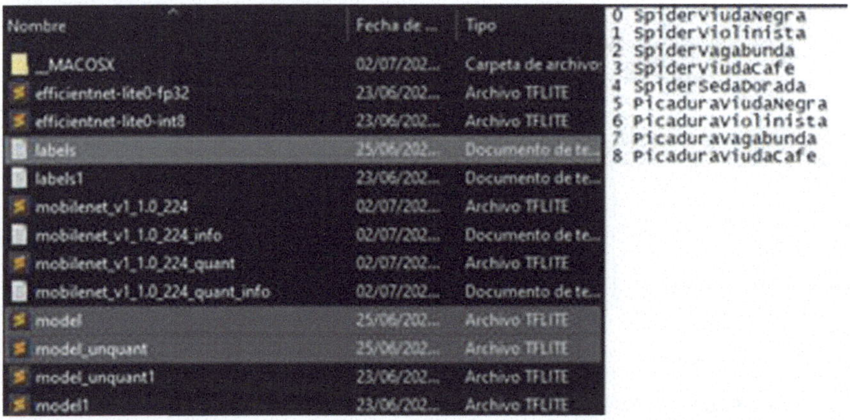

Fig. 9 Models exported and migrated to the Android Studio application for identification

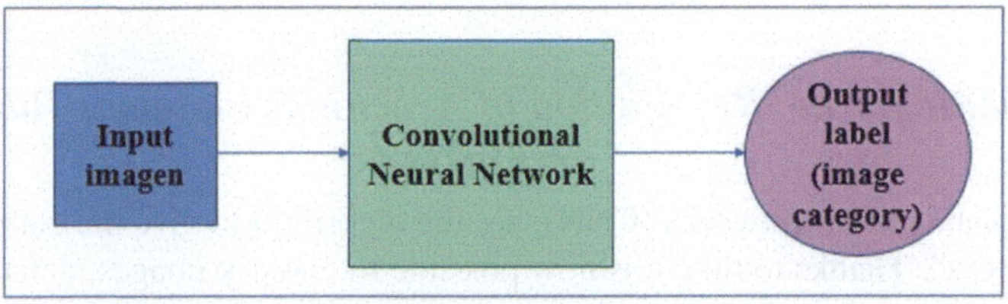

Fig. 10 Process performed by a convolutional neural network

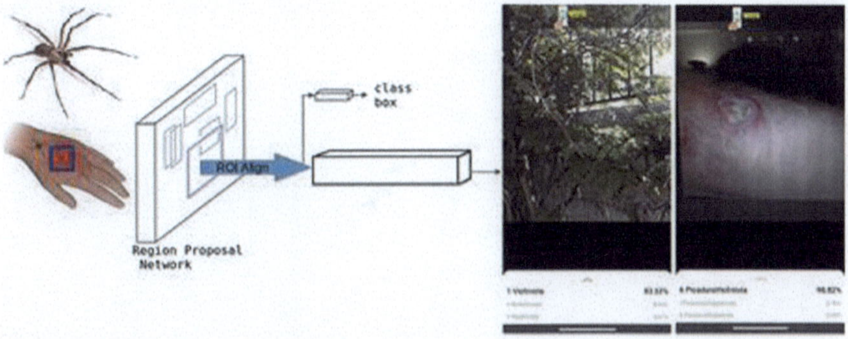

Fig. 11 Simplified diagram of the CNN used for the classification/segmentation of image. In the diagram, the network generates three types of output, the box where the arachnid is located, a binary mask that delimits the arachnid and the type (genus) of arachnids found or, failing that, the type of bite to which it corresponds

In order to make the convolutional neuronal network functional, a dataset of images of five different arachnid species, of spider webs and of the bites that each one of these species leaves after biting a person, was provided to it. Then, a new image that is not found in the dataset representing a bite or a spider was passed to it, so that the convolutional neuronal network was able to know which species corresponds to that image.

This module for the mobile application (see Fig. 11), has an exchange with the spider image recognition and the rest of the data. The arachnid identification stage consists of a deep learning architecture. An input image of the bite or arachnid is then segmented into several groups which are then classified as the recognition of the type of spider or the type of bite. The architecture is based on the convolutional neural network for the recognition.

In this paper, it was already proposed that the network will focus only on the recognition of arachnids and to identify what species the bite was from, simplifying the module to make it easier to train and less computationally burdensome. The proposed architecture consists of three convolutional channels, where each channel aims to select different feature sizes: a large channel with 11×11 size cores, a medium size channel with 4×4 size output filter and finally a small channel with 2×2 output filters. The outputs of the different channels are concatenated into a feature map that contains information at different scales of the input image, so for concatenation, the output of the medium and small channels is sampled so that the output of all three channels is the same size. Then the output is fed to a couple of convolution layers to recover the size of the original image and finally a 1×1 filter size convolution layer to have an image of the spiders or their bite effect.

The details of the architecture described are shown in Fig. 12.

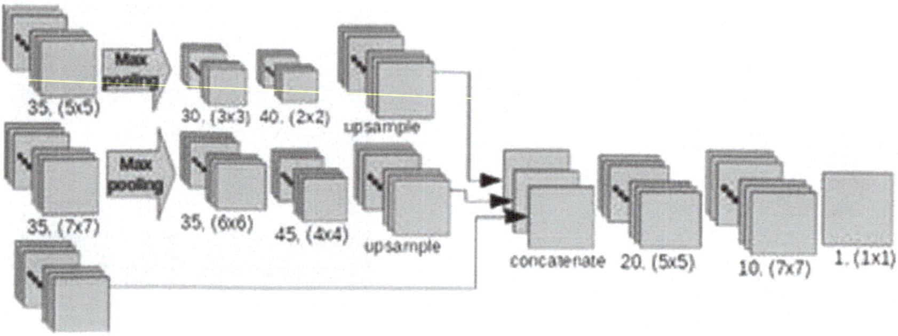

Fig. 12 Proposed architecture for the module. It consists of three channels of convolutional layers that are concatenated to have a good feature map with information at different scales

4.2 Results

The beta version of the application called Spider Bite, was developed based on the minimum requirements, which were the delivery of a very basic application with the ability to interact with the user using a simple but practical interface, where the user enters automatically without the need to log in and run it, This has the option of pro-cessing the image taken in real time in the case of seeing an arachnid species so that the application can define, by means of the camera, what type of species it is, or, in the case of an arachnid bite accident, suggest a nearby health center to go to immediately (see Fig. 13).

In Fig. 13, you can see how the application is composed, which has a dataset of the species that are being addressed, its operation is basic, then, the user enters, choose the option to upload an image from your gallery to process it and that the app throws the species or bite that is, or capture it in real time with the device's camera in the event of encountering the arachnid or having been stung at the time and thus

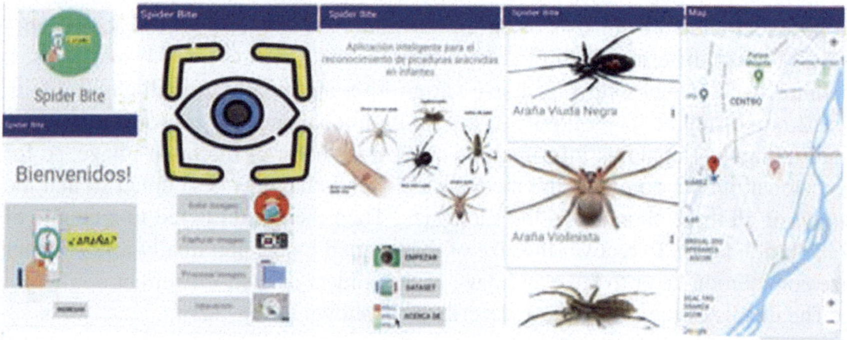

Fig. 13 Intelligent application developed to determine the most poisonous spider bite accidents in infants in the Misantla region

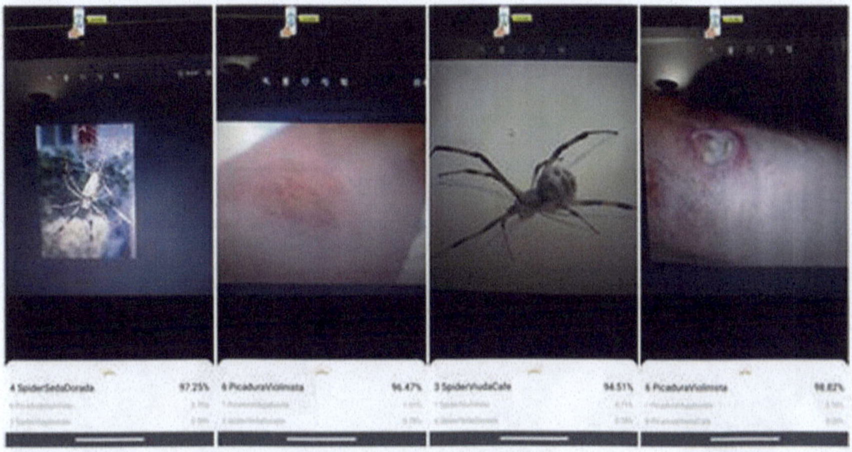

Fig. 14 Results obtained that are shown in the application once the generated TensorFlow Lite models are exported

this application gives us what we expect and also shows us the location of the person and the health center closest to him/her in Misantla.

The function that has the option of processing images within the application, is that, with the camera of the device, it shows an image and in automatic the application will say what is present in that image, the species of the arachnid, as well as the bite to which type of arachnid it corresponds (see Fig. 14).

One more example of the application's functionality is shown (see Fig. 15), in which the application was taken to field tests where it was presented with an arachnid species in a garden, which our intelligent application recognized as the next species.

Finally, the mobile application developed in the Android Studio platform gives feasible results of the model exported by TensorFlow Lite, which is called from the application, since it achieves 90% accuracy in terms of correct identification of arachnid species.

5 Conclusions and Future Work

It is through this whole validation process that we learned the potential of this intelligent application. According to the experiment, it is confirmed that it is possible to implement a technology platform for the recognition of arachnid stings in infants in real time using a mobile device that captures the part of the sting and this application will be able to identify which has been the arachnid to act immediately avoiding a misfortune in the affected part. Similarly, the lenses meet the following specifications: the use of a mobile device, Sony Xperia XA2 Ultra, with the 23Mpx rear

Fig. 15 Mobile application used in a garden for the identification of arachnid species

camera, f/2.3 with LED flash and 4 K recording. The Android 8.0 Oreo operating system and a Qualcomm Snapdragon 630 Octa-Core 2.2 GHz Cortex-A53 processor.

The result of this could be used, in this case, to focus strategic measures to prevent accidents in conflict areas. If the measures are implemented in areas where a higher probability of accidents is predicted, the impact of the measure will be greater. This provides a very good sense of how useful and practical the proposed methodology can be. With the use of this innovative application that combines Deep Learning and a mobile device-based model, it is possible to determine the places to go in the event of a child spider bite accident in the city of Misantla. The most important contribution is the possible prevention of future infant deaths in the city caused by spider sting accidents. We believe that this innovative technology has a promising application in other Latin American cities with similar problems of severe arachnid bites.

This project is very important for the city of Misantla because of the type of region, since it is mountainous and its climate is warm-humid-regular, and the children, being in the middle, are the most exposed to suffer from this problem because of the great diversity of species that live there. In addition, there is no intelligent system in place to help citizens with regard to an arachnid bite. And, until now, no work like the one presented above has been carried out in the country. On the other hand, the contribution that we have is the inclusion of image pre-processing techniques, in addition to the study in the fields of Deep Learning for the case study, as well as convolutional neural networks, the incorporation of three datasets into one and the creation of the classification model which uses images from three datasets, which are in one. The different methods and techniques used during the development of systems to recognize images have been explored. During the development several of the original decisions were alternated such as taking into account the recognition of objects within the image.

It was possible to confirm the great importance of the performance of a neural net-work, since it does not lie exclusively in its architecture but, in the set of data available it alters the final result. The two main complications that were presented more relevant in this work were: the model and image processing by how they are treated.

As future research to be incorporated into this project in terms of the mobile application, it is recommended that more existing arachnid species be added for recognition in terms of the arachnid bite. At the same time, this application should be implemented to detect and recognize other types of stings, such as those of poisonous snakes or even scorpions, and not only to recognize spider bites. That the application is capable of providing recommendations for the surveillance, prevention and care of accidents by the various types of bites according to the species to be detected. And, that it is developed in iOS devices, that it is multiplatform.

References

1. Alfonso, I.: Environment: Definition and importance of the environment around us (2019). [Online]. Available: https://cumbrepuebloscop20.org/medio-ambiente/. Last accessed 09 Oct 2019
2. Naturalista: Spider (Arachnid class). Naturalista (2019). [Online]. Available: https://www.naturalista.mx/taxa/47119-Arachnida. Last accessed 10 Oct 2019
3. Laura, O.C., Bastidas, M., Milton, E., Pino-Capcha, I.M.: SIG as a tool for the optimization in the attention of cases by bites or stings of arachnids in Colombia (2017)
4. Durán-Barrón, C.G.: Dangerous, only two homemade spiders. GACETA UNAM (2019). [Online]. Available: https://www.gaceta.unam.mx/solo-dos-aranas-de-casa-son-peligrosas/.
5. Experiment, G.A.: Teachable machine learning. Google (2017). [Online]. Available: https://teachablemachine.withgoogle.com/. Last accessed 07 July 2020
6. Roldan-Valadez, E., Favila, R., Martínez-López, M., Uribe, M., Méndez-Sánchez, N.: Imaging techniques for assessing hepatic fat content in nonalcoholic fatty liver disease. Ann. Hepatol. **7**(3), 212–220 (2014)
7. Padgalskas, V.: What is a Beta application?. Techlandia (2020). [Online]. Available: https://techlandia.com/aplicacion-beta-info_254507/. Last accessed 07 July 2020
8. Asunción, M., Fernández, C., Pomares, A., Galotto, M.: Distribution of smartphones worldwide. Scientific (2017). [Online]. Available: https://www.researchgate.net/figure/Distribucion-de-las-ventas-de-smartphones-en-todo-el-mundo-segun-su-sistema-operativo_fig3_319979830. Last accessed 10 July 2020
9. INaturalist: Misantla, MX, VER. NaturaLista. California Academy of Sciences y National Geographic Society (2019). [Online]. Available: https://www.naturalista.mx/places/misantla#q=arachnids. Last accessed 08 July 2020
10. de Roodt, A.R., Salomón, O.D., Lloveras, S.C., Orduna, T.A.: Poisoning by loxosceles spiders. Medicina (B. Aires). **62**(1), 83–94 (2002)
11. Molgó, M.N. et al.: Skin necrosis: A challenge for the doctor. Chil. Med. J. **142**(1), 118–124 (2014)
12. GK, I.: Data collection in clinical toxinology: debunking myths and developing diagnostic algorithms. J. Toxicol. Clin. Toxicol. **40**, 231–237 (2002)

13. Stoecker, G.J.: Diagnosis of loxoscelism in a child confirmed with an enzyme-linked immunosorbent assay and noninvasive tissue sampling. J. Am. Acad. Dermatol. **55**, 888–890 (2006)
14. Ministerio-Salud: Guidelines for the prevention, diagnosis, treatment and epidemiological surveillance of spider poisoning (2012)
15. Offerman, S.: The treatment of black widow spider envenomation with antivenin latrodectus mactans: A case series. Perm. J. **15**(3), 76–81 (2011)
16. Bagnato, J.I.: Convolutional neural networks: Machine learning and deep learning (2018). [Online]. Available: https://www.aprendemachinelearning.com/como-funcionan-las-convolutional-neural-networks-vision-por-ordenador/. Last accessed 15 July 2020

Evacuation Route Optimization in the Plaza de la Mexicanidad, Using Humanitarian Logistics

María Inés Borunda-Aguilar, Iván Juan Carlos Pérez-Olguín, Alberto Ochoa-Zezzatti, Erwin Adan Martinez-Gomez, and José Alberto Hernández

Abstract The article focuses on the Plaza de la Mexicanidad in Juarez City, Chihuahua. This plaza provides the town with an open area, where its inhabitants satisfy their commercial, social, cultural, recreational and family needs. Leading to large crowds of different features such as age, gender and distinct physical and emotional characteristics. Thus, providing the bases for a humanistic logistic simulation research topic for measuring the interrelated factors with the probability of occurrence of a potentially destructive phenomenon, based on the masses attitude during an event. Considering that Juarez City has a population of 1,428,508 according to the 2018 Juarez Strategic Plan report, being the largest city in the state of Chihuahua and the eight largest metropolitan area in Mexico.

Keywords Humanitarian logistics · Simulation · Route optimization

---

M. I. Borunda-Aguilar (✉) · I. J. C. Pérez-Olguín · A. Ochoa-Zezzatti · E. A. Martinez-Gomez
Autonomous University of Ciudad Juárez, Av. Hermanos Escobar, Omega, 32695 Ciudad Juárez, Chihuahua, México
e-mail: maria.borunda@uacj.mx

I. J. C. Pérez-Olguín
e-mail: ivan.perez@uacj.mx

A. Ochoa-Zezzatti
e-mail: alberto.ochoa@uacj.mx

E. A. Martinez-Gomez
e-mail: emartine@uacj.mx

J. A. Hernández
Autonomous University of Mexico State, Instituto Literario 100, Centro, Cuauhtémoc, 50000 Ciudad de México, México
e-mail: jose_hernandez@uaem.mx

A. Ochoa-Zezzatti et al. (eds.), *Technological and Industrial Applications Associated With Industry 4.0*, Studies in Systems, Decision and Control 347,
https://doi.org/10.1007/978-3-030-68663-5_3

1 Introduction

With the eminent technological development in software application, the use of crowd simulation programs [1] for urban development has been rapidly consolidated in both planning and designing of infrastructure and emergency exits. Thus, providing the bases for simulating alternative evacuation routes in case of possible accidents.

The article's objective is to design an evacuation route simulation for optimizing the key points to evacuate concertgoers in an event of an accident in the Plaza de la Mexicanidad, considering one factor such as an outdoor concert. The simulation will present agents embodied in real life behavior, representing groups of autonomous characters, called virtual agents. Among one of the objectives, is to generate an optimize algorithm capable of producing real-time visualization, which means that the frame rate must be at least 30 Hz [2]. Within the simulation design, was taken into account the monumental work of Enrique Carbajal (inaugurated on May 24th, 2013), a sculpture in the form of a colossal X with a height of 62 m.

The use of simulation can sharpen the planning of different scenarios in cities based on the design, development and implementation of emergency exit doors and infrastructure, for better evacuation. It can determine the reflected impact by designing more efficient routes and alternative routes for safeguarding the people's integrity within agglomerations in public events. The government can make use of crowd simulation to prevent disaster situations in public events, airports, and underpasses [3], leading to today's line of research regarding large crowds. However, other uses are the simulation of scenes in movies and video games as mentioned by Groenewegen [4].

2 Behavior of the Masses

There are two types of simulations (1) Microscopic, considered for individual or personal movements and (2) Macroscopic, regarding mass movements [5]. Macroscopic models are solved using analytical mathematical techniques, while microscopic models consider individual behavior and interactions between agents. However, as humans are rational beings, must of their behaviors are predictable; and this predictability allows analysts to observe people movements across the stage.

In the other hand, mass behavior presents a psychological speaking predictive behavior, as specific as the individual behavior. Such behavior determines how resilient and cooperative the crowds will be in each situation. For this, the culture of each crowd must be taken into account in order to determine behaviors that may lead to risker situations. In this type of cases, is essential not to introduce external agents, as they may trigger a catastrophic event by modifying the crowd's usual pattern. In other words, crowds move with predetermined patterns defined by culture, taste and interest, which help members not to create risk situations.

2.1 Stampedes of the Masses

The proper and accurate implementation of optimal contingency plans can safeguard the lives of people in a catastrophe situation, preventing irreversible damage caused by human stampedes. History shows evidence of endless dead, injured people and at high risk due to human stampedes.

One thing to improve, is to eradicate the lack of information and knowledge of the events' staff, who play a crucial role in mass mobilization. This can be achieved through constant training and education in contingency plans. These training should be based on activities that allow the individual to create a greater degree of resistance, avoiding negative situations towards the masses and thus causing accidents as losing someone life [6].

The first step towards identifying the potential and magnitude of a risk within a mass event, is to determine the initial assessment by asking: What is the purpose of knowing what is going on and what could go wrong? This, to determine the possible extent of the risk and how to minimize the possible damage, thus containing the negative impact. Next, the second step is the mobilization of resource, that is to say, to correctly organize the activities in the scene of the incident.

The fundamental thing is to identify and know the factors for determining the procedure to carry out in an effective manner to satisfy the alert needed for minimizing human damage and loss.

These factors are listed below:

- The first person responding to the alert must be trained to perform an initial evaluation and proceed according to the established evacuation methodology.
- After the initial evaluation, the person must determine the scope of the possible events. The initial evaluation is a sequence of activities aimed at obtaining the following information:
  - Precise location of the event
  - Time it occurred
  - Type of incident
  - Estimated number of victims
  - Potential risk
  - Exposed population
- The initial evaluation must be notified immediately. The bad decisions of the people who arrive at the place as a first instance could affect the mobilization.
- Information dissemination: To inform those involved and mobilize both resources and institutions that can provide timely help.

2.2 *Pre-disaster Planning*

To establish the procedures in case of an accident, based on the personnel training, it must incorporate the link between the assistance facilities and a multi-sectoral approach for stabilizing the victims.

Planning alternate evacuation routes through simulation, helps determine different behaviors of those involved in the incident, allowing for better preparation as to what to do, how to act, when and where to locate the personnel. With this, the target is to improve the timely information and decision-making. At this point, for the simulation process, past experiences can be taken into account allowing anticipation of unexpected events and minimizing its consequences.

Another factor to consider, is the event's staff as they are key role in the emergency's mobilization. Considering previous evacuation route scenarios, meeting points, space time coordination and assertiveness through their rapid response.

The goal of the evacuation planning is to minimize the total response and evacuation time, as well as minimizing any loss, especially human losses.

2.3 *Implementation of a Mass Victim Assistance System*

The Pan-American Health Organization considers five important points in its victim care model.

- Phase 1: Adoption of the system as a national policy.
- Phase 2: Sensitization.
- Phase 3: Training.
- Phase 4: Institutionalization of the system.
- Phase 5: Maintenance of the mass victim assistance system.

Considering the following codes:

- Red alert: Indicates fire.
- Yellow alert: Indicates earthquake or landslide.
- Blue alert: Indicates flood.
- White alert: Indicates chemical disaster.

3 Simulation Models

The model presented by Reynold, agents are considered to move individually and improvise, classifying this behavior as direction. Reynold mentions in previous research as the starting point of an investigation, triggering random movements in a symbolic way that reflects the external behavior on a reactive way, before a stimulus

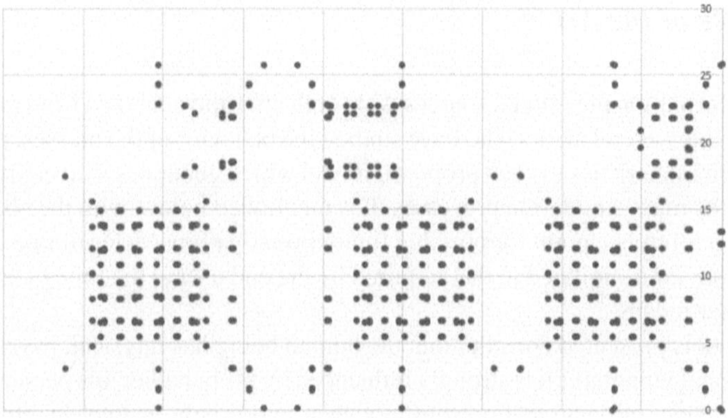

Fig. 1 Trajectory simulation

is provoke outside the environment causing the mass reaction [7]. Currently, space–time predictions, reactive behaviors, and direction movements have been combined on platforms dedicated to crowd simulation [8].

For this work, Menge simulation software is used where the agents' trajectories involved on stage are specified, placing them within specific mass and weight characteristics; and also considering velocity of slow or fast movements. Within the simulation of the scenario, the agents must find their way through the various obstacles that appear in the symbolic scenery of the Plaza de la Mexicanidad in Juarez City. Although, the agents must consider other agents in their path and avoid a collision (Fig. 1). In this work the hierarchical structures were considered, and the agents were placed in certain areas near the main stage [9], these by verifying their adjacent cells in search of their neighbors [10].

Using the Voronoi diagrams, calculations can be made by means of distance maps for the surfaces of each agent, coding the distances with the nearest neighbors [11], to determine the approximate ones with other individuals, obtaining the most optimal alternate evacuation routes.

In this simulation, an outdoor concert is represented, so most of the agents involved are close to the main stage. It is also considered the stimuli that the agents can receive, weather is known or new; in the same way, the agents' behavior can be altered in two ways, to act normal or to act in panic. In panic situations people tend to move faster and sharper and shorten the distances between individuals causing collapse between them. So, determining the agents' routes, factors and specifications is essential to the system's success or failure since their behavior triggers different decision-making and actions [4]. Without neglecting the existing obstacles in the simulation environment that reflect the behavior of new alternative evacuation routes of the masses.

3.1 State of the Art

For this information the official sources were collected and analyzed from the state of the art, finding several actors who have studied the behavior of the masses, providing symbolic characteristics to each proposed model which generates a diversity of alternatives in the mass simulation processes. It is mentioned below, with the objective of serving as a reference point for possible future investigations, determining the scope of each behavior a similated to that required by the end user at the time of simulation by computer means.

The models presented consider that the human being has physical, psychological and emotional vulnerabilities strongly influenced by his behavior, his personality and mood to reduce, mitigate and prevent in a clear and timely manner in situations of accident causing the eviction of the place (Table 1).

Table 1 State of the art models and features [2]

Articles	Model	Features
Stephen Chenney [12].	Representation and design of velocity fields using cellular automata	Agent flow movements
Shuai Zhang, Manchun Li, Feixue Li, Aili Liu, and Dong Cai [13].	Model in which the cells of a cellular automaton represent discrete positions in space	Allows agents to move from one cell to another
Hubert Klupfel [14].	Cellular models in the simulation of large events	Slots are implemented where agents can move through them
Adrien Treuille, Seth Cooper, and Zoran Popovi´c [15].	Model continuum crowds	Real time simulations, uniform movement and emerging behaviors are achieved
Yanbin Wang, Rohit Dubey, Nadia Magnenat-Thalmann, and Daniel Thalmann [16].	Interactive navigation for crowds	Allows a user to select a avatar from the crowd and control your movement through the virtual environment
Dirk Helbing and P´eter Moln´ar [17].	Social forces model	The psychological and mental process to decide Pedestrian speed is modified to reflect the decisions made during that stage
Olfa Beltaief, Sameh El Hadouaj, and Khaled Ghedira [18].	Psychological factors of perception and speed and objectives and preferences	Agents' behavior in three factors: interaction with the environment, collision avoidance interacting with each other and finally they are aware of their trajectory

(continued)

Table 1 (continued)

Articles	Model	Features
Nuria Pelechano, Kevin O'Brien, Barry Silverman, and Norman I. Badler [19].	Communication model and roles for agents	The following factors are considered: communication, stress, emotions, personality, and decisions

3.2 *Mathematical Model*

People are constantly being manage in groups without them realizing it. This can be observed in the main avenues, shopping centers, parks, among others. Therefore, more than 50% of the people, in small or large groups, move constantly within a direct or surrounding approach. The mathematical model below is presented by De Gyves from [20], which includes social forces to keep the group together:

$$f_i^{group} = f_i^{vis} + f_i^{atr} + f_i^{rep} \quad (1)$$

where:

f_i^{group} Represents the response of agent i to other group members.
f_i^{vis} It is the force that keeps the group members in the field of vision.
f_i^{atr} It is a term that directs agents to the center of the group.
f_i^{rep} It is a force of repulsion to keep group members apart.

Individuals in a crowd behave differently when they are alone or when they are part of a group. Within a multitude phenomenon such as polarization can occur, which occurs when two or more groups adopt divergent attitudes, opinions, or behaviors. In that case it is quite likely that the behavior among the groups of the crowd will result in confrontations if an agreement is not reached [21].

Being part of a crowd can be uncomfortable, but it can be even lethal in a possible accident.

3.3 *Simulated Case Studies and Comparison with Reality*

The simulation is the representation of the Plaza de la Mexicanidad maximum event of 10,000 people, in front of a concert. Menge was used to carry out the simulation process during the evacuation. To determine the behavior of the crowds in the eviction process using computer simulators.

This software allows you to configure the scenarios based on the number of agents involved, scenarios and obstacles. In Menge you can configure specific characteristics such as mass, weight, and speed of the simulated agents. Marking different assimilated behaviors based on their specific characteristics previously established in the program. All agents aim to reach the exit, traveling the most optimal route, minimizing the travel time, according to their physical and attitudinal response possibilities. The agglomerations that agents face puts them in a situation of waiting for the transfer process, which in this simulation we have in doors one and two corresponding to the areas that the agents selected as the nearest exit. Calculating the routes based on the specifications of the programmers, avoiding coalitions between those involved and the existing obstacles, calling it an evacuation plan.

Agents have a decision behavior like that of real people, considering their environment and objective. The environment or scenario involves both agents and obstacles within the virtual representation. The agents have a specified position at the beginning of the simulation (Fig. 2) that places them in the Cartesian plane in the dimensions of "x" and "y" establishing the speed and the structure parameters, such configuration can be altered to Programmer's disposition for a better simulation of the environment.

In this study, the scenario is made up of black rectangles that simulate the obstacles such as setting, stall selling food, or some merchandise and the points arranged by sections represent the agents involved. We can consider the agents as "i" to the wall as "w" and the obstacles as "x".

The simulation shows in Fig. 3 how the agents begin to move looking for the most optimal output.

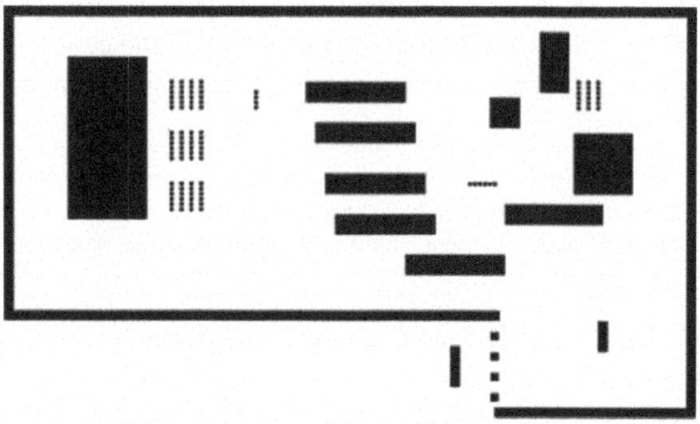

Fig. 2 Initial representation of agents, scenarios, and obstacles

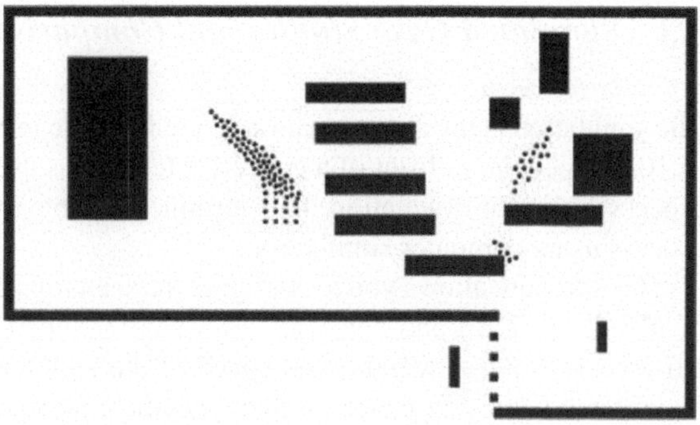

Fig. 3 Selection of the most optimal route

In Figs. 4 and 5 the agents present the speed difference according to their physical characteristics, the interesting thing is to see how regardless of the location on the stage, all the agents look for the same exit route, towards the nearest door.

Figure 6 shows us how the agents move around the stage, avoiding obstacles and forming a new tendency of movement assimilated and integrated by them.

In Fig. 7, the fastest agent passes through the first door, followed as shown in Fig. 8, of the other agents.

Figures 9 and 10 shows us how the agents are looking for a new exit, taking exit two as the access route.

However, it can be seen in Figs. 11, and 12 the beginning of the agglomeration of the masses, wanting to evict the place, it is here where situations that endanger the security of the agents can occur.

Figures 13 and 14 show us how to the last irrigation situations.

Finally, to Fig. 15, it shows the last agents leaving the place, however it is worth mentioning that these agents have characteristics of higher weight and lower speed, so it can be considered as the most vulnerable agents.

There is a considerable discrepancy in the amount of time needed for agents with more agile characteristics against agents with specifications that tend to limit their reaction capabilities.

Fig. 4 Agent displacement

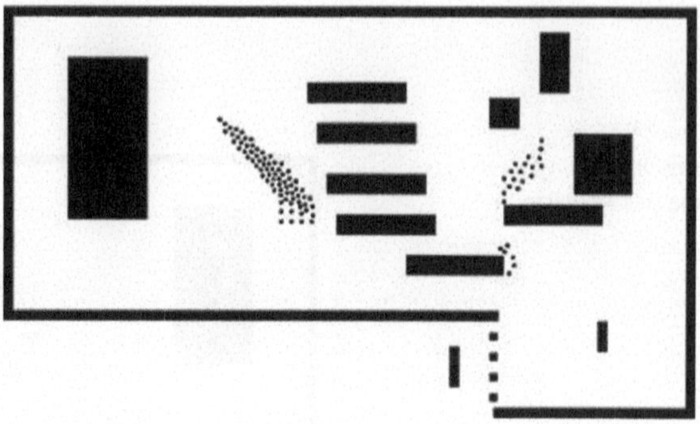

Fig. 5 Rearrangement of agents on the evacuation route

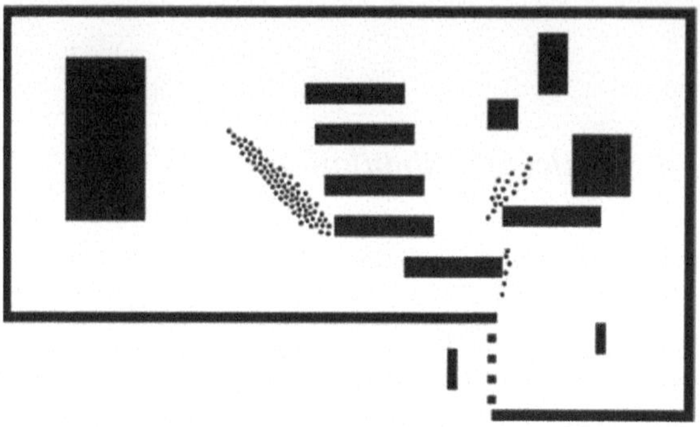

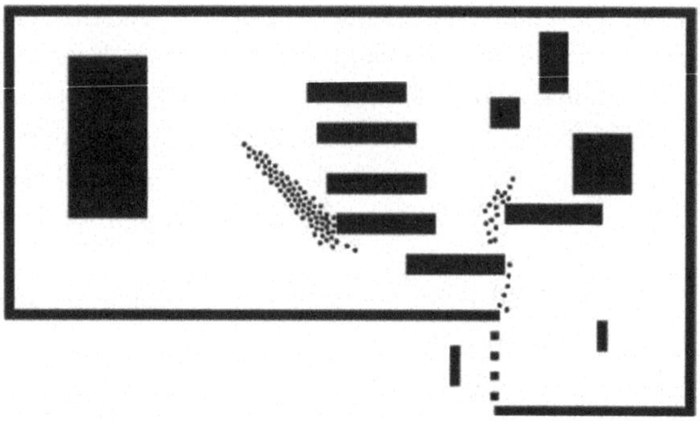

Fig. 6 Evasion of first obstacles

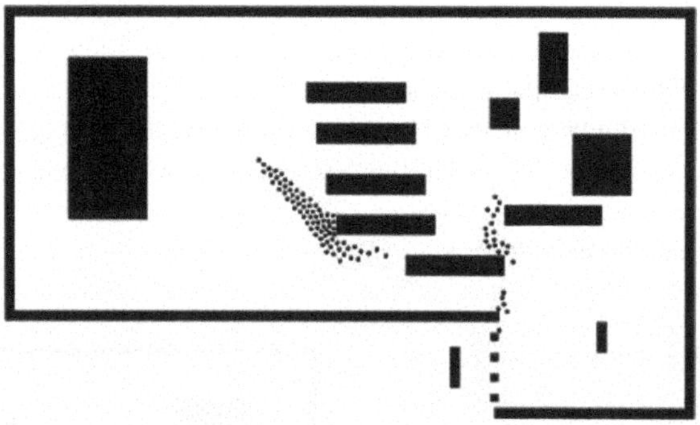

Fig. 7 Selection of the first exit door

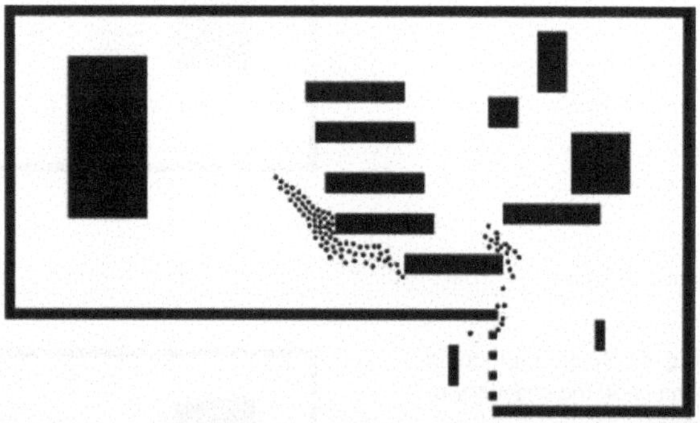

Fig. 8 First agent to leave the risk scenario

3.4 Trajectory Simulation

Figure 16 shows a cellular automaton model, simulating the evacuation and flow of agents during the incident in the Plaza de la Mexicanidad. Each box with arrow pointing indicates the position each agent can have at a given time, the black boxes indicate the obstacles that the agents have in their environment. The agents are autonomous and could navigate avoiding collisions with static or dynamic obstacles,

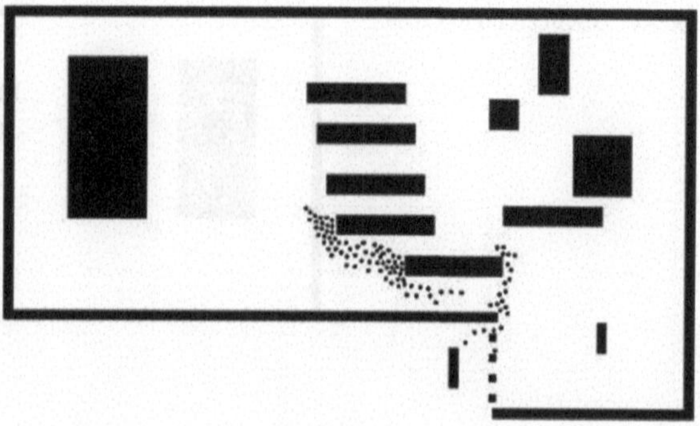

Fig. 9 First group of people to find the nearest exit door

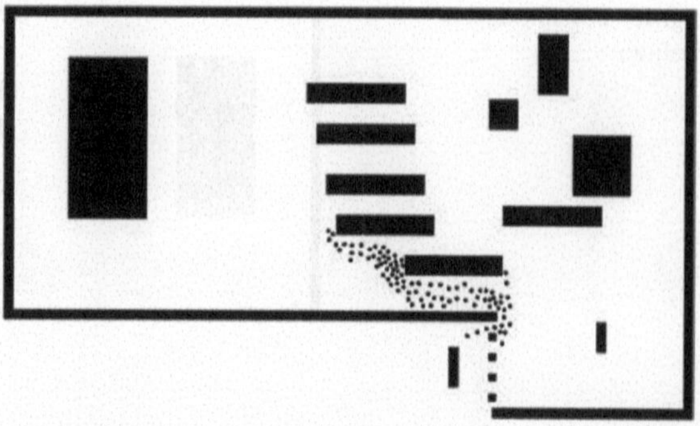

Fig. 10 First agglomeration at door one. Agents select door two

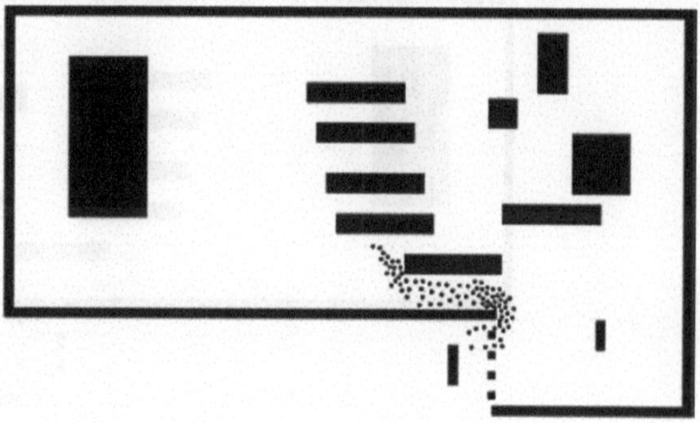

Fig. 11 Door agglomeration two

moving when the cells are free [11]. The cells can be occupied by different people at different times; however, the flow is the same with any member, determined by their escape decision and behaviors learned before.

The blue lines show us the best route estimating the time required for eviction, with the main objective of minimizing the number of victims during the displacement, as well as the evacuation time. It is considered an orderly behavior of the crowd, which minimizes evacuation times.

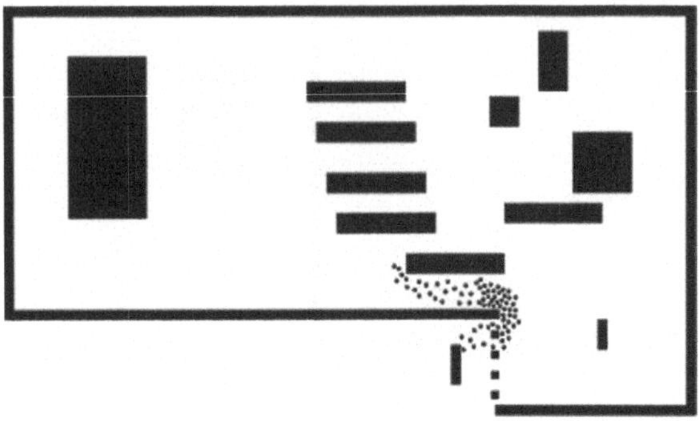

Fig. 12 Agents tend to cross

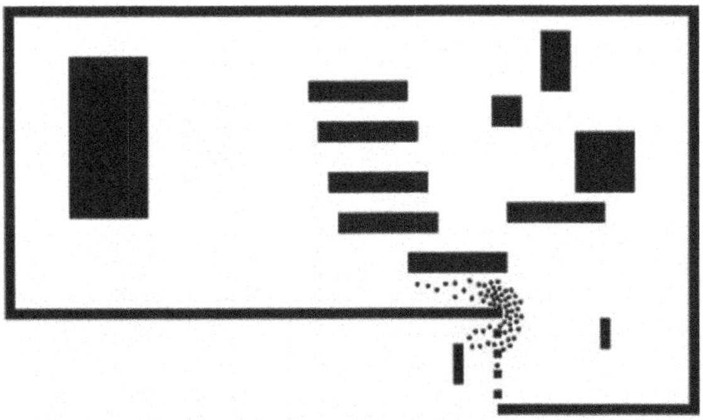

Fig. 13 Possible agents coalition

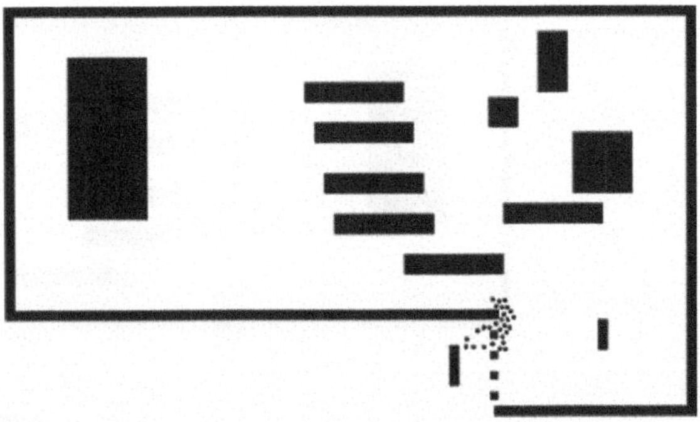

Fig. 14 Agents with greater weight and lower speed takes longer to leave

Agents can present two types of behavior: High level behavior: Agent decision making based on communication and learning [22] and Low-level movement: Based on the perception to move in a defined area avoiding collisions.

This leads to the factors that determine the processes to estimate the distribution of trained service personnel required in the affected area, to help the most vulnerable or at-risk people.

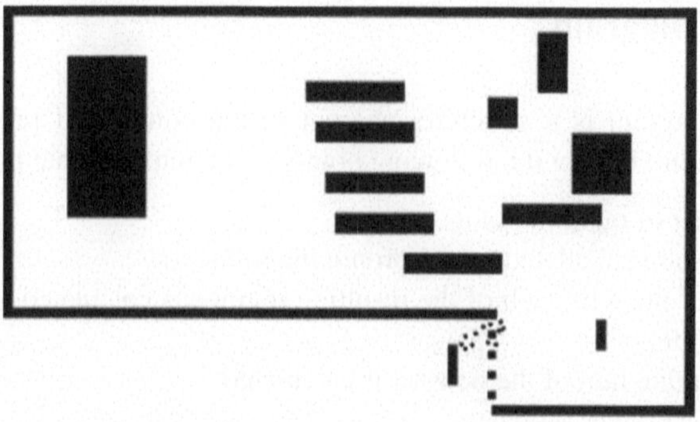

Fig. 15 Last agents to reach the goal

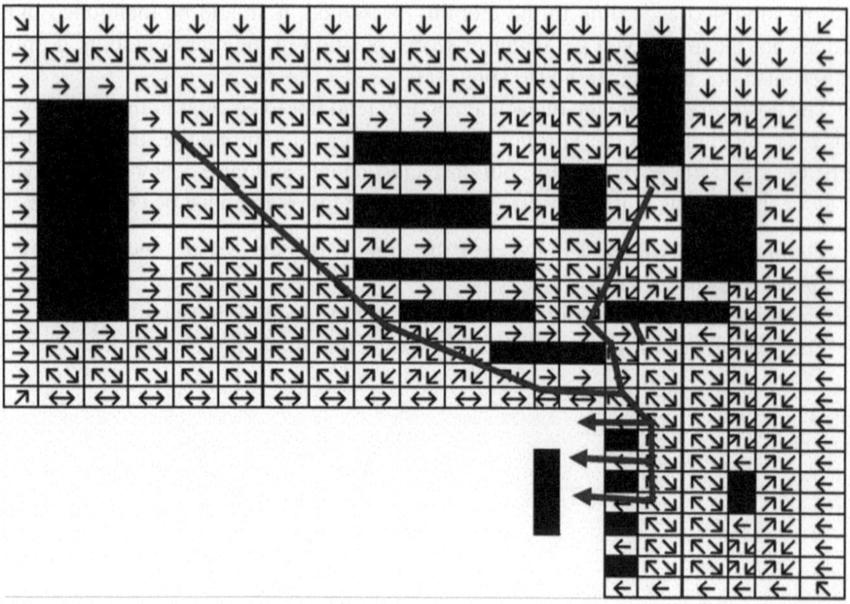

Fig. 16 Trajectory simulation

All this to safeguard the safety of people considering the generation of culture of prevention at spaces where the population acquires knowledge to form informed and aware citizens. *"Educate the population with planning to generate a culture of prevention"*.

4 Voronoi Diagram

The Voronoi diagram is a structure inherent to the concept of proximity and/or influence on nature having the following points as a computational procedure:

- Generate a set of random points.
- These points are calculated your Voronoi diagram.
- The center of mass of each of the resulting regions is calculated (this gives us a new set of points).
- The Voronoi diagram of the new set is calculated.

In the simulation the points of the model would be reflected for each agent, where each of them is assigned its Voronoi region that is formed by the coordinates of the location corresponding to each agent located on the stage. As they move the diagram will be modified according to the mobility presented, it should be mentioned that it does not mean that you get better or worse position in the representation, it only represents your new location [19] (Fig. 17).

The positional advantage of each agent may tend not to be clear to the naked eye, however, by capturing the Voronoi diagram, you can determine the space to which it belongs and the proximity of both obstacles, agents and emergency exits. Flavio et al. [23] mentions that to validate the model, the following must be considered:

- The numerical value averaged over the time of density, speed, and the flow of agents in each scenario.
- The relationship through the time of density and speed of the agent.
- The space–time profiles of the density, speed, and flow of agents.

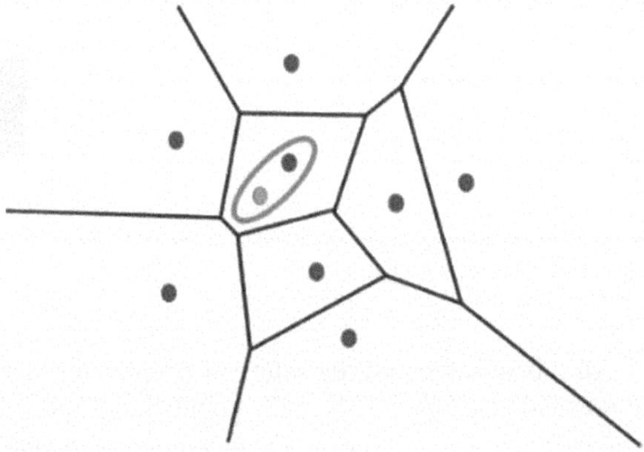

Fig. 17 Voronoi diagram [22]

5 Future Research and Works

Consolidating an education in prevention culture within the field of natural disasters, accidents and risk situations in places where crowds of people assist, can reduce the risks of discrepancies and threats that endanger human lives. Encouraging the values and attitudes towards providing efficient tools for the population's well-being, taking into account a risk management education focused on emergencies' prevention.

You can simulate new scenarios, with a greater number of agents involved, with different characteristics, and locate them in more distant areas, increasing the area of the door two and three to visualize if the displacement of the masses presents a greater mobility and also validate it with the Voronoi model.

6 Conclusions and Future Challenges

In conclusion, the behavior of the masses seems to move by disorderly and irrational impulses, however, the behavior of the multitudes responds to certain patterns, both psychological, emotional and attitudinal, not to mention previous events, which resemble the current one. Conditioning the mobility of the masses in an understandable, adaptable, and even possibly predictable way.

There are currently a large number of models that determine behaviors, patterns and routes with different impact factors that have been investigated and analyzed to achieve a better simulation of reality in an event that involves large numbers of people in a specific place, developing strategies that help minimize risks in catastrophic situations or natural disasters. For example, to develop mass simulations where the agents have a reactive behavior towards the other agents, when a similar pattern between the behavior of the agents is presented, a new binding group can be generated. Same that can tend to dominate the other groups or to follow.

Determining the interaction of different groups within the crowd, considering their behavior, emotions and aptitudes is essential in the investigation of mass simulation.

The simulation of the Plaza de la Mexicanidad, the behavior pattern of the agents that use the first two exit doors can be observed as the busiest option, which can cause bottlenecks. Thanks to this simulation, you can conclude that the first two doors would be essential to extend your space to be able to evict the crowd in a faster and more efficient way, since these are the busiest.

References

1. Wang, X., Zhang, J., Scalia, M.: Parallel motion simulation of large-scale real-time crowd in a hierarchical environmental model. Math. Probl. Eng. **1–15**, 2012 (2012)
2. De Gyves, O., Toledo, L., Rudom, I.: Comportamientos en simulaciόn de multitudes: revisiόn del estado del arte

3. Li, N., Zhang, Z.: Abnormal crowd behavior detection using topological methods. In: 2011 12th ACIS International Conference on Software Engineering, Artificial Intelligence, Networking and Parallel/Distributed Computing, pp. 13–18. IEEE, July 2011
4. Groenewegen, S.: Improving crowd behavior for games and virtual worlds. In: Proceedings of the Fifth International Conference on the Foundations of Digital Games—FDG '10, pp. 256–258, New York, New York, USA (2010)
5. Lemercier, S., Jelic, A., Kulpa, R., Hua, J., Fehrenbach, J., Degond, P., Appert-Rolland, C., Donikian, S., Pettr´e, J.: Realistic following behaviors for crowd simulation. Comput. Graph. Forum **31**(2pt2), 489–498 (2012)
6. Loera, A., Garcia, H., Rodríguez, J.: Optimization of shelter points and evacuation routes on a university campus, a case study for smart cities considering humanitarian logistics in case of flooding (2019)
7. Zezzatti, A.O., Contreras-Massé, R., Mejía, J.: Innovative data visualization of collisions in a human stampede occurred in a religious event using multiagent systems. In: 2019 23rd International Conference in Information Visualization–Part II, pp. 62–67. IEEE (2019). Sun, L., Qin, W.: Simulation of crowd behaviors based on event reaction. In: 2011 IEEE International Conference on Computer Science and Automation Engineering, pp. 163–167. IEEE, June 2011
8. Hadap, S., Eberle, D., Volino, P., Lin, M.C., Redon, S., Ericson, C.: Collision detection and proximity queries. In: Proceedings of the Conference on SIGGRAPH 2004 Course Notes—GRAPH '04, pp. 15–es, New York, New York, USA (2004)
9. Passos, E.B., Joselli, M., Zamith, M., Clua, E.W.G., Montenegro, A., Conci, A., Feijo, B.: A bidimensional data structure and spatial optimization for supermassive crowd simulation on GPU. Comput. Entertain. **7**(4), 1 (2009)
10. Sud, A., Govindaraju, N., Gayle, R., Andersen, E., Manocha, D.: Surface distance maps. In: Proceedings of Graphics Interface 2007 on—GI '07, New York, New York, USA, p. 35 (2007)
11. Pelechano, N., Badler, N.: Modeling crowd and trained leader behavior during building evacuation. IEEE Comput. Graph. Appl. **26**(6), 80–86 (2006)
12. Chenney, S. Flow tiles. In Proceedings of the 2004 ACM SIG-GRAPH/Eurographics symposium on computer animation—SCA'04, pp. 233–242, New York, USA, ACM Press (2004)
13. Zhang, S., Li, M., Li, F., Liu, A., Cai D.: A s141imulation model of pedestrian flow based on geographical cellular automata. In 2011 19th Int. Conf. Geoinformatics. **2**, 1–5. IEEE (2011)
14. Klupfel, H.: The simulation of crowds at very large events. In Traffic and Granular Flow'05, pp 341–346. Springer Berlin Heidelberg (2007)
15. Treuille, A., Cooper, S., Popovic, Z.: Continuum crowds. ACM Trans. Graph. **25**(3), 1160 (2006)
16. Wang, Y., Dubey, R., Magnenat-Thalmann N., Thalmann D.: An immersive multi-agent system for interactive applications. Vis. Comput. **29**(5), 323–332 (2012)
17. Helbing, D., Molnar, P.: Social force model for pedestrian dynamics. Phys. Rev. E, **51**(5), 4282–4286 (1995)
18. Beltaief, O., El Hadouaj, S., Ghedira K.: Multi-agent simulation model of pedestrian's crowd based on psychological theories. In 2011 4th Int. Conf. Logistics, 150–156. IEEE (2011)
19. Pelechano Gómez, N., O'Brien, K., Silverman, B.G., Badler, N.: Crowd simulation incorporating agent psychological models, roles and communication. First Int. Workshop Crowd Simul (2005)
20. De Gyves, O., Toledo, L., Rudomín I.: Comportamientos en simulaci´on de multitudes: revisi´on del estado del arte. ResearchGate, Enero (2013)
21. Loscos, C., Marchal, D. and Meyer, A.: Intuitive crowd behavior in dense urban environments using local laws. IEEE Theory and Pract. Comput. Graph. **122**, (2003)
22. Vogel, I., Van Der Ploeg, C.P., Brug, J., Raat, H.: Music venues and hearing loss: Opportunities for and barriers to improving environmental conditions. Int. J. Audiol. **48**(8), 531–536 (2009)
23. Mendes, F.S., Aude, J.S., Pinto P.C.V.: Construction of the voronoi diagram by a team of cooperative robots. (2004)

Automatic Fall Detection for the Care of Older Adults in Smart Cities

Sara Judith Ríos Dueñas, Jose Mejia, Alberto Ochoa, Jose Díaz, Lidia Rascon, Nelly Gordillo, and Eddy Sánchez-DelaCruz

Abstract As the number of elderly people increases, it is a necessity for smart cities to take care of elder special needs. As people age, the likelihood of accidents increases because of their motor skills decrease over time, this risk is not only latent when they live alone but also exists within the nursing homes. Because of this, constant care for older adults is a necessity for smart cities, this may not always be possible due to the lack of family members who can care for the elder or in the case of the nursing homes, the staff may not be enough to care for all adults. This leaves the need for systems that can constantly monitor older adults and respond and alert automatically in the event of accidents. In order to find a means to improve the quality of life of older adults who may suffer accidents, in this chapter it is presented an algorithm based on neural networks for the automatic fall detection of older adults in nursing homes. The implemented model was trained and tested on a database of video images containing fall situations.

Keywords CNNs architecture · Fall detection · Smart city

1 Introduction

One of the smart cities aims, is to provide more flexible, efficient, and sustainable services to its citizens. These services include (Mohanti et al. 2016) [24] smart infrastructure, smart transportation, smart energy, and smart healthcare. In this chapter we will focus on this last aspect of smart cities.

As life expectancy has increased the number of elderly people has also increased, this creates a necessity for smart cities to take care of elder special needs [8]. Older people are more prone to suffer accidents or put themselves in risk situations because

S. J. R. Dueñas · J. Mejia (✉) · A. Ochoa · J. Díaz · L. Rascon · N. Gordillo
Universidad Autónoma de Ciudad Juárez, Av. Hermanos Escobar, Ciudad Juárez, Mexico
e-mail: jose.mejia@uacj.mx

E. Sánchez-DelaCruz
Instituto Tecnológico Superior de Misantla, Veracruz, Mexico

A. Ochoa-Zezzatti et al. (eds.), *Technological and Industrial Applications Associated With Industry 4.0*, Studies in Systems, Decision and Control 347,
https://doi.org/10.1007/978-3-030-68663-5_4

their motor skills decrease over time, thus they needed a constant attention [5, 25, 26]. The risk is not only latent when they live alone but also exists within nursing homes. Older adults who are residents of the nursing homes may also suffer falls, among other accidents, or put themselves in risk situations, because they cannot be constantly taken care of. The absence of staff could be because they can be found attending to someone else patient, thus neglecting the other residents, or because insufficient personnel, therefore, it is important that those in charge of these residences improve their accident prevention and care system [7, 27].

These problems prompted for the incorporation of algorithms that can carry out automatic fall detection in older adults in real time, by means of portable devices. In recent times several proposals have opted for the use of neural networks to track people [1]. For example, structures were proposed to achieve the detection of falls using an Elman's neural network, that monitor older adults using portable devices [2]. Similarly, in [6] a system was designed and implemented for the fall detection in indoor spaces using WiFi devices. The main contribution is that it has greater comfort for users since they do not have to be charged with a device, which allows them to perform their daily activities with greater comfort and safety, in addition to that it was implemented in the design a base signal with greater sensitivity for the detection of activity, which allows to have a greater segmentation for the fall detection or activities similar to these.

Here, the design of an architecture, based on convolutional neural networks, is proposed for the detection of falls in older adults.

2 Theory

In this section is carry out a revision of the subjects of machine learning and specifically the topic of deep learning architectures. Additionally, a revision of the basic concepts of video processing is done.

2.1 Machine Learning

In the area of computer science, various methods have been created for the purpose of facilitate complex processes carried out by human beings. This area is called machine learning (ML), which is a set of techniques for training machines that are capable of learn to solve several problems. In Fig. 1, it is shown a graphic representation of the basic components of ML [3, 9].

Deep learning is a branch of the ML that has been developed for the design of algorithms based on artificial neural networks (ANN), in this branch different architectures have been de-signed that resemble biological ANN, thus creating a computational counterpart focused on the resolution of all kinds of problems, besides that this allows massive handling of information.

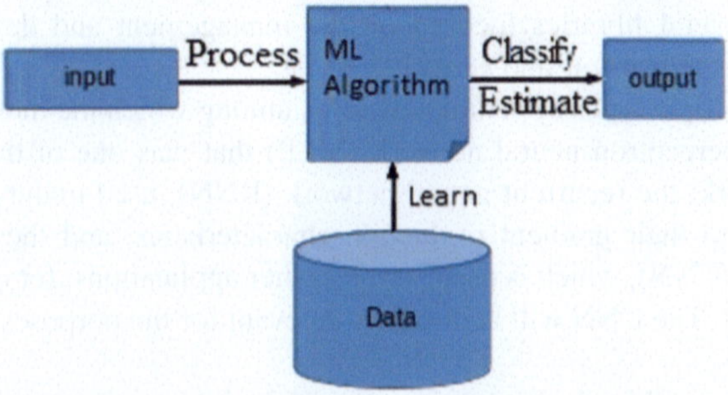

Fig. 1 Graphic representation of the basic components of ML

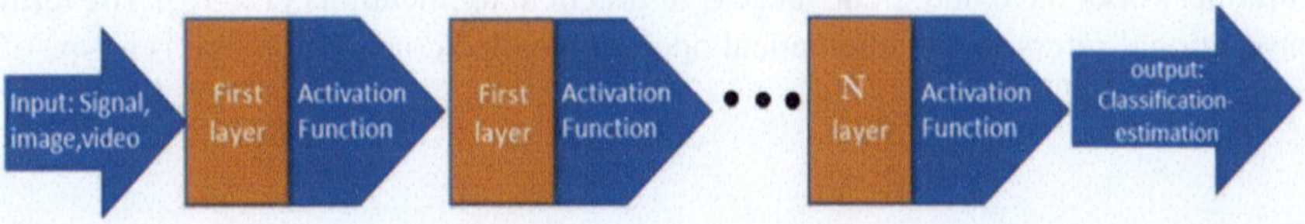

Fig. 2 Basic components of deep learning architectures, a "N" layers network

In deep learning, the word architecture refers to structures that have been designed in which neurons are connected within it, these structures also determine the behaviour of the network for carry out different tasks [9]. The general structures of the deep learning architectures are the layers, activations functions, and training parameters (Fig. 2).

2.2 *Layers*

The grouping of neurons is referred to as a layer and the number of layers used will determine the degree of an architecture [9, 10].

In [10] it is mentioned that the basic neural network structure must have three basic elements based on layers, which are: input layer, in which all the information from the external environment is received; the hidden layer, in which the information is processed and has no contact with the outside environment; and the output layer, which delivers a response from the neural network. The connections weighs of neurons within a network are changed during the learning stage to adapt to the problem at hand [10].

In [9] it is mentioned that the design of different neural networks architectures is important since the complexity of certain problems can vary. Neural networks are compared to Lego blocks since with these you can build almost any structure you want, like-wise with neural networks. The way in which a neural network is designed will depend on the skill and mastery of the designer.

There are several libraries focused on the management and design of neural networks such as *tensorflow* and *keras* [9].

There are multiple types of neural networks, among which the most essential are the multilayer perceptron neural network (MLP) that was one of the first models of neural network; the recurrent neural network (RNN), used mainly for language modelling due to their gradient re-duction characteristics; and the convolutional neural network (CNN), which is used, among other applications, for object tracking in videos [9, 11]. The CNN will be the most relevant for the porposes of the present chapter.

The CCNs are used for the processing of topology information in grid form, examples of these are time series data such as 1D cells, image data, such as 2D pixel cells, image classification, tracking and recognition of objects, the use of convolutional neural networks have had great success in practical applications [12, 13]. The term convolutional refers to a mathematical operation called convolution that is a type of linear operation [12].

3 Proposed Methodology

An architecture based on a convolutional neural network, consists of two main stages, the convolution stage and a second stage called pooling or reduction, which is used in most convolutional neural networks [12, 13].

The convolution function is an operation of the arguments of two functions, in the case of neural networks, the first argument would be the input layer (input) and the second argument a filter (kernel) between which they are applied. In the convolution operation, one of the advantages of this method is that the filter is used to obtain the same characteristics throughout the entire layer and this is of great help since it allows the reduction of training parameters and neural connections [12–14].

In [12] it is mentioned that convolution can help improve a neural network architecture by reducing interactions or reducing weights since the kernel is smaller than the input.

The pooling function consists of a reduction in size of the input matrix, this is done by means of a statistical extraction like the maximum of the blocks of the input matrix, or the average, or the minimum, among others. It is important mention that just as this technique has its advantages in reduction it also contains disadvantages such as loss of precision [14].

A typical convolutional layer is made up of several parallel convolutions that produce linear outputs, these are fed to a nonlinear input function, that optionally is followed by the application of the pooling function to modify the output in a layer [12].

3.1 *Activation Functions*

In [15] it is mentioned that the biological counterpart of artificial neural networks has states, when they are active or vice versa, likewise artificial ones have this characteristic. In the case of artificial neural networks those states are generated by "activation functions", these functions express when the state in which the architecture neuron is working is active or inactive. These functions can be linear or nonlinear such as the functions sigmoid or hyperbolic tangent. Linear: in the linear activation function, the output is a linear function of the input, while in the case of nonlinear functions, several types are used for example: the Sigmoid activation function where the output is mapped within the range of 0 to 1, so this affects the activation function on its slope [15]; The Hyperbolic tangent activation function where the output is within a range from −1 to 1 [15].

3.2 *Training Parameters of a Neural Network*

Neural networks have a great capacity for learning and to achieve their proper functioning must go through a stage called training patterns. The objective of the training is the adjustment of the network with the data and the proper development of the network to make it capable of performing a task [16, 17].

Training can be classified into two types, supervised training, unsupervised training.

- Unsupervised learning: In this type of training the weights of the network must be adjusted according to the correlation that exists in the input data, in this type of network training data is entered at the input, but it is not known what the output will be [17].
- Supervised learning: in this type of training the neural network is presented with certain data at the entrance, for training. This mode of training the network has the purpose of adjusting weights so that the output generated by the neural network is similar enough to the input. Supervised training in a few words performs the function of network supervision [16].

3.3 *Tools Used for the Development of ANN Architectures*

There exist several tools facilitating the programming of architectures (see Table 1), in the next section we describe some of the most common.

Pyhton Programming Language

In [21] it is mentioned that Python is a program for the design of algorithms and is a "high-level, interpreted and multipurpose programming language". This program has

Table 1 Available software to implement ANN

Software	Open source	Link
Computational	Yes	cntk.ai
Network toolkit		
Deeplearning4j	Yes	deeplearning4j.org
Caffe	Yes	caffe.berkeleyvision.org
Theano	Yes	deeplearning.net/software/theano
Torch	Yes	torch.ch
Python	Yes	python.org
TensoFlow/Keras	Yes	tensorflow.org
WEKA	Yes	cs.waikato.ac.nz/ml/weka/

great versatility because it can be operated on various operating systems or platforms. Python has several libraries, for the development of different types of algorithms, so it is important to have them to start a project as this will facilitate the design. This program has special libraries for the design of neural networks, for example: tensor flow; pytorch; and keras, which allow the management of highly complex architectures. In the following subtopics, some of the libraries used for the design of neural networks will be discussed.

Tensor Flow
It is a platform designed to work and design ML architectures, since it is an ecosystem with varied libraries and tools that facilitate the design of this type of algorithms. This platform is focused on solving real-world problems by applying deep learning [22].

Keras
It is a Python library designed for the management of neural networks, in addition that its main focus is to speed up the process of developing an algorithm by reducing the time to design and allows working on *tensorflow*. It is a highly friendly library for the design of neural networks and allows the management and creation of CNN and RNN type networks, including the combination of both. *Keras* has a special library for the design of convolutional neural networks called "keras api functional" [23].

3.4 Video Processing

There are several video formats and each one has different characteristics; it is important to know this information since if you want to work with video in neural networks you must have a library that has the ability to read the format that is being used. Below is a list of some common video formats:

- AVI: Audio Video Interleave, one of the advantages of this format is that it can be read by almost all decoders, however, the weight of the videos is high, occupying more memory space compared to other formats.
- WMV: this type of digital video format was created by the Microsoft company.
- FLV: this format was developed by the adobe flash player company and is of high quality, one of the most important advantages is that it is accepted by almost all operating systems.
- MOV: format developed by the Apple company, audio and video; this format is exclusive for devices belonging to the company

3.5 Action Recognition Through Video Sequences

The line of investigation of the recognition of actions in humans has focused on the processing of signals and images, in recent times the analysis of video images has been incorporated for the recognition of actions for the development and design of strategies to analyse images of video, in order to be able to detect threats or risk situations through video surveillance using cameras [18].

Strategies have been investigated that allow the automatic recognition of activities and patterns in which there is movement in order to track objects or people, for the detection or analysis of events [18]. The techniques frequently used for recognition of actions are, Fisher vectors, Hof system, neural networks and dense trajectories, among others [19].

The recognition of actions can be carried out in different ways, either by robot vision or a human computer interaction. Most of the studies that recognize actions, use video recordings in 2D cameras, even though, human actions are generally performed in a 3D space, this is mainly due because 2D video recording is pervasive, in this chapter we focus on this type of input (2D), since is the most commonly used in nursing homes.

For the analysis of actions, the human body must be recognized, the human body could be considered as a system in which limb and trunk movements are combined. Some methods that perform the recognition used the joints and trunk of the skeleton, such as the work of [6] which use Temporary Pyramids to organize and analyse joint data.

The recognition of actions has become a highly active, and most of the new developments are done using ANN, however, these developments have not had the same progress in recognition as the methods based on more classic methods. Part of the use of neural networks not having the same progress is because in some cases the databases used are too small for network train-ing or have noise [6].

The use of neural networks for the recognition of actions has had a greater challenge compared to the simple classification of images due to the variations of movement in the video, in addition to the need for a greater number of examples in the training of the network to extract as much information as possible and classify the actions. On most architectures, this has been solved by applying an architecture of two channels, one temporary and one spatial to separately train the networks in which the actions and appearances are separated, that is, the area where the action is taking place. In the temporal channel the network is classifying everything that entails the action, while in the spatial channel, the space where the action is performed is observed [19].

4 Methods

In this section, the scheme of the procedure for the development of a convolutional architecture for the aim of detect falls will be presented. The algorithm was trained with video images for the detection of falls of older adults.

The database was obtained from *Lei2- Laboratoire Electronique*, Informatique et image UMR CNRS 6306 Fall detection Dataset [20, 28]. This database shows different situations of daily life where accidents happen. It was designed for the prevention of reports of injuries, falls of older adults which are of the main causes that considerably reduce mobility and independence. The video is recorded at 25 frames/s and each with the resolution of 320 × 240 pixels, with different lengths of each video. The database consists of 191 videos in *Audio Video Interleave* (.avi) format. Also, note that for the making of this dataset, "videos are anonymous, and all those visible in the videos signed an authorization of diffusion and use of these videos for research purpose only" [20].

4.1 Fall and Non-fall Labelling

In this stage, the classification of the frames of each video was done manually. Each video was observed frame by frame until it is found the moment of the fall, then the video was di-vided into two parts (fall/no fall).

It is important to mention that in the videos provided by the database, after the fall, the actors who starred in the fall do not rise again, therefore, it is considered to be within the fall stage In Fig. 3, it is show the two parts (fall/no fall) of the same video.

4.2 Pre-processing

A maximum of 50 frames were set per video, this to reduce the computational cost and reduce training times, there were 206 videos of which only those that were more than 50 frames in length were selected Fig. 4.

Videos that had more than 50 frames were cut to have only 50 frames long. The next pre-processing step consisted in a resize of each frame. A 70 × 70 size change

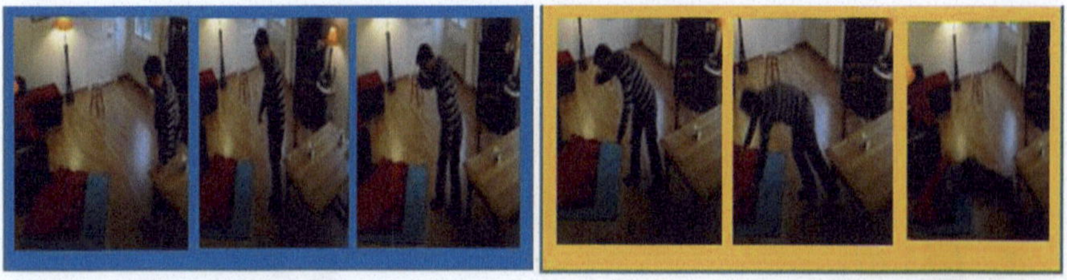

Fig. 3 Video classification into fall and not fall, the left image is a video segment with not fall, and the image on the right is a fall, images taken from videos in the dataset [20, 28]

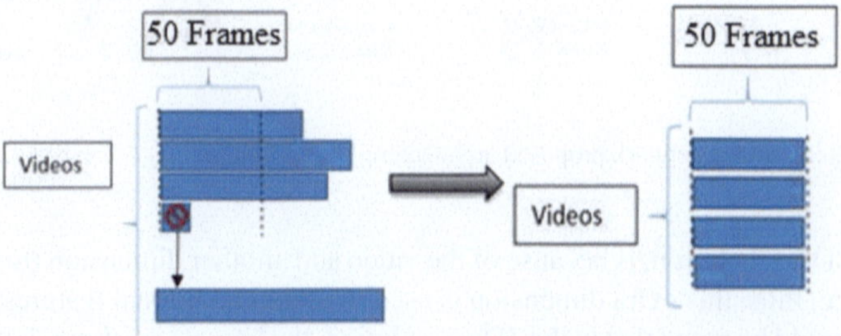

Fig. 4 From the database all videos were cut to 50 frames and videos with less than 50 frames were discarded

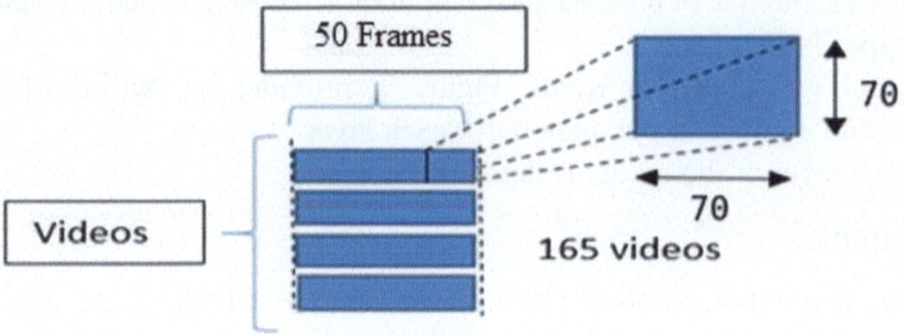

Fig. 5 Each video frame was reduced to a size of 70 × 70 pixels

was made to each frame of the videos (Fig. 5) this also was done to reduce the computational cost.

4.3 Architecture

A model was designed for the detection of falls, it was trained with video images from the database. The model has several layers which are composed of 3D convolutions, pooling stages, and dense layers. The use of 3D convolution instead of the more

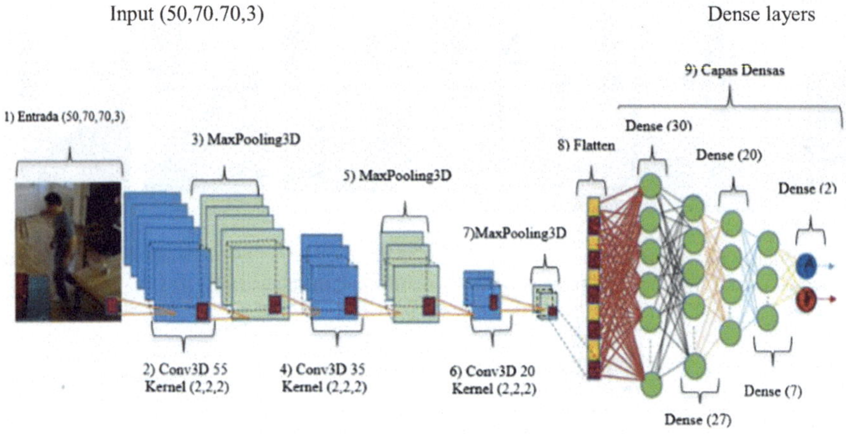

Fig. 6 Artificial neural network proposed architecture

common 2D convolution, is because of the video add another dimension (time) to the input image, thus, this extra dimension is used to obtain additional features from the video. It should be noted that in the 3D convolution, the kernels are three-dimensional structures. Each layer of the proposed architecture performs an important function for proper functioning of the model. The convolutional layers in the architecture, extract features in the video that help to classify the video, max-pooling layers are used to decrease the size of the data and computational cost, and dense layers classify the video type (fall/no fall).

The model was coded in Keras. Figure 7, provides an outline of the input dimensions and number of parameters for each layer.

Next, each layer of the model is described

Stage 1: Input

- The input has a dimension of [50 × 70 × 70 × 3] which means that the input consists of 50 frames of a size of [70 × 70] pixels × 3 channels corresponding to the RBG colors.

Stage 2: First convolutional layer (Conv3D_1)

- It is composed of 55 3D filters/kernels with each kernel of size [2 × 2 × 2].
- The output feature maps have dimension of [49 × 69 × 69, 55] obtained, which means that 55 layers of 49 frames with a size of [69 × 69] were obtained at the output, the input dimension was decreased because of the convolution operation. A linear activation was used.

Stage 3: first pooling layer (Maxpooling3D_1)

- The maxpooling layer, as shown in Fig. 6, decrease the size of the data by applying the maximum to blocks of 2 × 2 × 2 pixels, resulting in an output of [23, 34, 34,

```
[ ]  Layer (type)                 Output Shape              Param #
     =================================================================
     conv3d_1 (Conv3D)            (None, 49, 69, 69, 55)    1375
     _________________________________________________________________
     max_pooling3d_1 (MaxPooling3 (None, 24, 34, 34, 55)    0
     _________________________________________________________________
     conv3d_2 (Conv3D)            (None, 23, 33, 33, 35)    15435
     _________________________________________________________________
     max_pooling3d_2 (MaxPooling3 (None, 11, 16, 16, 35)    0
     _________________________________________________________________
     conv3d_3 (Conv3D)            (None, 10, 15, 15, 20)    5620
     _________________________________________________________________
     max_pooling3d_3 (MaxPooling3 (None, 5, 7, 7, 20)       0
     _________________________________________________________________
     flatten_1 (Flatten)          (None, 4900)              0
     _________________________________________________________________
     dense_1 (Dense)              (None, 30)                147030
     _________________________________________________________________
     dense_2 (Dense)              (None, 27)                837
     _________________________________________________________________
     dense_3 (Dense)              (None, 20)                560
     _________________________________________________________________
     dense_4 (Dense)              (None, 7)                 147
     _________________________________________________________________
     dense_5 (Dense)              (None, 2)                 16
     =================================================================
     Total params: 171,020
     Trainable params: 171,020
```

Fig. 7 Keras listing of the parameter for the proposed mode

55] which are 55 layers with 24 frames of a size of [34 × 34], which are half the size of each frame of the previous layer

Stage 4: second convolutional layer (Conv3D_2)

- As can be seen from Fig. 6, the second convolution layers has 35 filters of size [2 × 2 × 2], the output dimension is now of 23 frames with a size of [33 × 33].
- This layer has the ReLu activation function, since it has a good performance with convolutional networks. This is a nonlinear function that make zero all negative values [29].

Stage 5: Max pooling (Maxpooling_2)

- The pooling layer as an output of [11, 16, 16, 35], which are 35 layers with 11 frames of a size of [16 × 16] which are half the size of each frame of the previous layer.

Stage 6: third convolutional layer (Conv3D_3)

- In Fig. 6, it is shown that in the third convolution has an output of 20 layers each with 10 frames with a size of [15 × 15].

Stage 7: Max pooling (Maxpooling_2)

- In this pooling layer the is deceased to a size of [11, 16, 16, 35], which are 35 layers with 11 frames of a size of [16 × 16] which are half the size of each frame of the previous layer.

Stage 8: Flatten

- In this stage a grouping of the characteristics of the last layer is made as a single column vector, this is needed in *keras* to be able to make connections with the dense layer.

Stage 9: Dense layers: These are the classifying layers:

- Dense layer (30): 30 neurons, with *ReLu* activation function.
- Dense layer (27): 27 neurons with *ReLu* activation function.
- Dense layer (20): 20 neurons with *ReLu* activation function.
- Dense layer (7): 7 neurons with *ReLu* activation function.
- Dense layer (2): there are 2 neurons, in this layer it is classified whether an input is "fall" or "not fall", the *Softmax* activation function was used, which give a probabilistic output [29]. The training of the model was carried out within 10 epochs. The training was carried out using the *google Colab* platform and using a Tensor Processing Unit.

4.4 Evaluation Metrics

The Binary crossentropy metric was used to assess the model during training (loss function), this function gives the value of loss at each epoch, this to know the probability of the predictions of the model, that is to say how good or bad is the model. The predictions when having high loss values will mean that the prediction is bad otherwise if the values are low it will mean that the predictions are good [23, 30].

The Accuracy metric was also used as evaluation of the performance of a model to know the accuracy of a model [23].

Additionally, the training was carried out using the optimization method of adaptive moments, "Adam", which is a stochastic optimizer, that is, it optimizes randomly the parameters [23, 31].

To assess the model with the test data, the confusion matrix and the area under the curve (AUC) of the receiver operation characteristic (ROC) curve were used.

5 Results

This chapter will discuss the results obtained from the neural network model for the detection of falls. The first section presents the results of the training, accuracy obtained in the last peri-od of training, and the second section describes the results of the evaluation with the test set using the ROC curve and the confusion matrix [33, 34].

5.1 Training Results

- It was obtained as a result of 10 training periods with an accuracy of 0.971 in the last period of training.
- Training time was approximately 3 min per epoch.
- The training set was 178 videos among which there were 85 fall and 93 non-fall videos.
- Figure 8 presents the confusion matrix which shows the performance of the model using the 178 videos of the training set.

As a result, it was obtained that of the 178 videos evaluated:

True Positive (VP):
83 VP
False Negative (FN):
2 FN
False positive (FP):
3 FP
True Negative (VN):
90 VN

Fig. 8 confusion matrix of the training stage

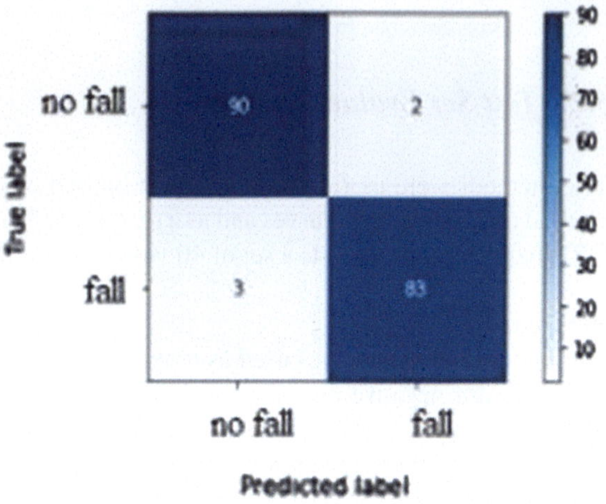

From the confusion matrix it is possible to calculate the following metrics:

$$\text{Sensitivity} = \text{VP/(Total Positives)} = 83/86 = 0.965$$
$$\text{Specificity} = \text{VN/(Total Negatives)} = 90/92 = 0.978$$
$$\text{Accuracy} = (\text{VN} + \text{VP})/\text{Total} = (90 + 83)/178 = 0.971$$

As training results, an accuracy of 0.971, a sensitivity of 0.965 and a specificity of 0.978 were obtained, as well as successfully achieving 173 videos of 178 and only err in 5, these results are indicative of how functional it can be the model created.

It is mentioned in [4] that the larger the training set, the greater the possibility of a neural network to learn, when a large amount of training data is not used, it may be that the model is over-adjusted or over-trained, that is, it will learn characteristics which are not the ones that the model wants to learn and when new scenarios are presented it does not achieve the detection objective.

In this model, an activation function called ReLu was used for training, which as mentioned earlier, this function helped to reduce training times [29], since preliminary tests were carried out with different models in this project without using ReLu activation function and training times ranged from 8 to 12 min.

The ReLu activation function helped the model to train since without this activation function the model stagnated at an accuracy of 0.55.

The results obtained from the confusion matrix of the training set are not the validation results of the model, this test was performed to know how the model behaved with the data used in the survey.

Of the 178 videos used in the training set, the model was successful in 175 videos and only failed in 5, this could be because the detection of people or actions implies a great challenge due to the variability in the funds of the scenarios, In addition to the lighting changes or if there are partial or total occlusions within the video images, all these may be factors that may affect the detection, causing the detection results to be incorrect [4].

5.2 *Test Set Evaluation*

Two methods were used to evaluate the proposed model with the test set. The first method used is the ROC curves and as a second method a confusion matrix was used [32]. To perform these tests a set of 40 videos was separated from the dataset.

ROC curve

This validation method was used as it tests the model by means of the true positive rate and false negative rate. As a training set, 40 videos were separated from the database, these videos were used only for the evaluation of the model's performance. Figure 9 shows the evaluation of the performance of the model evaluated with the ROC curve, the highest point of sensitivity is between 0.3 and 1, so it is an indicator

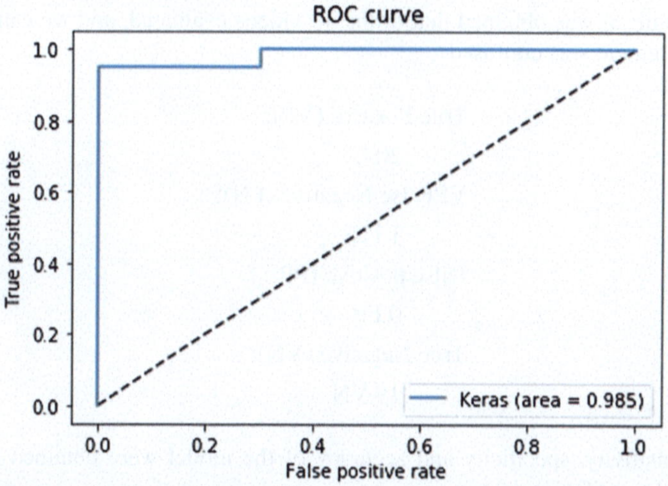

Fig. 9 ROC curve of the proposed model

that the model has good performance for the Fall detection also shows that this result shows that the model was correct in most of the videos in the test set.

Confusion Matrix
The confusion matrix validation method was performed in order to complement the validation of the model. The performance of the model is shown in Fig. 10.

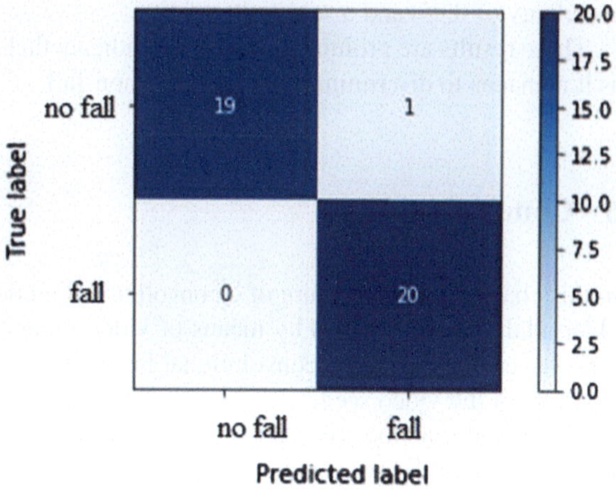

Fig. 10 Confusion matrix over the test set

As a result, it was obtained that of the 40 videos evaluated, and we can see that only one element was confused:

True Positive (VP):
20
VPFalse Negative (FN):
1 FN
False positive (FP):
0 FP
True Negative (VN):
19 VN

The sensitivity, specificity and accuracy of the model were obtained from the results of the confusion matrix.

$$\text{Sensitivity} = \text{VP/(Total Positives)} = 20/20 = 1$$
$$\text{Specificity} = \text{VN/(Total Negatives)} = 19/20 = 0.95$$
$$\text{Accuracy} = (\text{VN} + \text{VP})/\text{Total} = (20 + 19)/40 = 0.975$$

It was validated with 20% of the videos used in the training set since it gives us greater confidence in the operation of the model performed. Using a matrix of confusion allows us to see how many videos he evaluated as falling or not falling and in which ones, and in how many he failed, so in this way he gives us quantitative results. From the validation results it can be seen that from 40 videos, the model failed in 1, and correctly classify the others, thus the model has an Accuracy of 0.975, a specificity of 0.95, and a sensitivity of 1.

These results are promising since they indicate that the model works as expected as it manages to discriminate a fall from a non-fall.

6 Conclusions

In this chapter an architecture of a convolutional model for the detection of falls of older adults in smart cities by means of video images is presented. The Architecture was composed of 3D convolutional layers to take advantage of the volumetric structure of the video scenes.

The model function correctly in addition to achieving each of the objectives set at the beginning of the project, the model was trained with video images obtained from a database and an accuracy of 0.97 was obtained which indicates that the model is able to detect a fall, and at the time of validating the model, it was only wrong once, this is an indicator that the model implemented is highly competitive in relation to [2].

As a future work, the proposed architecture is planned that not only manages to detect the falls, but that it manages to track the test subjects through gait recognition [35–37] as well as distinguish faces, this would be a great contribution to the care of older adults. Also, it is planned to increase the number of videos in the test, to better assess the generalization of the proposed network.

References

1. Chen, Y., Yang, X., Zhong, B., Pan, S., Chen, D., Zhang, H.: CNNTracker: Online discriminative object tracking via deep convolutional neural network. Appl. Soft Comput. J. **38**, 1088–1098 (2016)
2. Chu, C., Chang, C., Chang, T., Liao, J.: Elman neural network identify elders fall signal base on second-order train method. In: 6th International Symposium on Next Generation Electronics (2017)
3. Schuh, A., Campos, M.D.B., Bez, M.: Universal access in human-computer interaction. Interact. Tech. Environ. **9738**(1), 92–101 (2016)
4. Nam, H.: Learning multi-domain convolutional neural networks for visual tracking hyeonseob nam. Comput. Vis. Pattern Recogn. 4293–4302 (2016)
5. Bian, Z., et al.: Fall detection based on body part tracking using a depth camera. IEEE J. Biomed. Heal. Informatics **19**(2), 430–439 (2015)
6. Wang, H., Zhang, D., Wang, Y., Ma, J., Wang, Y., Li, S.: RT-fall: a real-time and contactless fall detection system with commodity wifi devices. IEEE Trans. Mob. Comput. **16**(2), 511–526 (2017)
7. Barillaro, S., et al.: Diseño de sistema IoT de monitoreo y alarma para personas mayo-res Resumen Contexto Introducción Líneas de Investigación, Desarrollo e Innovación, pp. 712–716 (1980)
8. Mohanty, S.P., Choppali, U., Kougianos, E.: Everything you wanted to know about smart cities: The internet of things is the backbone. IEEE Consumer Electron. Mag. **5**(3), 60–70
9. Dipanjan Sarkar, T.G., Bali, R.: Hands-on transfer learning with Python, Primera ed. Pack Publishing Ltd, Birmingham (2018)
10. Martin del Brio, B., Molina, A.S.: Redes Neuronales y Sistemas Difusos, 2a edicion. Alfaomega, Bogota (2001)
11. López, R.F., Fernández, J.M.F.: Las Redes Neuronales Artificiales fundamientos teoricos y aplicaciones prácticas, 1ra edicio. Gesbiblo, S.L, Oleiros (2008)
12. Goodfellow, I., Bengio, Y., Courville, A.: Deep Learning, 1ra edicio. MIT, London (2016)
13. Long, J., Shelhamer, E., Darrell, T.: Fully Convolutional Networks for Semantic Segmentation. In: Proceedings of the IEEE Conference on Computer Vision and Pattern Recognition (2015)
14. Babaee, M., Dinh, D.T., Rigoll, G.: A deep convolutional neural network for video sequence background subtraction. Pattern Recogn. **76**, 635–649 (2018)
15. Matich, D.J.: Redes Neuronales: Conceptos Básicos y Aplicaciones. Universidad Tecnológica Nacional, México (2001)
16. Salas, R.: Redes Neuronales Artificiales. Departamento de Computación (2004)
17. Grado, T.D.E., Ingenier, E.N.: Entrenamiento de redes neuronales basado en algoritmos evolutivos (2005)
18. Lanzarini, L., Hasperué, W., Estrebou, C., Ronchetti, F., Monte, A.V.: Aprendizaje Automático aplicado a Reconocimiento de Patrones en Video y Minería de Datos. In: XVIII Workshop de Investigadores en Ciencias de la Computación (2016)

19. Feichtenhofer, C., Pinz, A., Zisserman, A.: Convolutional two-stream network fusion for video action recognition. In: Proceedings of the IEEE Conference on Computer Vision and Pattern Recognition (2016)
20. Charfi, I., Miteran, J., Dubois, J., Atri, M., Tourki R.: Optimized spatio-temporal descriptors for real-time fall detection: comparison of support vector machine and Adaboost-based classification. J. Electron. Imaging (2013)
21. Montoro, A.F.: Phyton 3 al descubierto, 1ra edicio. RC libros (2012)
22. Abadi, M., Barham, P., Chen, J., Chen, Z., Davis, A., Dean, J., Kudlur, M.: Tensor-flow: A system for large-scale machine learning. In: 12th {USENIX} Symposium on Operating Systems Design and Implementation ({OSDI} 16), pp. 265–283 (2016)
23. Gulli, A., Pal, S.: Deep learning with Keras. Packt Publishing Ltd (2017)
24. Silva, B.N., Khan, M., Han, K.: Towards sustainable smart cities: A review of trends, architectures, components, and open challenges in smart cities. Sustain. Cities Soc. **38**, 697–713 (2018)
25. Lemlouma, T., Laborie, S., Roose, P.: Toward a context-aware and automatic evaluation of elderly dependency in smart homes and cities. In: IEEE 14th International Symposium on A World of Wireless, Mobile and Multimedia Networks (WoWMoM), pp. 1–6. IEEE (2013)
26. Mirri, S., Prandi, C., Salomoni, P., Callegati, F., Campi, A.: On combining crowdsourcing, sensing and open data for an accessible smart city. In: Eighth International Conference on Next Generation Mobile Apps, Services and Technologies, pp. 294–299. IEEE (2014)
27. Xi, H., Ren, X., Zhai, S.G.: Smart pension: the elderly care service innovation with information technology. Sci. Res. Aging **2**(7), 12–20 (2014)
28. Et image, I.: U. C. 6306 Lei2 Laboratoire Electronique, Fall Detection Dataset. Fall Detection Dataset (2013)
29. Samarasinghe, S.: Neural Networks for Applied Sciences and Engineering: From Fundamentals to Complex Pattern Recognition. Auerbach Publications (2016)
30. Godoy, D.: NoComprensión de la entropía cruzada binaria/pérdida de registro: una explicación visual Title (2018). [Online]. Available: https://towardsdatascience.com/understanding-binary-cross-entropy-log-loss-a-visual-explanation-a3ac6025181a. Accessed 28 Oct 2019
31. Kingma, D.P., Jimmy, B.A.: Adam: a method for stochastic optimization. arXiv preprint arXiv: 1412.6980 (2014)
32. Lorca, J.C.: Uso de curvas ROC en investigación clínica. Aspectos teórico-prácticos (2011)
33. Hay, A.M.: The derivation of global estimates from a confusion matrix. Int. J. Remote Sens. **9**(8), 1395–1398 (1988)
34. Zelada, C.: Evaluación de modelos de clasificación (2017)
35. Figueiredo, J., Santos, C.P., Moreno, J.C.: Automatic recognition of gait patterns in human motor disorders using machine learning: a review. Med. Eng. Phys. **53**, 1–12 (2018)
36. Godfrey, A.: Wearables for independent living in older adults: gait and falls. Maturitas **100**, 16–26 (2017)
37. Harris, P., Garcia, M.B.: To fall is human: falls, gait, and balance in older adults. In: New Directions in Geriatric Medicine, pp. 71–90. Springer, Cham (2016)

Automatic Tumor Segmentation in Mammogram Images for Healthcare Systems in Smart Cities

Alberto Ochoa-Zezzatti and Jose Mejia

Abstract Breast cancer is one of the leading causes of cancer in women, thus this is an important issue for smart cities health systems. The aiming of smart cities is to offer a better quality of life of its citizens, by means of technology thus it is important to develop methods to diagnose breast tumors in shorter periods of time and with high degree of confidence. For this end, in this chapter, it is proposed to develop a new method for automatic segmentation of breast cancer tumors using deep learning, in order to be part of a modules focused on detection and automatic diagnosis or as an aid to medical staff as a support for diagnosis.

Keywords Breast cancer · Deep learning · Smart cities

1 Introduction

Breast cancer is one of the leading causes of cancer in women, approximately 15% of all cancer deaths among women is due to this ailment, according the World Health Organization [1]. One of the challenges in healthcare services in Smart Cities consist in give to the users a diagnostic in the shortest possible time and automatically, in the case of breast cancer detection, currently in some regions, this is not yet possible, with reporting waiting times of up to several months before having a diagnosis [2]. This is an important issue because an immediate attention increases the survival rate of patients.

The development of a breast cancer monitoring systems is a current effort toward an automatic system for health in smart cities [3–6]. With the advent of new technologies, powerful tools are provided in a smart healthcare framework. These new technologies aim to provide a seamless, usability, and accurate healthcare service to citizens of the smart cities [7]. One of these new enable technologies is Deep Learning.

A. Ochoa-Zezzatti · J. Mejia (✉)
Universidad Autónoma de Ciudad Juárez, Av. Hermanos Escobar, 32410 Cd Juárez, Chihuahua, Mexico
e-mail: jose.mejia@uacj.mx

A. Ochoa-Zezzatti et al. (eds.), *Technological and Industrial Applications Associated With Industry 4.0*, Studies in Systems, Decision and Control 347,
https://doi.org/10.1007/978-3-030-68663-5_5

The screening of mammographic images is a difficult process, and it is believed that around 10–30% of masses missed [8]. This is partially due because of the images acquired during the screening process are not perfect and usually have blur and low contrast. This may cause difficulty for the medical staff in classifying the different tissues. Additionally, in some cases the contrast be-tween normal and malignant tissues is minimal causing a false negatives diagnosis [9].

Another factor hinders the interpretation of a mammogram, is the experience of the radiologist at interpreting images, also, fatigue and routine, can reduce the effectiveness of manual diagnoses. It is estimated that approximately in 9% of the positive cancer detections, tumors were already visible on mammograms acquired from two years ago [10]. The problems outlined above, make a necessity to develop approaches to automatically segment malignant tissues, in order to help to reduce mistakes at interpreting images and to have a timely diagnosis. The use of computerized systems, for segmentation of tumors, is thus a key to en-able the automatic diagnosis and timely treatment of breast cancer.

One of the current trends in computer vision is the Deep Learning approach [11] which is used for almost all tasks in automatic image processing [12], these techniques are also applied to medical image processing and automatic diagnosis. The actual trend for the automatic segmentation or detections tumors is employing artificial neuronal networks [12] and deep learning networks [13, 14].

In this chapter we proposed to employ deep learning to segment mammary tumors in by using layers of deep convolutional networks. Our architecture is based on a two channel network and each channel includes convolutional depthwise layers to reduce the computational effort and non-local modules to capture long-range dependencies in the mammogram image. The proposed methodology is shown in Fig. 1.

2 Literature Review

Here we reviewed some of the most relevant algorithms for segmentation of mammogram tumours using deep learning.

In [15] they proposed a conditional Generative Adversarial Network for segmentation a breast tumor in a region of interest. The network is trained to recognize the tumor area and then creates a binary mask around it additionally it distinguishes between real and synthetic segmentations, thus it is capable of creating binary masks with realistic shapes. They noted that their network works even with a limited number of training samples.

In [13] proposed a pixel-based segmentation method using supervised and unsupervised approaches. Their semi supervised method is aimed to tumor segmentation, they used pixel information for the classification. As features for the classification they used the static and gray level run length matrix of each pixel and for feature reduction a Fisher discriminant analysis. Finally, they use a support vector machine with Bayes for tumor segmentation.

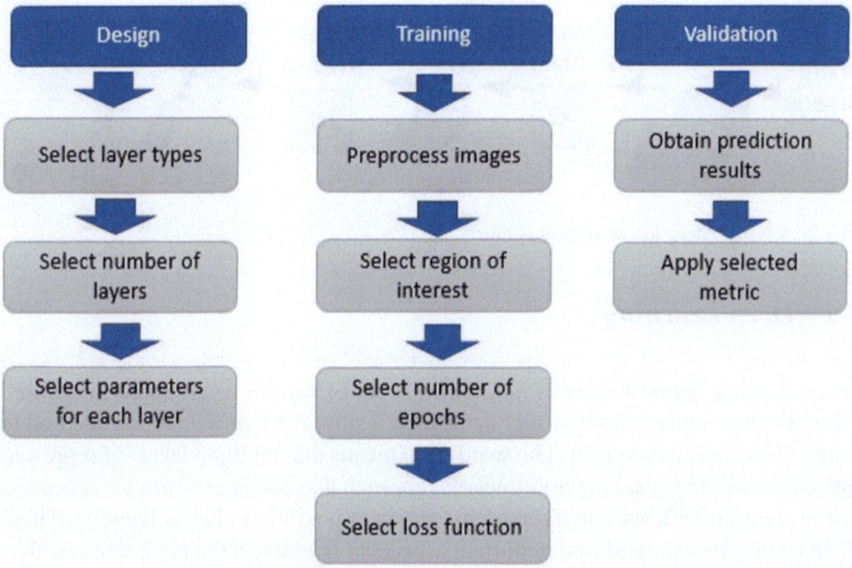

Fig. 1 Diagram of the methodology

In [16] it is developed an automatic image segmentation method based on deep networks for breast tumor segmentation, the method combines U-Net with attention gates, and a autoencoder scheme that contains a encoder-de-coder. The encoder is composed of a convolutional network while the decoder is a U-Net combined with the attention gates. In [17], they propose another attention-guided dense up-sampling network for breast tumor segmentation in mammograms images. In their proposed network they employ a structure they called asymmetrical encoder-decoder followed by an up-sampling block, and an attention-guided dense up-sampling block. This last block is designed to have three capabilities: compensates the information loss after a bilinear up-sampling, to fuse high- and low-level features, and finally to includes a channel-attention function in order to process different channels.

3 Theory Background

In this section, we described some of the building block used to develop the propose deep learning architecture, which are the depthwise layers of [18] and the non-local blocks of [19].

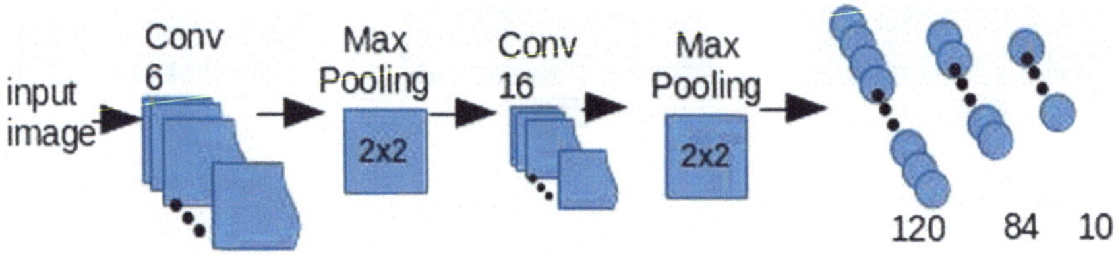

Fig. 2 A typical deep learning architecture

3.1 Deep Learning

Deep-learning neural networks that allows a set of neuron layers to be present-ed with raw data and to automatically determine features or representations need-ed to perform a classification [20]. The word deep means that multiple levels of representations are used by stacking non-linear layers such that each transform the representation at an initial level into a representation with possibly richer in features. These deep layers are supposed find important aspects or features of the input that could be determinant for discrimination tasks. Figure 2 shows a typical deep learning architecture, this network is composed of two convolutional layers (Conv), two maxpooling layers (Max Pooling) and three layers of dense networks of 120, 84, and 10 neurons, respectively.

The two layers of convolutional networks convert the input in a new representation with features that can help to classify the input. A key aspect of these convolutional layers that the features that they found are not designed by human engineers and instead are learned from the input data by training. The pooling layers are mainly for data reduction, generally are placed after each convolutional layer, and work by aggregating components by using a statistic like mean, variance or maximum value, the maximum is the most used statistic in the pooling operation. it is common to use the name of the statistic used as a prefix of the word pooling. Finally several layers of dense networks or multilayer perceptron are used to classify the input, these layers make use of features in the data operate, that is why convolutional layers placed before, in order to provide features to the dense layers.

There exist several improvements to the convolutional layers, next we revised two important types of recently suggested convolutional layers: deep wise layers and non-local layers.

3.2 Depthwise Layers

Depthwise convolution was introduced in [18] for use in the network MobileNet, which consist of convolutional layers with a more efficient model aimed to mobile and embedded vision applications.

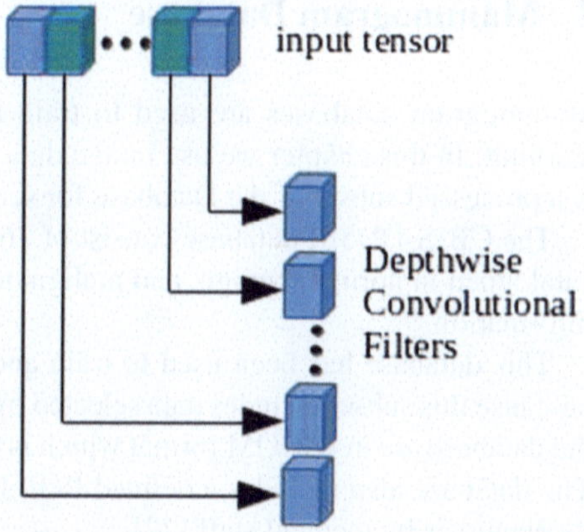

Fig. 3 Diagram for the depthwise convolution. Given an input tensor a filter is applied to each channel

The depthwise convolution uses separable convolutions which are factorized form of a standard convolution into a convolution for each channel and a 1 × 1 convolution called a pointwise convolution. The depthwise convolutions component ap-plies to each input channel a unique filter, this is show in Fig. 3.

After applying the filter, a pointwise convolution consisting of a 1 × 1 convolution is used to combine linearly the output of the depthwise layer. By using separable components, the depthwise convolution is very efficient as compared to standard convolution. The use depthwise convolution was tested in the MobileNet network by [18] demonstrating superior speed and accuracy when applied to a wide variety of tasks.

3.3 *Non-local Processing*

The non-local module for deep learning was developed by [19]. This block is inspired by non-local means algorithm introduced in the celebrated work of [21] where was proposed to capture the long-range dependencies of patches in an image and using these to average the real value of a pixel or path in a noise image. Similarly, this idea is applied to deep learning to extract features taking into account long-range dependencies between zones in the input image. The use of non-local in deep learning offers several advantages: non-local operations can capture long-range dependencies regardless of positional distance, non-local operations are efficient and are capable of obtain good result by using a few layers, and can be easily combined with other operations to increase the performance in any network.

4 Mammogram Database

Mammogram databases are used to train and initially test architectures of deep learning. In this chapter we use image data from the CBIS-DDSM database which is a processed subset of the Database for screening Mammography (DDSM).

The CBIS-DDSM database consist of 2620 mammography studies. Each case is catalogued in normal, benign, and malignant it also contains a verified pathological information.

This database has been used to train and test many algorithms of the literature be-cause this subset includes data selected by a trained medical doctor. The images in the database are in DICOM format which is usually used in biomedical applications. The database also includes a defined ROI delimitation of pathological mass-es and its diagnosis by medical staff [22].

Annotations for anomalies are given for each ROI and it is provided the exact position of the lesions by means of coordinates, this greatly eases the task of two generating the segmentation mask for each lesion.

5 Deep Learning Architecture for Breast Tumor Detection

In this section we describe in detail de steps outlined in Fig. 1 for the proposed method.

5.1 Architecture Design

The network architecture is based on [23] and [24]. However, in this work we modified each channel to include convolutional depthwise layers to reduce the computational effort and non-local modules to capture long-range dependencies in the mammogram image. As in [23] each channel obtains different kinds of features from the input image. The proposed architecture consists two channels (X1, X2):

- Channel X1: Consist of a depthwise input layer of kernel size (4 × 4) and depth multiplier of 3, followed by a maxpooling with kernel size of (2 × 2) and stride of 2. A second depthwise layer of kernel size (3 × 3) and depth multiplier of 35, and a third depthwise layer of kernel size (5 × 5) and depth multiplier of 65. Finally a Upsampling layer of (2 × 2) kernel.
- Channel X2: Consists of a convolutional layer of kernel size (2 × 2) with 10 ker-nels and activation ReLU, followed by a Non-Local block, a convolutional lay-er with 30 kernels of size (2 × 2) and a maxpooling with kernel size of (2 × 2) and stride of 2, then another convolutional layer with 35 kernels of size (2 × 2) and finally a convolutional transpose layer with 35 kernels of size (32 × 32).

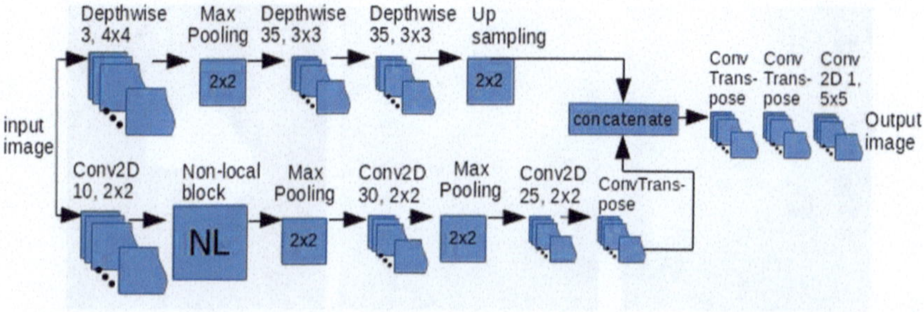

Fig. 4 Diagram of the proposed network

The two channels are concatenated and then the output goes through two convolutional transpose layers, with 5 kernels of (15 × 15) and 7 kernels of (7 × 7) respectively. Finally, as output a convolutional layer consisting of 1 filter with a size of 5 × 5 with sigmoidal activation. The network has as output, an image with a mask over the tumour. Figure 4 shows the architecture diagram.

5.2 *Training*

For training the architecture, first the images of the database were pre-processed before fed the network and then hyperparameter of the layer were selected, training set consisted of 800 images and a validation set of 200 images. The number of epochs require was 750. Since segmentation can be posed as a binary classification, the selected loss function was the Cross entropy. The training algorithm was Adaptive moments.

5.2.1 Prepossessing

The pre-processing consists of two steps: clipping the region of the tumor and decrease the image size. The size of the images from CBIS-DDSM database are 6511 × 4801 pixels and are images in grey scale. Also, de database provides a mask for the tumour. Using this information, the region containing the tumour, on both, image and mask were trimmed. The resulting region was reduced to a 60 × 60-pixel image. This was repeated for all images in the database. In Fig. 5, illustrated this process.

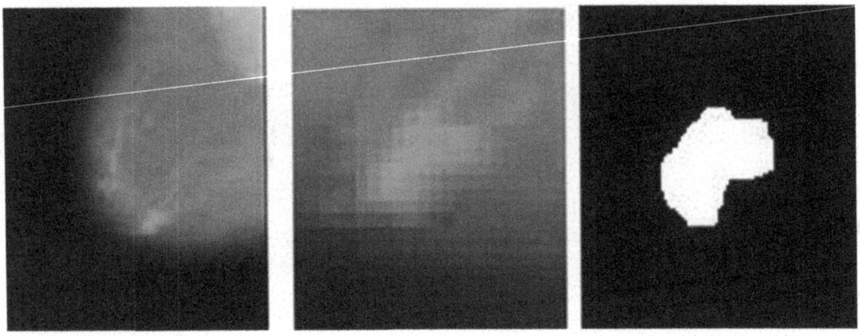

Fig. 5 Pre-processing of acquired images. Left, original mammographic image. Centre, trim delimiting the tumour area. Right, tumour mask located by the given coordinates

5.3 *Implementation*

The architecture was implemented in python employing the Keras library [25]. The module of depthwise layer of Keras was employed and the source code of [19] was used to implement the non-local module in the proposed architecture.

6 Validation

The results of the segmentation were evaluated using the Intersection over the Union Metric (IoU) metric [26]. This metric is frequently employed to measure the accuracy of segmentation task or object detection. The IoU is calculated dividing the intersection area between the segmented and true objects, by the area of its union [27]. The metric is commonly utilized to evaluate performance for the task on segmentation in convolutional neural network [28, 29]. The IoU is given by:

$$IoU = \frac{Intersection\ Area}{Union\ Area} \tag{1}$$

The images employed to validate consist of 200 mammograms. Figure 6 shows sample images of the test set and the result when it is processed by proposed architecture.

In Fig. 7, it is presented the results of the proposed network, In the last column, green is tumoral region correctly identified (true positive), blue is tumoral region no identified by the network (false negative), and red is false tumoral region identified by the network (false positive).

From Fig. 7, the green area, where tumoral region correctly identified is much greater than the areas with error, this could be sufficient for the medical staff to be aware of suspicious mass in the mammography.

Also, given that the network output is an area, this could be used to evaluate a quantitative measure in order to assess the response of the tumor to the treatment.

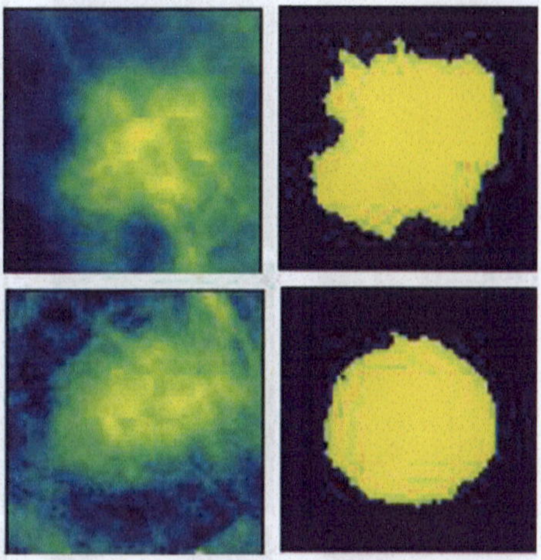

Fig. 6 Images of the test set, first column is the mammogram image and second column its associated mask

Figure 7 shows results from the proposed network. Each row shows the same region, first column is the region on the mammogram, second row is the ground truth segmentation, third column is the output segmentation from the network, and fourth column is a superimposed image from the first tree columns together. In the last column, green is tumoral region correctly identified (true positive), blue is tumoral region no identified by the network (false negative), and red is false tumoral region identified by the network (false positive).

Finally, in Fig. 8 we present the overall performance of the proposed network in the test set by using a boxplot. The method has a median of 0.89 in terms of IoU metric. Also, from the boxplot it can be seen that there is a small variance, this also correspond with the results show in Fig. 7 where the green area cover most of the tumor mass in the mammogram.

7 Conclusions

This research presents a new deep learning architecture for segmentation of masses in mammogram images. The proposed architecture consists of two channels, the first channel has several depthwise layers while the second it is designed for processing non-local features by using a non-local module. The performance of the proposed network is evaluated by using the IoU metric, results show a median of 0.89.7%. As a future research, it is planned to improve network performance adding more layers and channels.

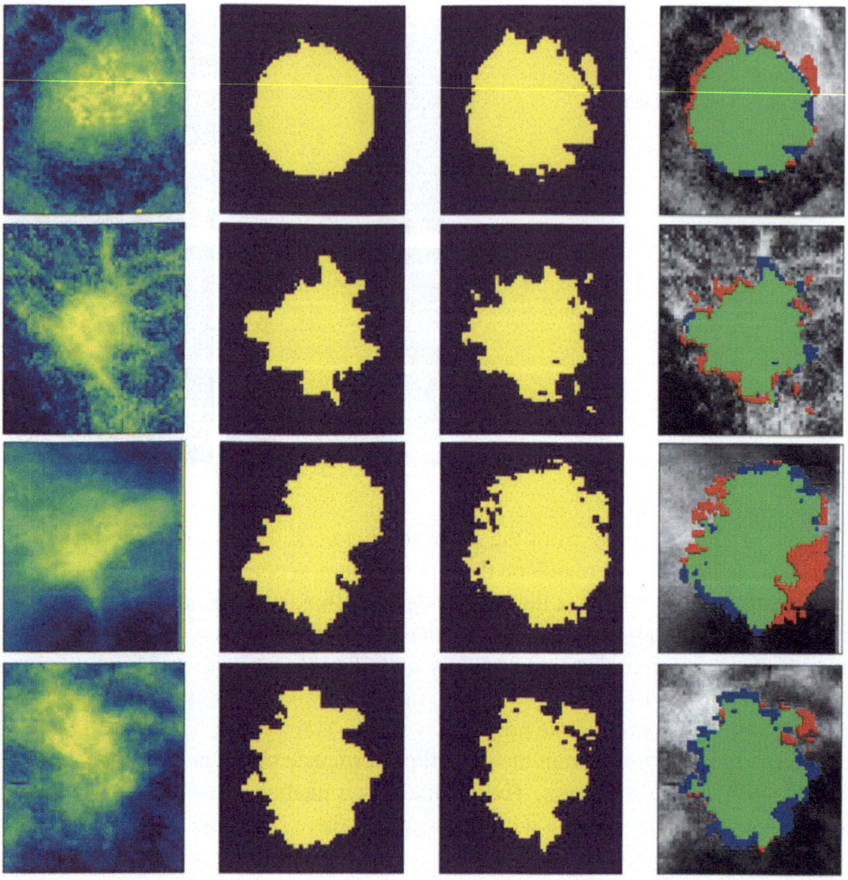

Fig. 7 Results from the proposed network. Each row shows the same region, first column is the region on the mammogram, second row is the ground truth segmentation, third column is the output segmentation from the network, and fourth column is a image from of the first tree columns together. In the last column, green is tumoral region correctly identified (true positive), blue is tumoral region no identified by the network (false negative), and red is false tumoral region identified by the network (false positive)

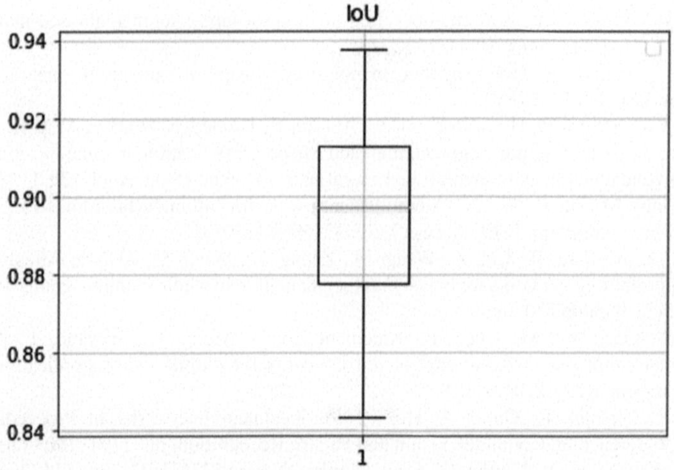

Fig. 8 Boxplot of the summarizing the performance of the network in the test set

References

1. WHO.: Cáncer de mama: prevención y control (2019). https://www.who.int/topics/cancer/breastcancer/es/
2. Avalos, M., Salazar, P., Larios, V.M., Durán-Limón, H.: Smart health methodology and services powered by leading edge cognitive services consumed in the cloud. In: 2016 IEEE International Smart Cities Conference (ISC2), pp. 1–6. IEEE (2016)
3. Alhamid, M.F.: Investigation of mammograms in the cloud for smart healthcare. Multimed Tools Appl **78**, 8997–9009 (2019). https://doi.org/10.1007/s11042-017-5239-z
4. Kashif, M., Malik, K.R., Jabbar, S., Chaudhry, J.: Application of machine learning and image processing for detection of breast cancer. In: Innovation in Health Informatics, pp. 145–162. Academic Press (2020)
5. Abdelaziz, A., Salama, A.S., Riad, A.M., Mahmoud, A.N.: A machine learning model for predicting of chronic kidney disease based internet of things and cloud computing in smart cities. In: Security in Smart Cities: Models, Applications, and Challenges, pp. 93–114. Springer, Cham (2019)
6. Rathee, D.S., Ahuja, K., Hailu, T.: Role of electronics devices for e-health in smart cities. In: Driving the Development, Management, and Sustainability of Cognitive Cities, pp. 212–233. IGI Global (2019)
7. Ghoneim, A., Muhammad, G., Amin, S.U., Gupta, B.: Medical image forgery detection for smart healthcare. IEEE Commun. Mag. **56**(4), 33–37 (2018)
8. Peng, W., Mayorga, R., Hussein, E.: An automated confirmatory system for analysis of mammograms. Comput. Methods Programs Biomed. (2015)
9. Vikhe, P.S., Thool, V.R.: Contrast enhancement in mammograms using homomorphic filter technique. In: 2016 International Conference on Signal and Information Processing (IConSIP), pp. 1–5. IEEE (2016)
10. Flores Gutiérrez, H., Flores, R., Benja, C., Benoso, L.: Redes Neuronales Artificiales apli-cadas a la detección de Cáncer de Mama (2015)
11. Goodfellow, I., Bengio, Y., Courville, A.: Deep Learning (2016)
12. Hamidinekoo, A., Denton, E., Rampun, A., Honnor, K., Zwiggelaar, R.: Deep learning in mammography and breast histology, an overview and future trends. Med. Image Anal. **47**, 45–67 (2018)

13. Azary, H., Abdoos, M.: A Semi-supervised method for tumor segmentation in mammogram images. J. Med. Sig. Sens. **10**(1), 12 (2020)
14. Rafegas, I., Vanrell, M.: Color encoding in biologically-inspired convolutional neural networks. Vis. Res. **151**, 7–17 (2018)
15. Singh, V.K., Rashwan, H.A., Romani, S., Akram, F., Pandey, N., Sarker, M.M.K., Torrents-Barrena, J.: Breast tumor segmentation and shape classification in mammo-grams using generative adversarial and convolutional neural network. Expert Syst. Appl. **139**, 112855 (2020)
16. Li, S., Dong, M., Du, G., Mu, X.: Attention dense-u-net for automatic breast mass segmentation in digital mammogram. IEEE Access **7**, 59037–59047 (2019)
17. Sun, H., Li, C., Liu, B., Liu, Z., Wang, M., Zheng, H., Wang, S.: AUNet: Attention-guided dense-upsampling networks for breast mass segmentation in whole mammo-grams. Phys. Med. Biol. **65**(5), 055005 (2020)
18. Howard, A.G., Zhu, M., Chen, B., Kalenichenko, D., Wang, W., Weyand, T., Adam, H.: Mobilenets: efficient convolutional neural networks for mobile vision applications (2017). arXiv preprint arXiv:1704.04861
19. Wang, X., Girshick, R., Gupta, A., He, K.: Non-local neural networks. In: Proceedings of the IEEE Conference on Computer Vision and Pattern Recognition, pp. 7794–7803 (2018)
20. LeCun, Y., Bengio, Y., Hinton, G.: Deep learning. Nature **521**(7553), 436–444 (2015)
21. Buades, A., Coll, B., Morel, J.M.: A non-local algorithm for image denoising. In: 2005 IEEE Computer Society Conference on Computer Vision and Pattern Recognition. (2005)
22. Lee, R.S., Gimenez, F., Hoogi, A., Miyake, K.K., Gorovoy, M., Rubin, D.L.: A curated mammography data set for use in computer-aided detection and diagnosis research. Sci Data **4**, 170177 (2017)
23. Jacobo, M.Z., Mejia, J.: Segmentation of brain tumor on magnetic resonance imaging using a convolutional architecture (2020). arXiv preprint arXiv:2003.07934
24. Guzman, M., Mejia, J., Moreno, N., Rodriguez, P.: Disparity map estimation with deep learning in stereo vision. CEUR (2018)
25. Chollet, F.: Keras, The python deep learning library. Astrophysics Source Code Library, ascl-1806 (2018)
26. Rahman, M.A., Wang, Y.: Optimizing intersection-over-union in deep neural networks for image segmentation. In: International Symposium on Visual Computing, pp. 234–244. Springer, Cham (2016)
27. Palomino, S., Concepción, L.P.: Watershed: un algoritmo eficiente y flexible para seg-mentación de imágenes de geles 2-DE **7**(2), 35–41 (2010)
28. Rezatofighi, H., Tsoi, N., Gwak, J., Reid, I., Savarese, S.: Generalized Intersection over Union : A Metric and A Loss for Bounding Box Regression, pp. 658–666 (2019)
29. Rosebrock, A.: Intersection over Union (IoU) for object detection. Mach. Learn. Object Detect. Tutor. (2016)

Impact of Industry 4.0: Improving Hybrid Laser-Arc Welding with Big Data for Subsequent Functionality in Underwater Welding

Alberto Ochoa-Zezzatti, Raúl Méndez, and Elías Carrum

Abstract This chapter shows the importance of different branches of Artificial intelligence (AI) in the development of the industry around the globe specifically in the welding processes. Making possible to take advantage of the humungous sets of data generated by the production lines, reducing costs in energy of raw material, thus increasing benefits.

Keywords Artificial intelligence · Industry 4.0 · Hybrid laser arc welding

1 Introduction

The artificial intelligence is growing faster, some of its applications are hidden in the shadows, while others are in front of us. The industry, being a bastion of the development, couldn't be left behind. We are already living the fourth industrial revolution; the automation is more present each day in the production lines. It is due to this that the different branches of AI (Big Data, Machine Learning, and Deep Learning) are taking more importance, improving the methods and technologies involved in the manufacture.

A. Ochoa-Zezzatti (✉)
Universidad Autónoma de Ciudad Juárez, Av. Hermanos Escobar, Omega, 32410 Cd Juárez, Chihuahua, México
e-mail: alberto.ochoa@uacj.mx

R. Méndez
Instituto Tecnológico de Parral, Avenida Tecnológico No. 57, Colonia Centro, Colonia Centro C.P. 33850 Hidalgo del Parral, Chihuahua, México

E. Carrum
COMIMSA Centro Conacyt, Saltillo, México

A. Ochoa-Zezzatti et al. (eds.), *Technological and Industrial Applications Associated With Industry 4.0*, Studies in Systems, Decision and Control 347,
https://doi.org/10.1007/978-3-030-68663-5_6

2 Industry 4.0

In this section, we are looking to introduce the main challenges that led to the creation of the idea of Industry 4.0, and what this involves.

2.1 The Origins

As mentioned by Roblek et al. industry 4.0 was first mentioned in 2011 in Germany as a proposal for the development of a new concept of German economic policy [1]. This way, the German industry was looking to introduce high-tech strategies into the manufacturing process. The main challenge the industries deal with is the rapid decision-making to improve productivity using different techniques of big data [2].

At this point, we could mention many of the innovations included in the improvements made when a production line is involved in the industry 4.0, but which care us the most is specifically one: how to optimize the processes: attempting to reduce costs and get the best final product possible.

2.2 Industry 4.0 Techniques

It is well-known that industry 4.0 has many techniques, the big data and data mining are some of them. The best definition of Big Data is "a comprehensive term for any collection of data sets so large and multifarious that it becomes difficult to process them using conventional data processing applications" [3].

The HACE theorem used to model the characteristics puts a problem explaining the difficult big data deals with and is shown in [4]: Imagine that several blind men are trying to size up a giant camel, which will be the Big Data in this context. The goal of each blind man is to draw a picture (or conclusion) of the camel according to the part of the information he collects during the process. Because each person's view is limited to his local region, it is not surprising that the blind men will each conclude independently that the camel "feels" like a rope, a hose, or a wall, depending on the region. To make the problem even more complicated, let us assume the camel is growing rapidly and its pose changes constantly. Just like an proof of how fast this "camel" can grow, Bernard Marr mentions in his note [5], there are 2.5 quintillion bytes of data created each day at our current pace, but that pace is only accelerating with the growth of the Internet of Things (IoT). Over the last two years alone 90% of the data in the world was generated.

3 Welding Technologies

There are so many things we can do with Big Data, but which would be the sense of investing in all this analysis if it won't be used practically. In this section, we introduce two welding techniques, the Laser Beam Welding (LBW) and the Gas Metal Arc Welding (GMAW).

3.1 *Laser Beam Welding (LBW)*

In the site of The Open University [6], is shown how an LBW works: A high energy laser beam is focused onto the surface to be welded, using mirrors and lenses (Fig. 1). The laser beam is coherent (in-phase waves), monochromatic (single wavelength) and highly collimated (parallel beam). The output can be continuous or pulsed. Solid-state lasers pulse by turning the flashlamps on and off or by "Q-switching" (Nd-YAG), where an optical shutter opens and closes to give a pulsed output (Fig. 2).

TWI, a laser provider for the industries, shows [7] that it can be used in the automotive industry as well. This kind of weld is used in the manufacture of components that cover engine parts, transmission parts, alternators, solenoids, fuel injectors, fuel filters, air conditioning equipment, and airbags.

From another point of view, MECCO, also a provider of laser technology says in [8] that even when the processes and applications of laser welding are most prominent in the automotive industry, where lasers boosts productivity at a low cost, these are also used in the jewelry and medical industries to put together metals on a smaller level. Any material with a high heat conductivity can be laser welded.

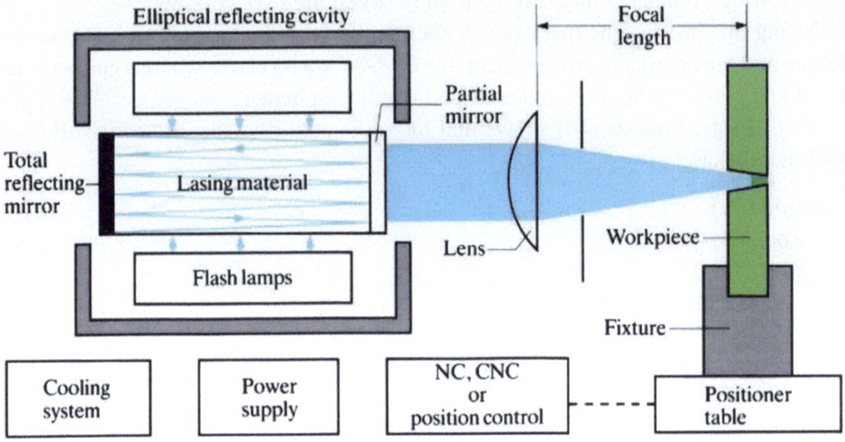

Fig. 1 LBW structure

NUMBER	LASER POWER (KW)	WELDING SPEED (mm/s)	WIRE FEED RATE (m/min)	CURRENT (A)	VOLTAGE (V)	STICK OUT (mm)	DEFOCUS (mm)	GAP (mm)	Penetration %	Mechanical properties UTS MPa
1	8.5	23	13	204	31.8	17	0	0.5	100	477
2	8	23	13	206	31.7	17	0	0.5	85	420
3	9.2	23	14	236	40	17	0	0.0	85	435
4	9	22	14	238	40	16	0	0.0	84	418
5	9	21	12	197	38	16	0	0.0	91.33	581
6	8	20	13	209	39	16	0	0.5	100	701
7	8.5	23	14	228	39	16	0	0.5	86	559
8	8.3	23	14	221	39	16	0	0.5	85	410
9	8.75	22	13	209	39	16	0	0.5	100	516
10	9	20	14	229	40	16	0	0.0	100	585
11	8	20	13	205	39	16	0	0.0	84	433
12	9.5	20	14	234	40	16	0	0.0	88	575
13	9	25	12	124	38	17	0	0.5	86	563
14	9	22	14	211	40	18	0	0	100	591
15	8	22	14	228	40	18	0	0	100	580
16	8	22	10	148	38	18	0	0	100	602
17	7	22	10	149	38	18	0	0	73	390
18	8	30	10	163	38	18	0	0	100	581
19	8.5	30	10	151	38	18	0	0	100	583
20	8	30	10	163	39	18	-3	0	100	582
21	8	30	11	209	39	18	-3	0	100	535
22	8	30	11	205	39	18	-3	0	100	534
23	8	30	11.5	208	35	18	-3	0	84	584
24	8.5	30	12	208	36	18	-3	0	100	588
25	9	18	13	203	36	16	-3	0.5	92	581
26	9	18	13	203	36	16	-5	0.5	97	585

Fig. 2 Data generated by Medina

3.2 *Gas Metal Arc Welding (GMAW)*

Before defining GMAW, it is important to define Arc Welding (AW), which is the base of the first one. TWI establish in [9] that AW as a welding process using an electric arc to create heat to melt and join metals. A power supply creates an electric arc between a consumable or non-consumable electrode and the base material using either direct (DC) or alternating (AC) currents.

The electric arc from the power supply creates an intense heat of around 6500 °F which melts the metal at the joint between the two workpieces.

During this process the metals react chemically with the gases in the air (mainly oxygen and nitrogen), due to this effect, the GMAW was created, where a gas behaves as a shield between the molten metals and the atmosphere.

In his thesis, Liratzis [10] shows that the most common gases used in different combinations are:

- Argon (Ar).
- Carbon dioxide (CO_2).
- Oxygen (O_2).

3.3 *Hybrid Welding*

Medina proposes in [11] an analysis of the behavior of the mechanical properties and depth penetration using hybrid welding which combines both LBW and GMAW. In the same article, he made a simulation for which he left an opportunity to improve the work by looking for a tool that allows adjusting the parameters and getting the best results.

4 Optimization

In this section, the different behavior of the methods tested by Medina with his provided data [11] will be shown. Although they are few to make a Big Data analysis, this processing will show some ways in which the data could be handed. To simplify the process of data we used Weka.

4.1 Database

In the data used as training set we have as independents variables:

- Laser power.
- Welding speed.
- Wire feed rate.
- Current.
- Voltage.
- Stick out.
- Defocus.
- GAP.

And the result variable is mainly the Ultimate Tensile Strength (UTS).
Due to the small dataset, was chosen to split it in 66% to train and 34% to test.

4.2 Multilayer Perceptron (MLP)

A multilayer perceptron is one of the basic deep learning algorithms. In a single layer perceptron, the input variables are multiplied with weights and add Bias. The weights are updated when an error is detected in the output. In a multilayer, the process of multiply is made sometimes. For example, a MLP with 3 layers could be represented as $f(x) = f(3)(f(2)(f(1)(x)))$. Just like Kain shown in his publication [12], each layer is represented as $f(Wx + b)$, where f is the activation function, W is the set of parameters or weights in the layer, x is the input vector, which can also be the output of the previous layer and b is the bias vector.

After applying a MLP to process the existing data we have the results shown in Fig. 3.

```
=== Summary ===

Correlation coefficient                  -0.0749
Mean absolute error                      85.6295
Root mean squared error                 119.6287
Relative absolute error                 157.8281 %
Root relative squared error             183.4381 %
Total Number of Instances                 9
```

Fig. 3 MLP results

```
=== Summary ===

Correlation coefficient                   0.224
Mean absolute error                      55.9417
Root mean squared error                  68.5264
Relative absolute error                 103.109  %
Root relative squared error             105.0781 %
Total Number of Instances                 9
```

Fig. 4 SMO results

4.3 Sequential Minimal Optimization (SMO)

The sequential minimal optimization surged as a solution to support vector machines (SVM). Platt describes in [13] that SVMs have empirically been shown to give good generalization performance on a wide variety of problems such as handwritten character recognition, face detection, pedestrian detection, and text categorization.

SMO decomposes the overall QP problem into QP sub-problems, using Osuna's theorem to ensure convergence [13].

After defining superficially what's and how works SMO, the result of evaluating the dataset output the results of Fig. 4.

4.4 Other Techniques

Even if artificial intelligence (AI) provides many techniques to calculate values from the variables known, as was said before, it is an important part of the engineer to analyze the problem and to choose the best one, either by experience or by testing the candidates (just like before).

(a) (b)

Fig. 5 Virtual training using an intelligent system **a** and real practice in shallow waters for equipment of a deep-draft ship

5 Conclusions and Future Challenges

In this paper, the main demonstration is how much the AI impacts the industry, giving rise to the fourth revolution in the industry, where it starts taking advantage of the large amount of information generated each minute by the machinery around the globe. Making these huge sets of data in an industry more sustainable, fast and overall, with bigger profits.

Also, it is presented how a relatively simple problem like the hybrid welding that has "just" eight input variables is out of the range of human ability. It is here where the machine processing of information demonstrates its utility.

In contrast, we have the popular fear that in the future the industry starts requiring fewer and fewer employees, which is false. The truth is that what will be needed is an even more skilled workforce. It is of utmost importance to perform a Virtual Model of Underwater Welding to achieve cognitive skills as part of a Virtual Training in workers who require welding in shallow waters and for processes of the Industry 4.0 and to improve the competitiveness of organizations, as is shown in Fig. 5.

References

1. Roblek, V., Meško, M., Krapež, A.: A complex view of industry 4.0. SAGE Open **6**(2), 1–11 (2016)
2. Lee, J., Kao, H.-A., Yang, S.: Service innovation and smart analytics for industry 4.0 and big data environment. Procedia **16**, 3–8 (2014)
3. Tamhane, D.S., Sayyad, S.N.: Big data analysis using HACE theorem. Int. J. Adv. Res. Comput. Eng. Technol. **4**(1) (2015)
4. Khan, S.: Hace theorem based data mining using big data. Res. Inventy Int. J. Eng. Sci. **6**(5), 83–87 (2016)
5. Marr, B.: Forbes. 21 May 2018 [En línea]. https://www.forbes.com/sites/bernardmarr/2018/05/21/how-much-data-do-we-create-every-day-the-mind-blowing-stats-everyone-should-read/#c2b6ac060ba9 [Último acceso: 21 December 2019]

6. The Open University.: OpenLearn. The Open University, 6 March 2018 [En línea]. https://www.open.edu/openlearn/science-maths-technology/engineering-technology/manupedia/laser-beam-welding [Último acceso: 21 December 2019]
7. TWI Global.: Industrial Lasers and Applications in Automotive Welding (n.d.) [En línea]. https://www.twi-global.com/technical-knowledge/published-papers/industrial-lasers-and-applications-in-automotive-welding [Último acceso: 21 December 2019]
8. MECCO.: Laser Welding Applications MECCO (n.d.) [En línea]. https://www.mecco.com/application-laser-welding [Último acceso: 21 December 2019]
9. TWI Global.: What is Arc Welding?—Definition and Process Types (n.d.) [En línea]. https://www.twi-global.com/technical-knowledge/faqs/what-is-arc-welding [Último acceso: 21 December 2019]
10. Liratzis, T.: Tandem Gas Metal Arc Pipeline Welding, Cranfield. Cranfield University, United Kingdom (2007)
11. Medina, G.Y.M., Siller, E.G.C., Pérez, A.F.M., Garces, R.S.: Mechanical Properties and Depth Penetration Optimization Using NSGA-III in Hybrid Laser Arc Welding. Cambridge University Press (2019)
12. Kain, N.K.: Understanding of Multilayer Perceptron (MLP). 21 November 2018 [En línea]. https://medium.com/@AI_with_Kain/understanding-of-multilayer-perceptron-mlp-8f179c4a135f [Último acceso: 21 December 2019]
13. Platt, J.C.: Sequential minimal optimization: a fast algorithm for training support vector machines. Microsoft (1998)

Interpersonal Relationships and Reciprocity: Their Influence in Knowledge Transfer Inside of Mexican Hotels

Aurora Máynez, Hilda Zorrilla-Nuñez, Alberto Ochoa-Zezzatti, and Andres Hernández Gómez

Abstract Knowledge stands out as a strategic asset that has been employed from past times up to the present age. Its management began in recent decades, with an increasing business interest, except in the case of the tourism and hospitality sector. Knowledge transfer is one of the vital areas of this field. Interpersonal relationships, reciprocity and the intention to share are among the factors that affect it. To advance knowledge, in the context of hotels located on the island of Cozumel, Mexico, this work aims to establish whether the first two parameters influence the transfer of intra-hotel knowledge, through the mediation of the intention to share. The research design was quantitative, non-experimental, cross-sectional, correlational-causal, with an intentional non-probabilistic sample. Surveys were employed as the preferred data collection technique, applied through a self-administered questionnaire. The variables were measured with scale items available in literature, translated into Spanish, and adapted to the context. They were valued with a Likert-like measurement scale. Modeling structural equations (SEM) with partial least squares (PLS) was used as a statistical analysis technique. Interpersonal relationships have a significant impact on the intention to share and the expectation of reciprocity, but do not directly affect knowledge transfer. The expectation of reciprocity significantly influences the transfer and the intention to share. The intention to share knowledge impacts knowledge transfer. The strongest relationship arises between interpersonal relationships and the intention to share knowledge, mediated by an expectation of reciprocity. Interpersonal relationships influence the expectation of reciprocity both in a direct and indirect manner. Sharing knowledge is not only essential to improving the performance of individuals and businesses, it is also a moral challenge for organizations. Our findings show that, because of sharing knowledge, new services and work activities are created, and that knowledge becomes a part of normal work tasks in Mexican hotels. Moreover, we agree that interpersonal relationships favor knowledge transfer. On the expectation of reciprocity, we identify that these variable influences knowledge transfer. In this context, the variable was conveyed through people's trust to share their knowledge with their co-workers in their certainty that, in doing so,

A. Máynez (✉) · H. Zorrilla-Nuñez · A. Ochoa-Zezzatti · A. H. Gómez
Doctorado en Tecnología, Universidad Autónoma de Ciudad Juárez, Ciudad Juárez, Mexico
e-mail: amaynez@uacj.mx

A. Ochoa-Zezzatti et al. (eds.), *Technological and Industrial Applications Associated With Industry 4.0*, Studies in Systems, Decision and Control 347,
https://doi.org/10.1007/978-3-030-68663-5_7

they strengthen their relationship with them. Finally, the intention to share was the main direct antecedent of knowledge transfer inside hotels; is further reflected in the confidence of employees on the importance of sharing their knowledge and in their certainty that they can provide important knowledge for the hotel.

Keywords Competitive advantage · Corporative strategies · Smart business in a smart city

1 Introduction

To compete in the markets, for decades the most used assets were material, physical and tangible. However, for some time now, organizations have been increasingly using intangible resources for their ability to create value and to promote sustainable competitive advantage [63, 65]. Today, emerging technologies like internet of things (IoT), cloud computing, big-data, and cyber-physical systems, is radically changing the conventional business processes [61]. Therefore, the ability of organizations to promote knowledge transfer is crucial for their adaptation to industry 4.0 to transform their activities into smart business [33]. Undoubtedly, social capital is a relevant factor for knowledge sharing, also for this intangible asset to be stored in electronic repositories, that is, social capital is useful to explain the use of technology for knowledge management [53]. Thus, the hotel sector cannot ignore the application of knowledge transfer processes to generate competitive strengths.

This intangible good, created within firms, is used in particular contexts, at specific moments of time and space [63], which strengthens its contribution in strategic terms, since knowledge asymmetries derive in performance differences between firms [5, 80]. Knowledge management always entails challenges. On the one hand, poorly shared and inadequately managed knowledge is easily eroded [6]. Moreover, a significant component of this asset is people [5]: sharing knowledge depends on the will of everyone [81]. On the other hand, when employees leave (by personal decision or corporate strategies such as downsizing or automation) [47, 62], knowledge is lost, which affects the quality of the goods or services produced and the renewal of organizational capacities [34, 78]. Consequently, there is a clear need for means to protect and preserve this intangible asset [15]. One alternative for this is knowledge transfer.

Knowledge transfer is one of the vital areas of knowledge management [6]. This academic field was created in recent decades [44, 68, 85] and there is a growing business interest in it [62], because of its impact on the performance and competitiveness of firms [10, 85]. This discipline has grown rapidly in most industries (both in theoretical and empirical terms), for example Guofeng et al. [33] highlighting importance of the transfer knowledge for the construction join ventures context with the exception of the tourism and hospitality sector, despite the fact that, by the nature of its services, this sector is knowledge-intensive: in order to compete in it, firms must be able to generate new knowledge that will enable them to meet the needs of their clients [63].

Among the factors that influence the transfer of knowledge are the interpersonal relations between the sender and the receiver, the expectations of reciprocity and the intention to share this intellectual asset. On the one hand, it is argued that reciprocal relationships can facilitate the exchange of knowledge [14, 49]. In addition, it is indicated that individuals share more with those with whom they have personal relationships, with those united by friendly ties [42]. Finally, another immediate antecedent to transference is intention: people must be willing and motivated to share their knowledge [13, 28, 54].

In the context of hotels located on the island of Cozumel, Mexico, the objective of this research was to investigate the influence of individual aspects on knowledge transfer. Particularly, the work tries to answer if the interpersonal relations and the expectation of reciprocity influence the transfer of knowledge within the hotels, through the mediation of the intention to share. The document is organized as follows: in the first part, the theoretical framework of the variables studied is presented; in the second part, the methodology used in the research work is presented; in the third part, the findings are shown and finally the discussion and conclusions are presented.

2 Literature Review

Organizations should encourage and facilitate knowledge sharing activities, thereby seeking to improve their performance [46]. Several authors acknowledge that many of the initiatives to manage knowledge fail due to the lack of willingness of employees to share knowledge, both in technological means [53] or in a tacit way. Although there are efforts and progress in identifying the factors associated with intra-organizational knowledge transfer, there is still a need to continue with their research [18, 67, 76], since in some of them little is known about the topic [51]. Specifically, and because of its practical importance, it is essential to investigate why employees are willing and motivated to share their knowledge with their colleagues [9, 14].

Within the organization there are different types of knowledge, but due to its strategic attributes, the tacit knowledge stands out. Organizational knowledge is an integral part of the activities carried out within firms and comes from the integration of individual knowledge of its members [12, 32, 58]. Although for taxonomic purposes it is classified both tacitly and explicitly, both types coexist and interact continuously in different situations [12, 79].

Tacit knowledge is distinguished by its strategic contribution. It is an asset that is difficult to transfer or imitate, valuable, unique, non-substitutable, imperfectly movable, and causally ambiguous and is therefore considered a highly valuable organizational resource for building and maintaining sustainable competitive advantage [3, 26, 28, 58]. Compared to the explicit, the tacit one is more difficult to express, as it is intuitive and cannot be verbalized; it consists of mental models, beliefs and perspectives that are difficult to transmit [70]. Moreover, it is impossible to find it in databases, textbooks, or manuals and therefore it is internalized in people's minds and souls [36]. In this research work, this type of knowledge is analyzed. This is because it is inferred that explicit knowledge is documented and available to employees, in manuals and other types of instruments.

Knowledge transfer is defined as the process through which organizational actors (senders and recipients) exchange, apply and are influenced by the experience and knowledge of others [58, 82]. Although it is carried out by individuals, the implications are greater for the groups involved [5]. Thanks to the transfer, knowledge is shared, disseminated and accumulated and, consequently, productive capacities are renewed, which in turn allows tasks to be carried out in more effective and efficient ways [58, 60, 78, 82]. It should be noted that in order to be classified as a value-adding activity, it must bring about changes in existing practices, policies and behavior, as well as develop new processes, ideas, practices and policies [28].

The intention to share knowledge is a fundamental condition for the successful transfer of knowledge and is conceptualized as the willingness of the sender, individually, to share this intellectual asset; this is because a necessary condition is to reveal parts of one's knowledge [28, 37]. Originally, the process is complicated by the characteristics of tacit knowledge, but the situation becomes even more complex when the person refuses to share [51]. Therefore, although there are many advantages to sharing knowledge, it must be clear that the intention to share is not a pre-existing condition in the organizational environment but is associated with the will of the people. Other works, postulate variables such as intra-organizational navigation of employees (understood as a proactive behavior that affects job performance and that implies the discovery and use of other employees, resources and organizational skills), highlighting the need for proactive interactions to occur with different people, groups or functional areas [72]; that is, the individual performance of people depends on their ability to relate to others.

There are several barriers to individual knowledge sharing. On the one hand, people do not always share what they know, because they are not willing to do so [51]. In addition, sometimes employees do not share because they think their knowledge is at risk, or they fear that it will be considered irrelevant or unimportant [14]. Moreover, because of the intrinsic value of knowledge, people avoid sharing it and thus generate a hoarding of experience [31, 39]. In short, knowledge sharing depends on the intrinsic motivation of the sender to share, it cannot be forced [45].

Motivation plays a fundamental role when it comes to sharing knowledge. In this sense, it is necessary to understand why individuals choose to share in certain cases and avoid it in others [74]. On the one hand, if employees believe that their knowledge will contribute to organizational performance, their willingness to share will be greater [59]. In addition, if the parties to the exchange have an expectation that value will be created, their willingness to participate will be greater, so they should anticipate that interacting, exchanging and combining knowledge will be worthwhile, even if they are not certain [69]. On the other hand, by sharing knowledge, the sender enjoys thinking that the receiver values what is offered [25, 38]. Finally, another relevant aspect is the perception of competence or self-efficacy [59]: if the issuer feels that his or her ability will contribute to improved organizational performance, his or her attitude toward sharing will be more positive [13]. In summary, the intention to share predicts knowledge transfer [13,50]. Therefore, it is proposed that:

H1. The intention to share knowledge positively and significantly influences the transfer of knowledge.

Knowledge transfer is a complex activity, product of prerequisites and contextual factors [60]. One of the factors that influences it is reciprocity, which is defined as the practice of giving and receiving something that is valuable to the parties [73]. This concept refers to the mutual reinforcement between two parties, regarding the actions carried out between them [27]. Socially, it is a component of the person's cognitive system (values, ideas and experiences), which collects information, facts and feelings associated with past exchanges; based on it, the expected value of current and future decisions is evaluated and commitments are determined [71, 77]. In short, reciprocal relationships are personal ties that can be improved through contributions or support, from which help is expected in the future [53].

Reciprocity is a fundamental aspect of human activities. On the one hand, individuals engage in social exchanges, anticipating that their efforts will encounter reciprocity [7]. Moreover, they think that in return for their actions, they will receive a reward that justifies the costs they incur [57]. In this type of exchange one party benefits another and although there is a general expectation of reciprocity, the exact nature of the return is not specified, nor is there an exact price, so it is impossible to negotiate about it [11, 19].

The expectation of reciprocity influences the intention to share and the transfer of knowledge. The relationship of giving and receiving encourages the sharing of knowledge between people [2], and influences the quantity and quality of shared knowledge [55]. By transferring knowledge, the issuer is confident that its conduct will be reciprocated through a benefit in the future [17]. This is because reciprocal relationships between co-workers increase communication and the exchange of this intangible asset [2] and can facilitate exchange [14]. Therefore, the attitude to exchange knowledge will be more positive in people who seek more and better relationships with their colleagues and in those who trust their colleagues and are interested in reciprocity (i.e., expect to receive something from others in the future) [13, 14, 54]. Then, it is considered that:

H2. The expectation of reciprocity influences knowledge transfer positively and significantly.

On the other hand, the willingness to share also depends on reciprocity [64]. When employees believe that they are working in an equitable environment, and that their reciprocal relationships with others can be improved through their sharing of knowledge, they will exhibit a positive attitude towards that activity [48]. Finally, it has been observed that reciprocal benefits can be an effective motivator to facilitate the intention to share knowledge and thus achieve mutual cooperation: if employees believe that by sharing their knowledge they can obtain reciprocal benefits from other colleagues, they will be more likely to perceive that activity as favorable and positive so motivational factors such as self-efficacy of knowledge, reciprocal benefits and enjoyment or joy in helping others, are those that drive the employee to have the intention of sharing what he knows with his colleagues [59]. Therefore, it is suggested that:

H3. The expectation of reciprocity has a positive and significant impact on the intention to share knowledge.

Interpersonal relationships impact the achievement of the firm's goals. In first instance, they constitute a valuable organizational resource for the conduct in social affairs since they provide to their members a capital of collective property, which is embedded within the networks of knowledge and mutual recognition [16]. On the other hand, they motivate people to act in a way that generates mutual benefits [6]. Although formal and informal relationships coexist in the organization, the latter has attracted greater interest from researchers. This type of interpersonal relations is understood as any kind of contacts, voluntarily initiated by the actors [1]; the original purpose is to strengthen bonds of empathy and solidarity in the environment of work.

3 Methodology

The research design was quantitative, non-experimental, cross-sectional, correlational-causal in scope. Quantitative, because latent variables were measured and hypotheses were tested with inferential statistical techniques; non-experimental, because no changes were provoked in the independent variables, but only existing situations were observed; correlational, because we sought to determine and measure the degree of relationship between variables; and finally, transversal, because data were collected at a moment in time [43].

The sample was intentional non-probabilistic. The field work was carried out from September 26 to 30, 2016. Voluntary collaboration was sought from hotel employees who allowed access to carry out the survey. A total of 187 people participated, of which 1 questionnaire was eliminated, as it was incomplete; therefore, 186 useful responses were available, representing 99.5% of the responses received. This sample size was adequate, since it met the minimum recommended sample for the proposed model (the highest number of structural relations received by an endogenous variable was multiplied by 10, in this case 3) [21].

As a data collection technique, the survey was used, through a self-administered questionnaire, with three sections: informed consent, assessment of the variables studied and demographic data. In the first section, the objective of the work was communicated, and the participant was asked for his or her consent to use the information in an aggregate form. In the second section, the variables to be studied were presented, which were measured through statements that were assessed with a Likert-type measurement scale, with five points of response assignment, from totally disagree to totally agree. Finally, items were presented with demographic aspects regarding sex, age, schooling, position in the hotel and age of the participant, and category and type of hotel chain. The variables of the study were measured with items from scales available in the literature, translated into Spanish, and adapted to the context. Table 1 shows the items used to measure the variables. Twenty-four items were included in the initial measurement model: the expectation of reciprocity was measured with 5 items from the Kankanhalli et al. [53] and Lin [59] scales; interpersonal relationships were assessed with 6 items built from the Plouffe and Grégoire [72], Vaux and Harrison [83], Requena Santos [75], Chua [23] and Chow and Chan [22] scales; intention to share knowledge, with 6 items from the Lin [59] and Maynez Guaderrama [66] scales; and knowledge transfer with 7 items from the Maynez Guaderrama [66] scale.

As a statistical analysis technique, structural equation modeling (SEM) with partial least squares (PLS) was used. Among the main reasons are its ability to model multiple relationships simultaneously and its ability to control the endogeneity problem. PLS is also a preferred technique to model relationships when one variable is dependent and independent at the same time. Also, this statistical technique is considered highly robust against non-normal data [61]. The analysis was performed in two stages: in the first stage, the (external) measurement model was reviewed to ensure the reliability and validity of the constructs (variables); then, in the second stage, the (internal) structural model was evaluated to assess the predictive relevance of the model and thus test the research hypotheses [4, 35]. Smart PLS 3 statistical software was used. As a first step, the convergent validity of the proposed measurement model was reviewed. Because it exhibited low factorial loads or generated discriminant validity problems, five items were eliminated: one from the expectation of reciprocity (reci1); two from interpersonal relationships (rel4 and rel5); one from intention to share knowledge (int3), and one from knowledge transfer (tran1). Then, the final measurement model had 19 items. Table 2 shows the final statistics of the refined measurement model; in all cases, the statistics met the recommended cut-off criteria.

In addition to the above, the discriminant validity of the measurement model was evaluated. The Fornell-Larcker criterion and the Heterotrait-Monotrait ratio were used for this purpose. The results of both tests are shown in Table 3.

In the first section of the table, the value of the square root of the Analysis of Variance Extracted (AVE) is shown; this value was compared with the correlations between constructs (shown under the diagonal of the first section of Table 3). In none of the cases did the construct correlation exceed the AVE square root, indicating discriminant validity. In addition, the value of the heterotrait-monotrait ratios was checked to ensure that they did not exceed the established cut-off point of 0.85. In all cases the values of the ratios were lower than the cut-off point. According to this statistical test, it was proved that the measurement model has discriminant validity.

4 Results

The demographic statistics of the sample are the following: the majority of the participants were men (62.3%); 77.8% of the participants were between 20 and 40 years old; 38.2% had the equivalent of high school education or a professional career that had

Table 1 Operationalization of variables

Construct	Items
Expectation of reciprocity	When I share my knowledge, I expect my co-workers to help me when I need it. (reci1)
	I share my knowledge with my co-workers with confidence (reci2)
	When I share my knowledge with colleagues, my relationship with them becomes stronger (reci3)
	When I share my knowledge, I think I will get an answer when I have doubts (reci4)
	When I share my knowledge with colleagues, I expect them to share what they know, when I need it (reci5)
Interpersonal relations	I am good at relating to the people who make decisions in the hotel (rel1)
	I have a good relationship with the decision makers (rel2)
	I take time to develop relationships with others (rel3)
	I use my personal relationships to sole work situations (rel4)
	I have a good relationship with the people I work with (rel5)
	I often communicate with my co-workers (rel6)
Intent to share knowledge	When I am asked to share my knowledge, I am generally willing to help and to teach (int1)
	I am willing to help any colleague, as I am sure he appreciates the value of what I share with him (int2)
	I am confident in my ability to share knowledge with my co-workers (int3)
	I have the experience to share valuable knowledge (int4)
	I am sure it is important to share my knowledge (int5)
Knowledge transfer	I am sure I can share important knowledge for this hotel (int6)
	It is common for employees to share their knowledge (tran1)
	Thanks to the shared knowledge, the activities that are developed are increasingly efficient (tran2)
	Based on shared knowledge, we make changes in the way we work (tran3)
	Shared knowledge becomes part of normal work activities (tran4)
	By sharing new knowledge, we are able to create new services (tran5)
	By sharing knowledge, we are able to create new work activities (tran6)
	By sharing knowledge, we are able to better perform the services (tran7)

Source Own elaboration

not been completed; 13.5% worked as waiters, 13.0% as managers or area managers, and 10.3% as administrative assistants; 31.1% had less than one year's seniority and 19.7% had 1 or 2 years working in the hotel. 84.9% of participants reported working in Mexican chain hotels and 52.2% indicated that the hotel category was 5 stars.

Figure 1 shows the magnitudes of the structural relationships described in the model. According to the results, five of the six proposed scenarios were not rejected. This implies that in the context of hotels in Cozumel Island, interpersonal relationships influence the intention to share knowledge (path = 0.459, t = 7.067) and the expectation of reciprocity (path = 0.469, t = 6.969), but do not directly affect knowledge transfer (path = 0.145, t = 1.758). That is, interpersonal relationships do influence the transfer of knowledge, but this occurs through the expectation of reciprocity and the intention to share knowledge, which are derived from relationships of trust and empathy between subjects. The expectation of reciprocity directly influences the transfer of knowledge (path = 0.225, t = 2.520, $p < 0.05$) therefore H2 was supported. Moreover, the structural link between the expectation of reciprocity and the intention to share knowledge was significant at a level of $p < 0.001$ (path = 2.30, t = 3.321), which implies that H3 was supported; that is, this expectation

Table 2 Convergent validity of the measurement model

Construct	Average factorial load	Cronbach's Alpha	Composite reliability	Analysis of variance extracted
Expectation of reciprocity	0.736	0.723	0.826	0.544
Interpersonal relations	0.730	0.711	0.821	0.536
Intention to share knowledge	0.733	0.786	0.855	0.544
Knowledge transfer	0.765	0.859	0.895	0.587

Source Prepared by the authors based on Smart PLS results

Table 3 Discriminant validity of the measurement model

Construct	Fornell-Larcker criterion				Heterotrait-Monotrait ratio		
	IC	ER	RI	TC	IC	ER	RI
Intention to share knowledge (IC)	0.737						
Expectation of reciprocity (ER)	0.445	0.737			0.569		
Interpersonal relations (RI)	0.567	0.469	0.732		0.756	0.629	
Knowledge transfer (TC)	0.481	0.426	0.420	0.766	0.567	0.537	0.523

Source Own elaboration based on results of Smart PLS 3.0

directly and indirectly affects the transfer of knowledge, The aim of the project is to promote the development of the community, through the mediation of the intention to share knowledge. Finally, the intention to share knowledge has a positive and direct impact on the transfer of knowledge within the hotels (path = 0.299, t = 2.955, $p < 0.05$) thus H1 was supported.

According to coefficient R^2, the structural relations of the model explain 0.220 of the variances of the expectation of reciprocity, 0.363 of the variances of the intention to share knowledge, and 0.301 of the variance of knowledge transfer within hotels. In other words, interpersonal relationships explain in a way: weak to the expectation of reciprocity; moderate to the intention to share knowledge, together with the expectation of reciprocity; moderate to the transfer of knowledge within the hotels, simultaneously with the expectation of reciprocity and the intention to share.

To determine the predictive validity of the model, the Stone Geisser Q^2 indicator was reviewed and calculated through the blindfolding procedure in the Smart PLS 3 software [20, 41]. The results indicate the existence of predictive validity since the Q^2 values (Table 4) of the endogenous variables of the model were greater than zero [8, 20].

Table 5 shows the direct, indirect, and total effects of the proposed relationships. The strongest relationship is between interpersonal relationships and the intention to share knowledge (0.567), through the mediation of the expectation of reciprocity (0.108). In addition, interpersonal relationships also directly and indirectly influence the expectation of reciprocity (0.469). Finally, interpersonal relationships, through the mediation of the two previous relationships, indirectly impact the transfer of knowledge (0.275).

Our results are in line with the approaches that emphasize that knowledge is not a commodity, and in order to use it in problem solving, it needs to be shared among colleagues, friends and experts [40]. Likewise, the findings show the need for business support, if what is sought is collaboration and work in informal networks [24]. Thus, it is reiterated that social interaction and close, cooperative, reliable, reciprocal and

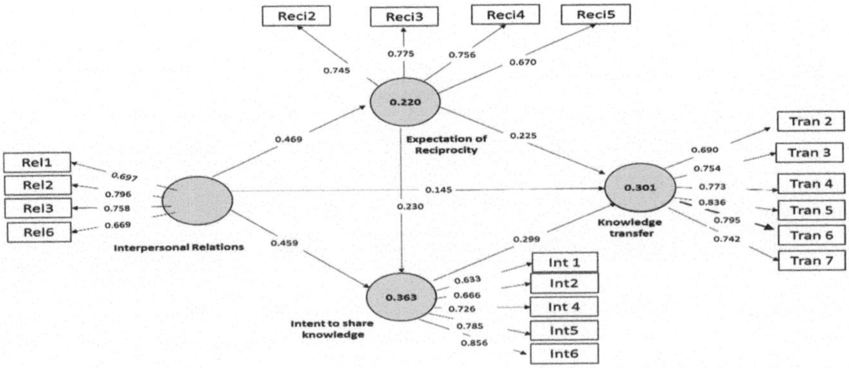

Fig. 1 Contrasted model. *Source* Smart PLS

Table 4 Indicator Q^2 of the endogenous variables of the model

Construct	Sum of squares of prediction error (SSO)	Sum of squares of the error using the average for the prediction (SSE)	Cross validation of Redundancy construct $Q^2 = 1 - (SSE/SSO)$
Intention to share knowledge	930,000	764,483	0.178
Expectation of reciprocity	744,000	665,303	0.106
Knowledge transfer	1,116,000	944,297	0.154
Intention to share knowledge	930,000	764,483	0.178

Source Own elaboration based on results of Smart PLS 3.0

Table 5 Direct, indirect, and total effects of the relationships proposed

Structural path	Direct effect	Indirect effect	Total effect
Intention to share knowledge → Transfer of knowledge	0.299 (2.955)*		0.299 (2.955)*
Expectation of reciprocity → Knowledge transfer	0.225 (2.520)*	0.069 (2.127)*	0.293 (3.562)*
Expectation of reciprocity intention → Share knowledge	0.230 (3.321)*		0.230 (3.321)*
Interpersonal relationships Intention → Share knowledge	0.459 (7.067)**	0.108 (2.719)*	0.567 (10.844)**
Interpersonal relations → Knowledge transfer	0.145 (1.758)	0.275 (4.512)**	0.420 (6.395)**
Interpersonal relations → Expectation of reciprocity	0.469 (6.969)**		0.469 (6.969)**

*$p < 0.05$; **$p < 0.01$
Source Own elaboration

healthy interpersonal relationships favors knowledge transfer, which coincides with various studies [6, 22, 29, 30, 37, 52, 56, 74, 81].

On the other hand, it is worth noting the direct and indirect influence of the expectation of reciprocity on the transfer of knowledge (0.293), through the mediation of the intention to share this intangible asset (0.069). These results converge with academic precedents published in the early works of Han and Pashouwers [37] and later of Wills and DePaulo [84], which indicate that informal networks involve frequent and lasting interpersonal relationships where it is possible to find reciprocity. In other words, people seek out their friends when they need help and share their knowledge with them as this strengthens their expectation of future benefits [29, 37, 52, 84]. In terms of intent to share and transfer knowledge, reciprocity is relevant because the sender believes that personal ties with the recipient can be enhanced by contributions or support that occur in the present and expects that their conduct will be reciprocated in the future. In this sense, some studies have offered evidence [17, 53]. Our findings support the postulates that the tacit culture of 'give and take' encourages the sharing of knowledge and influences the quantity and quality of knowledge shared [2, 14, 55]; it enables employees to have a more positive attitude towards sharing what they know, because they believe that their relationships of solidarity can be improved through the sharing of knowledge [48, 59].

To conclude, it only remains to indicate that the variable with the greatest direct influence on knowledge transfer was the intention to share. These results coincide with previous work indicating that workers will be more likely to transfer, when they think they are competent and effective and consider that it will be worthwhile, since they perceive that their knowledge will favors the performance of the firm [13, 59, 69].

5 Conclusions

Sharing knowledge is not only essential for improving the performance of individuals and companies, it is a moral challenge for organizations; it is also a pro-social activity that benefits cooperation among employees and group effectiveness [9]. When individual knowledge is communicated, social acceptance and friendship is encouraged and promoted. In this research work, we investigated, in the context of hotels located on the island of Cozumel, whether interpersonal relationships and the expectation of reciprocity influence the transfer of knowledge, through the mediation of the intention to share this intangible asset. The findings show that, because of sharing knowledge, services and work activities are standardized and create a new added value to the client. Then, from this shared knowledge of new ways of doing things, productive processes are transformed, and these renewed activities become part of everyday life. In the hotel context, this can happen in the improvement of customer check-in or check-out times or in follow-up programs to customer complaints.

In this research, the intention to share was the main direct antecedent of knowledge transfer within hotels. This variable was mostly reflected in two aspects: the employees' confidence in the importance of sharing their knowledge and their certainty that they can contribute important knowledge to the hotel in which they work. On the other hand, the greatest effect on knowledge transfer is caused by interpersonal relationships. This is mediated by the intention to share. This construct was reflected, to a greater extent, both in the favorable perception of workers of their relationship with decision-makers and in the time, they spend on developing relationships with their colleagues. In that sense, these findings lead to emphasize the importance of this type of relationship in the organizational context.

Finally, reciprocity also directly and indirectly influences the transfer of knowledge. This variable is expressed in people's confidence in sharing their knowledge with their co-workers and in their certainty that, in doing so, they are strengthening their relationship with them. In this paper, it became clear that interpersonal relationships impact on reciprocity. As with other research, this effort has limitations. This is a transversal analysis, in a reduced time horizon, carried out in the Mexican tourism sector located in Cozumel Island, Mexico. In addition, people were accessed in a non-probabilistic way. This limits the possibility of generalizing the results to other contexts and these could change over time. Therefore, it is recommended to replicate the work in other contexts, in a probabilistic sample and try to carry out the analysis with a longitudinal perspective.

References

1. Aalbers, R., Dolfsma, W., Koppius, O.: Rich ties and innovative knowledge transfer within a firm. Br. J. Manage. **25**(4), 833–848 (2014). https://doi.org/10.1111/1467-8551.12040
2. Abdelwhab Ali, A., Panneer Selvam, D.D.D., Paris, L., Gunasekaran, A.: Key factors influencing knowledge sharing practices and its relationship with organizational performance within the oil and gas industry. J. Knowl. Manage. (2019). https://doi.org/10.1108/JKM-06-2018-0394
3. Ambrosini, V., Bowman, C.: Tacit knowledge: some suggestions for operationalizacion. J. Manage. Stud. **38**(6), 811–829 (2001). https://doi.org/10.1111/1467-6486.00260
4. Anderson, J.C., Gerbing, D.W.: Some methods for respecifying measurement models to obtain unidimensional construct measurement. J. Market. Res. **19**(4), 453–460 (1982). https://doi.org/10.1177/002224378201900407
5. Argote, L., Ingram, P.: Knowledge transfer: a basis for competitive advantage in firms. Organizat. Behav. Human Dec. Process. **82**(1), 150–169 (2000). https://doi.org/10.1006/obhd.2000.2893
6. Asrar-ul-Haq, M., Anwar, S.: A systematic review of knowledge management and knowledge sharing: trends, issues, and challenges. Cogent Bus. Manage. **3**(1), 1127744 (2016). https://doi.org/10.1080/23311975.2015.1127744
7. Bal, P.M., Jansen, P.G.W., van der Velde, M.E.G., de Lange, A.H., Rousseau, D.M.: The role of future time perspective in psychological contracts: a study among older workers. J. Vocat. Behav. **76**(3), 474–486 (2010)
8. Barroso, C., Carrión, G.C., Roldán, J.L.: Applying maximum likelihood and PLS on different sample sizes: studies on SERVQUAL model and employee behavior model. In: Vinzi, V.E., Chin, W.W., Henseler, J., Wang, H. (eds.) Handbook of Partial Least Squares: Concepts, Methods and Applications, pp. 427–447. Springer Berlin Heidelberg, Berlin, Heidelberg (2010)
9. Bavik, Y.L., Tang, P.M., Shao, R., Lam, L.W.: Ethical leadership and employee knowledge sharing: exploring dual-mediation paths. Leader. Quart. **29**(2), 322–332 (2018). https://doi.org/10.1016/j.leaqua.2017.05.006
10. Berends, H., Vanhaverbeke, W., Kirschbaum, R.: Knowledge management challenges in new business development: case study observations. J. Eng. Technol. Manag. Jet-M **24**(4), 314–328 (2007). https://doi.org/10.1016/j.jengtecman.2007.09.006
11. Blau, P.M.: On the Nature of Organizations. Wiley, New York (1974)
12. Bloodgood, J.M., Morrow Jr., J.L.: Strategic organizational change: exploring the roles of environmental structure, internal conscious awareness and knowledge. J. Manage. Stud. **40**(7), 1761–1782 (2003). https://doi.org/10.1111/1467-6486.00399
13. Bock, G.-W., Kim, Y.-G.: Breaking the myths of rewards: an exploratory study of attitudes about knowledge sharing. Inf. Res. Manage. J. **15**(2), 14–21 (2002). https://doi.org/10.4018/irmj.2002040102

14. Bock, G.-W., Zmud, R.W., Kim, Y.-G., Lee, J.-N.: Behavioral intention formation in knowledge sharing: examining the roles of extrinsic motivators, social-psychological forces, and organizational climate. MIS Q. **29**(1), 87–111 (2005)
15. Bollinger, A.S., Smith, R.D.: Managing organizational knowledge as a strategic asset. J. Knowl. Manage. **5**(1), 8–18 (2001). https://doi.org/10.1108/13673270110384365
16. Bourdieu, P.: The forms of capital. In: Handbook of Theory and Research for the Sociology of Education, pp. 241–258. Greenwood, New York (1986)
17. Cabrera, E.F., Cabrera, A.: Fostering knowledge sharing through people management practices. Int. J. Hum. Res. Manage. **16**(5), 720–735 (2005). https://doi.org/10.1080/09585190500083020
18. Castro, C., Fontela, E.: Transfer of socially complex knowledge in mergers and acquisitions. J. Knowl. Manage. **11**(4), 58–71 (2007). https://doi.org/10.1108/13673270710762710
19. Chadwick-Jones, J.K.: Social Exchange Theory: Its Structure and Influence in Social Psychology. Academic Press, New York (1976)
20. Chin, W.W.: How to write up and report PLS analyses. In: Vinzi, V.E., Chin, W.W., Henseler, J., Wang, H. (eds.) Handbook of Partial Least Squares: Concepts, Methods and Applications. pp. 655–690. Springer Berlin Heidelberg, Berlin, Heidelberg (2010)
21. Chin, W.W., Newsted, P.R.: Structural equation modeling analysis with small samples using partial least squares. In: Hoyle, R.H. (ed.) Statistical Strategies for Small Sample Research, vol. 1, pp. 307–341. Sage Publications, Thousand Oaks, California (1999)
22. Chow, W.S., Chan, L.S.: Social network, social trust and shared goals in organizational knowledge sharing. Inf. Manag. **45**(7), 458–465 (2008). https://doi.org/10.1016/j.im.2008.06.007
23. Chua, A.: The influence of social interaction on knowledge creation. J. Intell. Capit. **3**(4), 375–392 (2002). https://doi.org/10.1108/14691930210448297
24. Cross, R., Borgatti, S.P., Parker, A.: Making invisible work visible: using social network analysis to support strategic collaboration. Calif. Manag. Rev. **44**(2), 25–46 (2002). https://doi.org/10.2307/41166121
25. Dixon, N.: The neglected receiver of knowledge sharing. Ivey Bus. J. **66**(4), 35–40 (2002)
26. Dyer, J.H., Hatch, N.W.: Using supplier networks to learn faster. MIT Sloan Manage. Rev. **45**(3), 57–63 (2004)
27. Ekeh, P.P.: Social Exchange Theory: The Two Traditions. Harvard University Press, Cambridge (1974)
28. Evanschitzky, H., Ahlert, D., Blaich, G., Kenning, P.: Knowledge management in knowledge-intensive service networks: a strategic management approach. Manage. Dec. **45**(2), 265–283 (2007). https://doi.org/10.1108/00251740710727287
29. Fullwood, R., Rowley, J., Delbridge, R.: Knowledge sharing amongst academics in UK universities. J. Knowl. Manage. **17**(1), 123–136 (2013). https://doi.org/10.1108/13673271311300831
30. Ghobadi, S., D'Ambra, J.: Knowledge sharing in cross-functional teams: a coopetitive model. J. Knowl. Manage. **16**(2), 285–301 (2012). https://doi.org/10.1108/13673271211218889
31. Granovetter, M.S.: The strength of weak ties. Am. J. Sociol. **78**(6), 1360–1380 (1973)
32. Grant, R.M.: Toward a knowledge-based theory of the firm. Strat. Manage. J. **17**(Winter Special Issue), 109–122 (1996). https://doi.org/10.1002/smj.4250171110
33. Guofeng, M., Jianyao, J., Shan, J., Zhijiang, W.: Incentives and contract design for knowledge sharing in construction joint ventures. Automat. Const. **119**, 1–11 (2020). https://doi.org/10.1016/j.autcon.2020.103343
34. Hafeez, K., Abdelmeguid, H.: Dynamics of human resource and knowledge management. J. Oper. Res. Soc. **54**(2), 153–164 (2003). https://doi.org/10.1057/palgrave.jors.2601513
35. Hair Jr., J.F., Hult, G.T.M., Ringle, C.M., Sarstedt, M.: A Primer on Partial Least Squares Structural Equation Modeling (PLS-SEM) (2nd ed.). Sage Publications, Inc. (2016)
36. Haldin-Herrgard, T.: Difficulties in diffusion of tacit knowledge in organizations. J. Intell. Capit. **1**(4), 357–365 (2000). https://doi.org/10.1108/14691930010359252

37. Han, J., Pashouwers, R.: Willingness to share knowledge in healthcare organisations: the role of relational perception. Knowl. Manage. Res. Pract. **16**(1), 42–50 (2018). https://doi.org/10.1080/14778238.2017.1405144
38. Hansen, M.T.: The search-transfer problem: the role of weak ties in sharing knowledge across organization subunits. Adm. Sci. Q. **44**(1), 82–111 (1999). https://doi.org/10.2307/2667032
39. Hansen, M.T., Nohria, N.: How to build collaborative advantage. MIT Sloan Manage. Rev. **46**(1), 22–30 (2004)
40. Henriksen, L.: Knowledge management and engineering practices: the case of knowledge management, problem solving and engineering practices. Technovation **21**(9), 595–603 (2001). https://doi.org/10.1016/S0166-4972(01)00023-2
41. Henseler, J., Ringle, C.M., Sinkovics, R.R.: The use of partial least squares path modeling in international marketing. In: Sinkovics, R.R. Ghauri, P.N. (eds.) New Challenges to International Marketing Advances in International Marketing, vol. 20, pp. 277–319. Emerald Group Publishing Limited (2009)
42. Henttonen, K., Janhonen, M., Johanson, J.-E.: Internal social networks in work teams: structure, knowledge sharing and performance. Int. J. Manpower **34**(6), 616–634 (2013). https://doi.org/10.1108/IJM-06-2013-0148
43. Hernández Sampieri, R., Fernández Collado, C., Baptista Lucio, M.d.P.: Metodología de la investigación (6a. ed.). Mc Graw Hill Education, México (2014)
44. Herschel, R.T., Nemati, D.R.: Chief knowledge officer: critical success factors for knowledge management. Inf. Strat. Executive's J. **16**(4), 37–45 (2000). https://doi.org/10.1080/07438613.2000.10744623
45. Hooff, B.v.d.: What one feels and what one knows: the influence of emotions on attitudes and intentions towards knowledge sharing. J. Knowl. Manage. **16**(1), 148–158 (2012). https://doi.org/10.1108/13673271211198990
46. Hsu, I.-C.: Knowledge sharing practices as a facilitating factor for improving organizacional performance through human capital: a preliminary test. Expert Syst. Appl. **35**(3), 1316–1326 (2008). https://doi.org/10.1016/j.eswa.2007.08.012
47. Hua, T.K., Platts, K.: A connectance-based approach for managing manufacturing knowledge. Ind. Manage. Data Syst. **104**(2), 158–168 (2004). https://doi.org/10.1108/02635570410522134
48. Huber, G.P.: Transfer of knowledge in knowledge management systems: unexplored issues and suggested studies. Eur. J. Inf. Syst. **10**(2), 72–79 (2001)
49. Hung, Y.-C., Chuang, Y.-H.: Factors affecting knowledge sharing behavior: a content analysis of empirical findings. Paper presented at the International Conference of Pacific Rim Management, July (2009)
50. Israilidis, J., Siachou, E., Cooke, L., Lock, R.: Individual variables with an impact on knowledge sharing: the critical role of employees' ignorance. J. Knowl. Manage. **19**(6), 1109–1123 (2015). https://doi.org/10.1108/JKM-04-2015-0153
51. Jolaee, A., Md Nor, K., Khani, N., Md Yusoff, R.: Factors affecting knowledge sharing intention among academic staff. Int. J. Educat. Manage. **28**(4), 413–431 (2014). https://doi.org/10.1108/IJEM-03-2013-0041
52. Kachra, A., White, R.E.: Know-how transfer: the role of social, economic/competitive, and firm boundary factors. Strateg. Manag. J. **29**(4), 425–445 (2008). https://doi.org/10.1002/smj.668
53. Kankanhalli, A., Tan, B.C., Wei, K.-K.: Contributing knowledge to electronic knowledge repositories: an empirical investigation. MIS Q. **29**(1), 113–143 (2005)
54. Kwon, K., Mai, L., Peng, N.: Determinants of consumers' intentions to share knowledge and intentions to purchase on s-commerce sites: incorporating attitudes toward persuasion attempts into a social exchange model. Euras. Bus. Rev. **10**, 157–183 (2020). https://doi.org/10.1007/s40821-019-00146-5
55. Lai, H.-M., Hsieh, P.-J.: The effects of the norm of reciprocity and interpersonal trust on knowledge-sharing behavior: a longitudinal analysis. Paper presented at the Pacific Asia Conference on Information Systems. PACIS 2018 Proceedings (2018)

56. Levin, D.Z., Walter, J., Appleyard, M.M., Cross, R.: Relational enhancement: how the relational dimension of social capital unlocks the value of network-bridging ties. Group Org. Manage. **41**(4), 415–457 (2015). https://doi.org/10.1177/1059601115574429
57. Liao, L.-F.: Knowledge-sharing in RyD departments: a social power and social exchange theory perspective. Int. J. Hum. Res. Manage. **19**(10), 1881–1895 (2008). https://doi.org/10.1080/09585190802324072
58. Liao, S.-H., Hu, T.-C.: Knowledge transfer and competitive advantage on environmental uncertainty: an empirical study of the Taiwan semiconductor industry. Technovation **27**(6–78), 402–411 (2007). https://doi.org/10.1016/j.technovation.2007.02.005
59. Lin, H.-F.: Effects of extrinsic and intrinsic motivation on employee knowledge sharing intentions. J. Inf. Sci. **33**(2), 135–149 (2007). https://doi.org/10.1177/0165551506068174
60. Liyanage, C., Elhag, T., Ballal, T., Li, Q.: Knowledge communication and translation—a knowledge transfer model. J. Knowl. Manage. **13**(3), 118–131 (2009). https://doi.org/10.1108/13673270910962914
61. Mahmood, T., Mubarik, M.S.: Balancing innovation and exploitation in the fourth industrial revolution: role of intellectual capital and technology absorptive capacity. Technol. Forecast. Soc. Chang. **160**, 1–9 (2020). https://doi.org/10.1016/j.techfore.2020.120248
62. Malone, D.: Knowledge management. A model for organizational learning. Int. J. Account. Inf. Syst. **3**(2), 111–123 (2002). https://doi.org/10.1016/S1467-0895(02)00039-8
63. Marco-Lajara, B., Zaragoza-Sáez, P.d.C., Claver-Cortés, E., Úbeda-García, M.: Hotel performance and knowledge sources in Spanish tourism districts. Curr. Issues Tour. **21**(17), 1988–2013 (2018). https://doi.org/10.1080/13683500.2016.1225696
64. Martín, N., Martín, V., Trevilla, C.: The influence of employee motivation on knowledge transfer. J. Knowl. Manage. **13**(6), 478–490 (2009). https://doi.org/10.1108/13673270910997132
65. Massingham, P.: An evaluation of knowledge management tools: part 1—managing knowledge resources. J. Knowl. Manage. **18**(6), 1075–1100 (2014). https://doi.org/10.1108/JKM-11-2013-0449
66. Maynez Guaderrama, A.I.: La Transferencia de Conocimiento Organizacional como Fuente de Ventaja Competitiva Sostenible: Modelo Integrador de Factores y Estrategias (Doctorado Documento), Universidad Popular Autónoma del Estado de Puebla, Puebla, Mexico (2011)
67. Mazloomi, H., Jolly, D.R.: Knowledge transfer in alliances: determinant factors. J. Knowl. Manage. **12**(1), 37–50 (2008). https://doi.org/10.1108/13673270810852377
68. Moffett, S., McAdam, R., Parkinson, S.: Developing a model for technology and cultural factors in knowledge management: a factor analysis. Knowl. Process Manage. **9**(4), 237–255 (2002). https://doi.org/10.1002/kpm.152
69. Nahapiet, J., Ghosal, S.: Social capital, intellectual capital, and the organizational advantage. Acad. Manag. Rev. **23**(2), 242–266 (1998). https://doi.org/10.5465/amr.1998.533225
70. Nonaka, I., Takeuchi, H.: The Knowledge-Creating Company: How Japanese Companies Create the Dynamics of Innovation. Oxford University Press (1995)
71. Pesämaa, O., Pieper, T., da Silva, R.V., Black, W.C., Hair, J.F.: Trust and reciprocity in building inter-personal and inter-organizational commitment in small business co-operatives. J. Cooperat. Organizat. Manage. **1**(2), 81–92 (2013). https://doi.org/10.1016/j.jcom.2013.10.003
72. Plouffe, C.R., Grégoire, Y.: Intraorganizational employee navigation and socially derived outcomes: conceptualization, validation, and effects on overall performance. Person. Psychol. **64**(3), 693–738 (2011). https://doi.org/10.1111/j.1744-6570.2011.01223.x
73. Portes, A.: Social capital: its origins and applications in modern sociology. In: LESSER, Eric L. Knowledge and Social Capital, pp. 43–67. Butterworth-Heinemann, Boston (2000). https://doi.org/10.1146/annurev.soc.24.1.1
74. Reagans, R., McEvily, B.: Network structure and knowledge transfer: the effects of cohesion and range. Administ. Sci. Quart. **48**(2), 240–267 (2003). https://doi.org/10.2307/3556658
75. Requena Santos, F.: Análisis de redes sociales: Orígenes, teorías y aplicaciones, 497 pp. Centro de Investigaciones Sociológicas, Madrid

76. Rhodes, J., Hung, R., Lok, P., Ya-Hui Lien, B., Wu, C.-M.: Factors influencing organizational knowledge transfer: implication for corporate performance. J. Knowl. Manage. **12**(3), 84–100 (2008).https://doi.org/10.1108/13673270810875886
77. Settoon, R.P., Bennett, N., Liden, R.C.: Social exchange in organizations: perceived organizational support, leader–member exchange, and employee reciprocity. J. Appl. Psychol. **81**(3), 219–227 (1996). https://doi.org/10.1037/0021-9010.81.3.219
78. Shamim, S., Cang, S., Yu, H.: Supervisory orientation, employee goal orientation, and knowledge management among front line hotel employees. Int. J. Hospit. Manage. **62**(Supplement C), 21–32 (2017). https://doi.org/10.1016/j.ijhm.2016.11.013
79. Sternberg, R.J.: What do we know about tacit knowledge? Making the tacit become explicit. In: Sternberg, R.J., Horvath, J.A. (eds.) Tacit Knowledge in Professional Practice. Researcher and Practitioner Perspectives. Lawrence Erlbaum Associates, Mahwah, New Jersey (1999)
80. Thomas, J., Watts, S., Henderson, J.: Understanding "strategic learning": linking organizational learning, knowledge management, and sensemaking. Organ. Sci. **12**(3), 331–345 (2001). https://doi.org/10.1287/orsc.12.3.331.10105
81. Titi Amayah, A.: Determinants of knowledge sharing in a public sector organization. J Knowl Manage **17**(3), 454–471 (2013). https://doi.org/10.1108/JKM-11-2012-0369
82. Van Wijk, R., Jansen, J.J., Lyles, M.A.: Inter-and intra-organizational knowledge transfer: a meta-analytic review and assessment of its antecedents and consequences. J. Manage. Stud. **45**(4), 830–853 (2008). https://doi.org/10.1111/j.1467-6486.2008.00771.x
83. Vaux, A., Harrison, D.: Support network characteristics associated with support satisfaction and perceived support. Am. J. Community Psychol. **13**(3), 245–265 (1985). https://doi.org/10.1007/BF00914932
84. Wills, T.A., DePaulo, B.M.: Interpersonal analysis of the help-seeking process. In: Snyder, C.R., Forsyth, D.R. (eds.) Handbook of Social and Clinical Psychology: The Health Perspective, vol. 162, pp. 350–375. Pergamon Press, New York (1991)
85. Wong, K.Y., Aspinwall, E.: An empirical study of the important factors for knowledge-management adoption in the SME sector. J Knowl Manage **9**(3), 64–82 (2005). https://doi.org/10.1108/13673270510602773

Modern Technology Applications Including Metaheuristics and Artificial Intelligence Based Applications for Industry 4.0

Brainwaves Behavior During the Learning Curve Associated with the Manufacturing of a Product with Legos

Félix Lira-Casas, Ana García-Acosta, Jorge de la Riva-Rodríguez, and Marco Gallo

Abstract Within the workplace, it is essential to have prior training for the making of a product, either for the training of new staff or improvement in a process. The only disadvantage is that the learning task needs repetition several times until the worker can carry out the assigned task efficiently and within the set time. Unfortunately, in the literature review performed in the area of neuroergonomics, information has not been found regarding the workload effects within the brainwaves during the attainment of a task learning curve. Therefore, this work focused on a learning curve experiment that consists of the manufacture of a product with Legos in an assembly line involving seven people. A wireless Emotiv headband comprised of fourteen Electroencephalogram (EEG) electrodes along with the Emotiv PRO software were used for the brainwaves recording. After performing the statistical analysis, it was found that the presence of more significant brainwave changes occurred during the first ten minutes of the product manufacturing, and these changes were located in the frontal region of the cerebral cortex. The results of this study also indicated higher activity within the Low Beta, High Beta, and Gamma brainwaves than with respect to other brainwaves. This research will assist in the future implementation of more efficient processes based on repetitive tasks

Keywords Brainwaves · Learning curve · EEG · Repetitive tasks

1 Introduction

It is well-known that nonstop improvement is essential nowadays to remain competitive; it is, therefore, necessary for companies to develop strategies and techniques that focus on competitiveness trough process optimization and innovation.

F. Lira-Casas · A. García-Acosta (✉) · J. de la Riva-Rodríguez · M. Gallo
Tecnológico Nacional de México, Instituto Tecnológico de Ciudad Juárez, Ciudad Juárez, Mexico
e-mail: agarcia@itcj.edu.mx

A. Ochoa-Zezzatti et al. (eds.), *Technological and Industrial Applications Associated With Industry 4.0*, Studies in Systems, Decision and Control 347,
https://doi.org/10.1007/978-3-030-68663-5_8

A fundamental feature associated with process optimization involves the manufacture learning curve, in other words, the human repetition of task during the apprenticeship stage associated with the standard time attainment in the assembly line. This accurate timing allows costs minimization and improves productivity.

The learning curve for a manufacturing process, very often involves long times, and also several repetitive tasks before reaching the standard time, creating a monotonous, tired and stressful environment for the workers.

A study conducted by the University of Shanghai, by Ndaro and Wang [1], concentrated on the fatigue displayed during the training of a laparoscopic surgical simulation, and the learning curve was monitored using an electroencephalogram system. This experiment, involved twenty students in the range of 24–26 years of age, each person lacking emotional problems, alcoholism, drug abuse, etc. Besides, none of these students had any experience in the use of surgical simulations, during four continuous hours at room temperature conditions; the students performed the simulation for 20 times, intending to elicit fatigue. At the end of the experiment it was concluded that there was significant learning and fatigue effects when the training was repeated for a series of tests. However, for the training to be effective and efficient, continuous monitoring is necessary, to identify the point within the learning curve of maximum apprenticeship. In addition, the presence of fatigue is a significant signal for the efficiency in terms of the necessary time taken to complete the laparoscopic task and the accuracy obtained in terms of the errors displayed while performing the task. Although these results concentrated only in the laparoscopic surgical training, these principles also apply to doctors involved in patient surgeries. The results of this experiment can also be extrapolated to the manufacturing workplace, to identify the presence of fatigue associated to the learning curve.

The goal of this work is the study of the brainwaves behavior associated with the learning curve of a series of people in the assembly of a product with legos until the standard time attainment.

1.1 Theoretical Framework

In the industry sector, the product needs to comply with the appropriate measurements to fulfill the quality and customer requirements; this goal is obtainable when the work is performed efficiently.

When a new operator is hired or new techniques are implemented within the assembly line, it is necessary to determine the amount of time required, even for the assembly of only part of the product. The obtainment of the standard time can take hours or even weeks; This time is measured using a learning curve [2]. However, the quality of the tasks performed might deteriorate if the human factor is not taken into account; therefore, it is crucial to know to what extent the complexity of the work learned can lead to exhaustion on the workers, its effect on the brainwaves [3].

The human brain in any state produces electrical impulses traveling through neurons. These impulses are called brainwaves, displaying different patterns that can be read and interpreted using an electroencephalogram and are involved with different states of consciousness. These signals are captured using electrodes that are located in specific places on the scalp. [4], and divided into five different frequency periods: delta brainwaves, these signals have the lowest frequency between 0.5 Hz and 3 Hz; theta is frequency is comprised in the 4–7 Hz; alpha between 8 and 13 Hz; beta, being one of the fastest brainwave, ranges between 14 and 30 Hz; gamma brainwaves are associated with frequencies higher than 30 Hz [5].

Also, the cerebral cortex presents five divisions classified into different parts or lobes; These lobes are associated with the performance of certain daily basis functions, such as The Frontal lobe, region located in front of the brain, is function involves reasoning, decision making and its related with the work processing speed at any moment. The Temporal lobe is located on the bottom section of the brain, possesses the function of memorization, auditory senses, as well as the sense of balance, which allows a person to walk appropriately [6].

The Parietal lobe is located behind the frontal lobe, involves the function of sensory perception, such as touch, it is involved with pain and discomfort. The Occipital lobe is related to the temporal lobe and is responsible for visual sense [6].

The study of the effect of brainwaves in an assembly line can be carried in a laboratory setting instead of an industrial site, using legos for the making of a product. The use of Lego blocks for research and education purposes has been well-documented. There exist studies that use Legos in educational and research settings for the simulation of manufacturing processes, creation of virtual prototypes and as test components [7]. The use of Lego blocks is called MIB Dynamics (Interactive Assembly of Blocks), and it has been applied at the Federal University of Itajubá since 2003, using Toyota's production system [8].

2 Methodology

In this work seven students between 19 and 25 years old participated, they did not present any addictions, such as drug usage, and did not consume any alcoholic beverage during the past 48 h. All people slept an average of eight hours before experimenting. The learning curve method was performed separately in each of the participants while assembling 40 pieces of legos in a workstation, and their amount of time was measured using a chronometer.

During the assembly process, in each of the participants the Neuroheadest Emotiv EPOC [9] was placed, consisting of 14 electrodes and using the Emotiv PRO software [10]. the brain signals were monitored and recorded until each person attained its standard time.

2.1 Standard Time Determination

To determine the standard time that all participants have to attain, a test subject was trained until he arrived at a constant pace in the construction of several pieces, then video recording was taken to be able to time each of the hand movements performed during the assembly of each Lego car. In order to determine the sample size necessary for each time (Fig. 1) the following formula was used [Reference [11] and the following reference http://egyankosh.ac.in//handle/123456789/31711]:

$$n = \left(\frac{40\sqrt{n' \sum x^2 - \sum(x)^2}}{\sum x} \right)^2 \tag{1}$$

where:

n = Sample size (number of observations),
n' = Number of preliminary observations
x = value of each observation
40 = constant for a 95% confidence interval.

Fig. 1 Sample size of every element during the assembly

Element	Sample Size	Element	Sample Size
1	2.56877465	13	0.83931935
2	0.34814917	14	0.48500639
3	0.38137431	15	0.5969151
4	0.23067728	16	4.11393746
5	0.16471228	17	0.10906495
6	1.36301561	18	2.66642804
7	0.20593005	19	0.89384592
8	0.48166578	20	0.1354259
9	0.7749821	21	0.6507919
10	0.68276765	22	2.62566446
11	2.7335267	23	1.06641776
12	0.96370503	24	2.64480291
		25	2.18445709

Fig. 2 Normal time in seconds associated to each elememt

Element	Normal Time	Element	Normal Time
1	6.014735	13	6.3175
2	6.9033365	14	6.175
3	3.3572335	15	1.659935
4	4.11407	16	2.6847
5	5.380895	17	5.2955945
6	8.237545	18	2.8979655
7	6.051215	19	5.80602
8	3.290895	20	6.61352
9	4.82049	21	5.417185
10	6.5436	22	7.190645
11	4.87027	23	6.47159
12	3.09358	24	7.29847
		25	7.666405

Once the sample size is determined, the normal time of each hand movement is obtained (Fig. 2) and finally the standard time is calculated, resulting in 2 min with 28 s.

2.2 *Participants Involved in the Study*

Once the standard time was determined, participants were recruited for the experiment; Seven students were recruited in the range of 19 and 25 years old. Each participant had to fill a form (Fig. 3), registering the following information: sleep hours, smoking habits, alcoholic beverage consumption, and they were also asked about any drug of medicine consumption, for allergies or other diseases. After completing the form, the Emotiv EPOC system was placed in their head for EEG recording (Fig. 4).

TECNOLÓGICO NACIONAL DE MÉXICO		TEC DE JUAREZ Forjando el futuro...
Name:	Age:	Gender:
Hours of sleep:	Smoke:	Alcohol:
Other medications:		
		Standard Time: 2min 28 seconds
Car	Time	Pulse

Fig. 3 Registration form for each person involved in the experiment

Fig. 4 Photo of a person perfiorming the experiment, and using the Emotiv EPOC Neurohead band

2.3 Experiment and Data Processing

During the experiment, each participant data corresponding to the assembly times were recorded until the standard time established previously was reached. A stopwatch was employed during the recording and it was monitored that each of the finished cars contained all of the lego pieces in their respective place in the assembly.

In the next step, all the data obtained during the experiment were saved and exported to a European Data Format (EDF), for its subsequent use with the Matlab

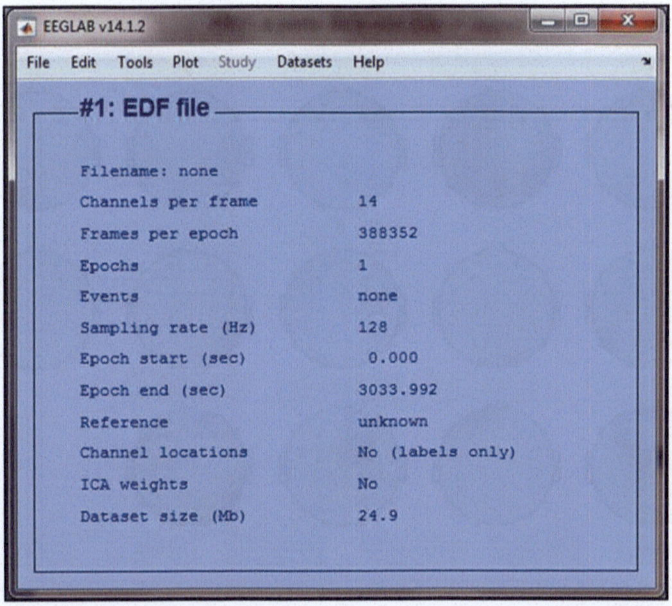

Fig. 5 Summary of imported data using the EEGLAB graphical user interface

software. Reference [12] and the EEGLAB toolbox [13]. Within this application, the data extracted from the Emotiv PRO software is imported, and in the graphical user interface the following data are displayed: the number of channels, frames per epoch, the frequency of the sample, the recording time, etc., (Fig. 5).

The data obtained during each test presents external artifacts not related to brain signals, such as muscular, respiratory and even cardiac. It is important to eliminate the noise associated with these artifacts using different techniques such as the ICA (Independent Component Analysis Method) [13] and ASR (Artifact Subspace Reconstruction Method) [14]. The ICA method analyzes the different neuronal components in the brain, from the signals recorded by the electrodes on the surface of the head (Fig. 6). In order to eliminate the noise components from the data analysis, it is necessary to detect the corresponding activity, either the ocular movement of component one or the irregular behavior that component number 14 presents. For signal elimination, different judgments are explored using either the spectrogram display of each of the components (Fig. 7) or through statistical analysis (Fig. 8).

The ASR method continuously detects irregular signals throughout the study according to specific conditions, for example, if the signal displays a flat behavior during a certain amount of time, are poorly correlated or present excessive amplitude, they will be eliminated. At the end of the filtering process, a clean EEG is obtained for subsequent statistical analysis (Fig. 9).

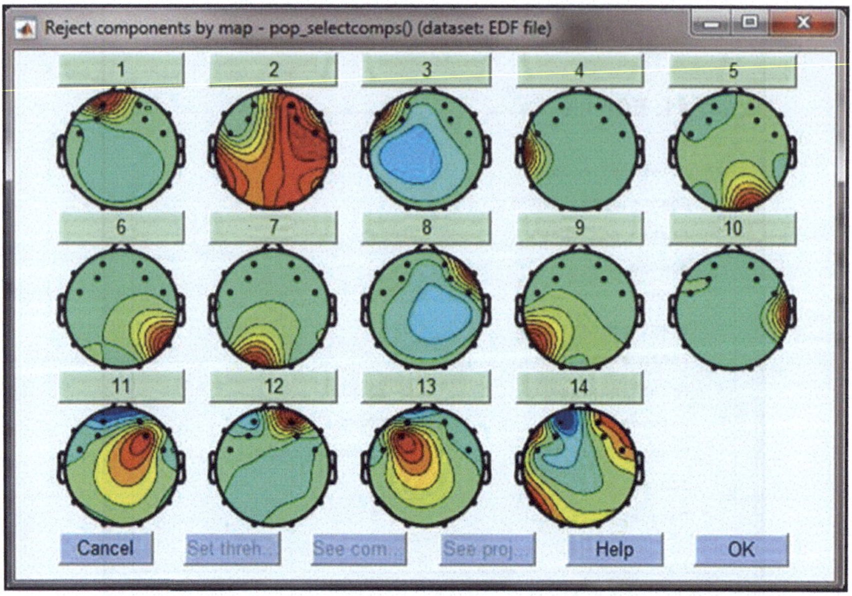

Fig. 6 Active components during the experiment and noise elimination with the ICA method

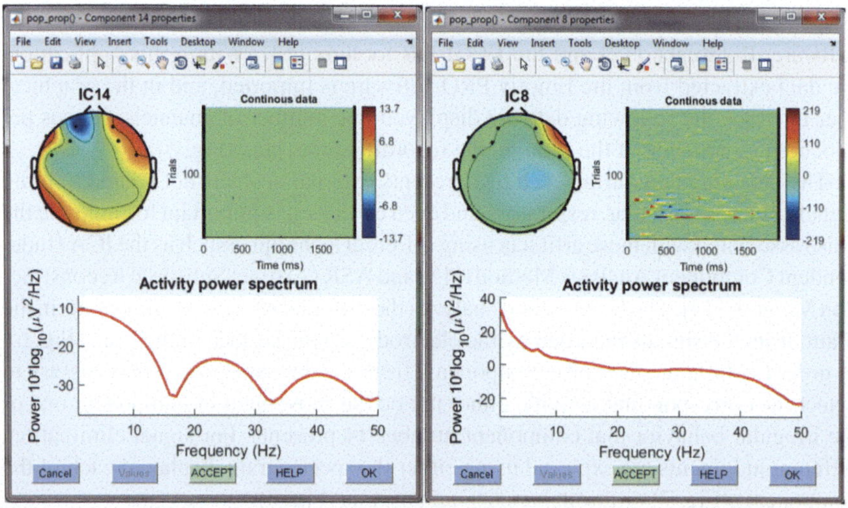

Fig. 7 Power versus Frequency plot of extracted components using the ICA method

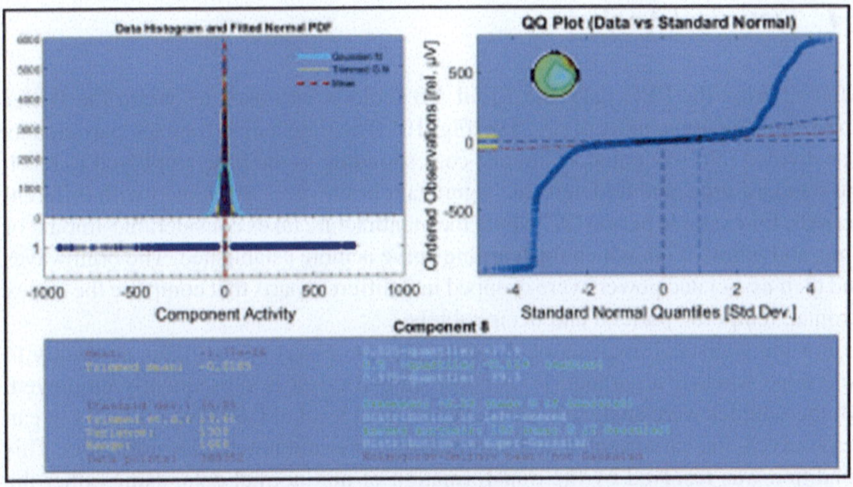

Fig. 8 Statistical information of an extracted component using the ICA method

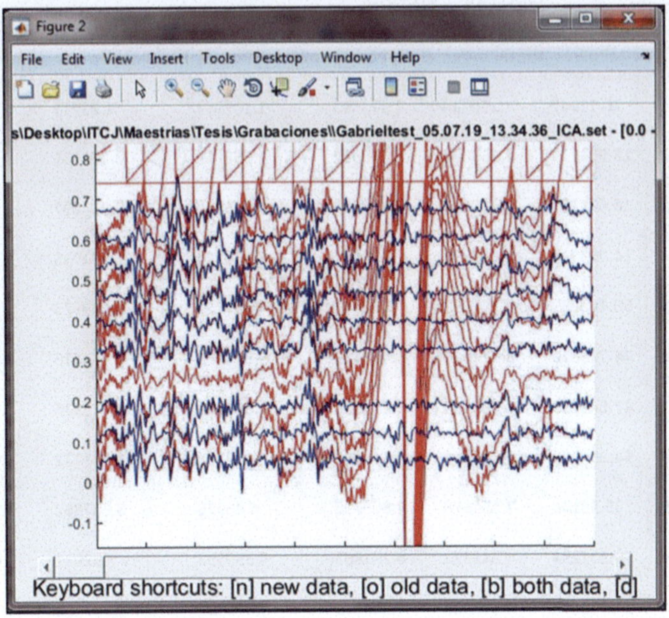

Fig. 9 Noise signal removal using the ASR method

2.4 Statistical Analysis

After filtering the EEG data, the EEGLAB toolbox was used to obtain the Power Spectrum Density in units of μv^2/Hz (Fig. 10). The time length for each participants was divided into five equal segments, corresponding to the time employed to reach the standard time, and to determine significant brainwaves associated with different periods, for example period "A" where the manufacture takes considerable amount of time, and point "B" in which the learning curve is more established. The brainwaves and their associated power, were grouped into different parts that comprise the cortex (frontal, temporal, parietal and occipital lobe).

For the statistical analysis of the data, it is necessary to test their normality to determine whether a parametric or non-parametric test is subsequently employed. All the samples were analyzed with a significance level of 0.05; From Fig. 11, it can be observed, the variance and and brainwaves power did not display normality. This condition was resolved by the transformation of non-normal data, using either the

absolute power		microvolt	ts^2/Hz			
Channel	deltaPower	thetaPower	alphaPower	lowbeta Power	Highbeta Power	gammaPower
1	10.11947	1.50714	1.373393	1.102591	1.23735	1.453848
2	15.933253	2.353081	2.038887	1.658162	2.174285	2.895409
3	16.691701	1.413546	0.539515	0.306926	0.214897	0.167636
4	14.243728	2.023995	1.486021	1.160744	1.532935	2.097023
5	50.768089	7.342993	5.52377	4.299757	5.579923	7.567645
6	44.715819	8.422081	6.462422	5.393183	7.084416	9.884966
7	47.547232	6.839969	5.281006	3.866641	4.775826	6.205105
8	54.850963	9.813191	6.515782	5.088768	7.185722	9.966414
9	35.53144	7.551678	4.803796	4.46978	8.607882	12.183478
10	50.332254	4.345622	4.281865	4.359552	7.92031	10.701096
11	14.762992	4.347525	4.674179	3.923802	5.667132	7.52723
12	21.363526	6.607091	6.702556	5.18632	6.15083	7.757984
13	22.125996	4.671816	5.701186	5.127747	7.419951	9.728686
14	30.07133	8.187083	9.728738	8.46098	9.356206	10.840871

Fig. 10 Power μv^2/Hz and frequency values during a period of time associated with the learning curve

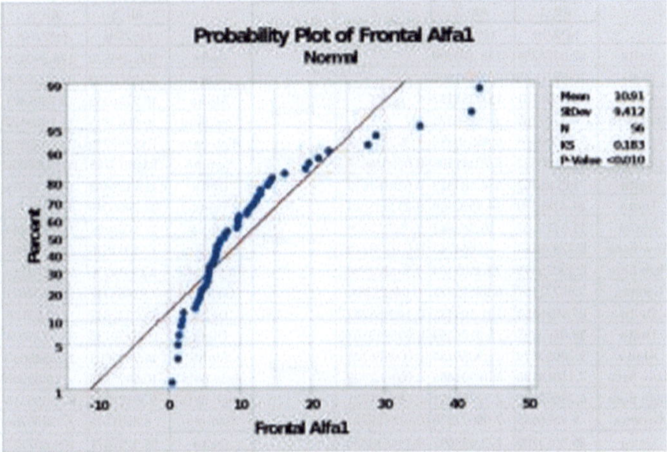

Fig. 11 Non-normality plot for the alpha brainwave in the frontal lobe

square root or the natural logarithm of the data. Reference [15]. It should be taken into account that the data resulting from this transform (α of 0.146) were used in the statistical analysis (Fig. 12).

To determine the significance between brainwaves from different periods along the learning curve, the paired t student was employed and shown in (Fig. 13). In this figure brainwave powers (real data without transformation) are shown for each period within the different lobes. The p-values, from the paired t-student method, are also presented in Fig. 13, where the values with significant change are highlighted

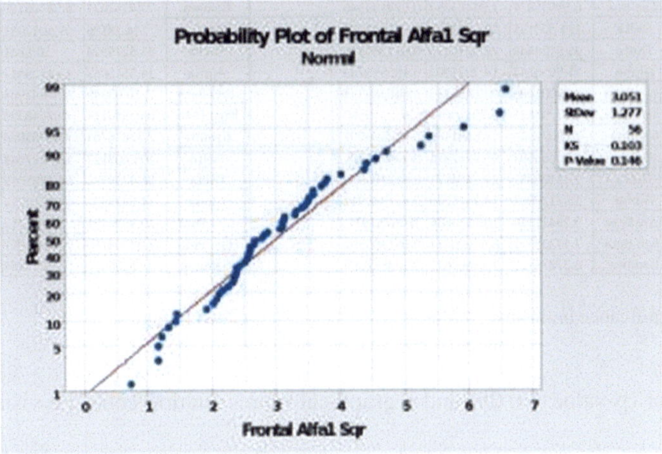

Fig. 12 Normality plot, obtained after data transformation for the alpha brainwave in the frontal lobe

Regions		All_1	All_2	
		ts^2/Hz	ts^2/Hz	pvalue
Frontal	Delta	97.0278291	146.160309	-
	Theta	20.6873449	33.2781966	-
	Alpha	10.908731	15.3351132	-
	Low Beta	8.1737265	9.46196375	-
	High Beta	5.20500668	6.45510389	7.99935E-07
	Gamma	4.3800672	5.20341186	0.001624438
Temporal	Delta	115.86362	154.583513	0.685361622
	Theta	24.4182339	34.3300409	0.293325864
	Alpha	11.5715102	16.039642	0.003918279
	Low Beta	8.22896164	9.964743	0.001834229
	High Beta	6.19075579	7.83505164	0.001243878
	Gamma	5.69771157	6.99535871	0.01427364
Parietal	Delta	97.5915153	121.503131	0.149965242
	Theta	19.0522295	25.3108444	0.095969567
	Alpha	9.22091879	11.9056144	0.004906225
	Low Beta	7.11014136	8.28901486	0.002862222
	High Beta	6.23098686	7.52793421	0.001331771
	Gamma	5.9506645	7.09672336	0.016725223
Occipital	Delta	80.8053709	123.633692	0.194830833
	Theta	17.0486079	27.3188381	0.191186207
	Alpha	10.1418247	13.7127903	0.002384347
	Low Beta	7.74915757	9.29421314	0.007758175
	High Beta	5.86153143	7.17308921	0.000291548
	Gamma	5.33783586	6.37873729	0.046694164

Regions		All_3	All_4	
		ts^2/Hz	ts^2/Hz	pvalue
Frontal	Delta	106.235576	128.390879	-
	Theta	32.0586402	42.0825298	-
	Alpha	16.2716353	16.7995471	-
	Low Beta	10.9069815	11.1969522	-
	High Beta	7.15342779	7.54269113	-
	Gamma	5.68013205	5.65066852	-
Temporal	Delta	132.199498	145.933531	0.6193379
	Theta	37.7466049	47.2204736	0.20454781
	Alpha	19.1069204	18.1179844	0.65373035
	Low Beta	12.3931576	12.2710383	0.80040856
	High Beta	9.23939236	9.06523114	0.8105738
	Gamma	7.91463357	7.44152621	0.24171976
Parietal	Delta	107.232847	114.23536	0.63202645
	Theta	29.5230133	34.5279974	0.69364994
	Alpha	14.7921003	14.5545067	0.52742505
	Low Beta	10.5495293	10.63949	0.45492572
	High Beta	9.39354014	8.63550314	0.34116566
	Gamma	8.8691145	7.73836764	0.07539656
Occipital	Delta	95.3262871	103.607122	0.19740282
	Theta	27.7404571	36.4798988	0.2358536
	Alpha	14.7525226	14.8425302	0.57947455
	Low Beta	10.7564656	10.9906461	0.61245092
	High Beta	8.17920471	7.98394286	0.81668357
	Gamma	7.14665679	6.56306529	0.253716

Regions		All_2	All_3	
		ts^2/Hz	ts^2/Hz	pvalue
Frontal	Delta	146.160309	106.235576	-
	Theta	33.2781966	32.0586402	-
	Alpha	15.3351132	16.2716353	-
	Low Beta	9.46196375	10.9069815	-
	High Beta	6.45510389	7.15342779	-
	Gamma	5.20341186	5.68013205	-
Temporal	Delta	154.583513	132.199498	0.733354056
	Theta	34.3300409	37.7466049	0.125371506
	Alpha	16.039642	19.1069204	0.29297038
	Low Beta	9.964743	12.3931576	0.091287127
	High Beta	7.83505164	9.23939236	0.018043586
	Gamma	6.99535871	7.91463357	0.074149346
Parietal	Delta	121.503131	107.232847	0.337152292
	Theta	25.3108444	29.5230133	0.521260929
	Alpha	11.9056144	14.7921003	0.185595754
	Low Beta	8.28901486	10.5495293	0.046449387
	High Beta	7.52793421	9.39354014	0.031279569
	Gamma	7.09672336	8.8691145	0.077526001
Occipital	Delta	123.633692	95.3262871	0.224228285
	Theta	27.3188381	27.7404571	0.512166929
	Alpha	13.7127903	14.7525226	0.411348917
	Low Beta	9.29421314	10.7564656	0.05752971
	High Beta	7.17308921	8.17920471	0.051306591
	Gamma	6.37873729	7.14665679	0.178522009

Regions		All_4	All_5	
		ts^2/Hz	ts^2/Hz	pvalue
Frontal	Delta	128.390879	110.396245	-
	Theta	42.0825298	85.2836183	-
	Alpha	16.7995471	15.0195854	-
	Low Beta	11.1969522	9.20050789	-
	High Beta	7.54269113	6.87981721	0.06586345
	Gamma	5.65066852	4.91210541	-
Temporal	Delta	145.933531	122.17958	0.59736725
	Theta	47.2204736	87.1829761	0.13192705
	Alpha	18.1179844	16.9198837	0.7355829
	Low Beta	12.2710383	10.1131778	0.09052647
	High Beta	9.06523114	8.14444929	0.08416076
	Gamma	7.44152621	6.59405193	0.06132754
Parietal	Delta	114.23536	95.5017031	0.13009827
	Theta	34.5279974	50.5841	0.90466218
	Alpha	14.5545067	13.4128817	0.36367613
	Low Beta	10.63949	8.57842964	0.0438192
	High Beta	8.63550314	7.97648393	0.12862092
	Gamma	7.73836764	7.06353864	0.10691085
Occipital	Delta	103.607122	102.297943	0.35705282
	Theta	36.4798988	54.4295019	0.19493732
	Alpha	14.8425302	14.1612024	0.93310026
	Low Beta	10.9906461	9.181213	0.05088012
	High Beta	7.98394286	7.07539657	0.03297549
	Gamma	6.56306529	5.75615564	0.05179011

Fig. 13 Significance brainwave values (red) and their associated p-values

in red color (p-value < 0.05) and a graphical representation could be visualized in Fig. 14.

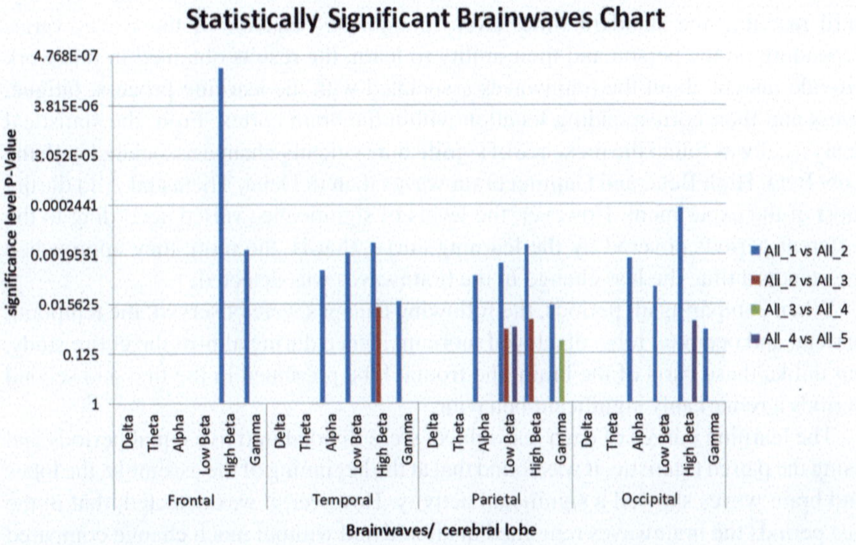

Fig. 14 Chart graph of brainwaves p-values during the manufacture learning curve

3 Results

A comparative analysis of the most significant p-values in all brain regions (frontal, temporal, parietal and occipital) was performed. From the values determined, it was found that the brain waves that predominates over all the values in the frontal region are the High Beta and Gamma.

4 Discussion

A large sustained activity with Beta waves can be detrimental for a worker, either in his normal day or when being trained with a new product or industrial process; One of the findings for the participants in this experiment was the difficulty associated during the assembling of the first LEGO cars, but as time passed the subjects rapidly became familiar with the location of each piece within the assembly and their timing improved considerably.

5 Conclusions and Future Challenges

Throughout this study, brainwaves were analyzed during the manufacture learning curve, by developing a lego product from zero knowledge at the start of the training until reaching the standard time. Even though the behavior of the waves varies depending on the person and their ability to learn, the results obtained in this work provide insight about the brainwaves associated with the learning process, fatigue, stress and their corresponding location within the brain cortex. From the statistical analysis, it was found the presence of significant statically change associated with the Low Beta, High Beta, and Gamma brain waves than in Delta, Theta and Alfa during most of the experiment. However, the levels of significance varied according to the different periods covered by the learning curve, that is, the more they approached the standard time the less change in the brainwaves was detected.

When comparing all periods, the following changes were observed: the temporal, parietal and occipital lobe, displayed more incidence during almost the entire study, but unlike these parts of the brain, the frontal lobe presented in the first and second periods a remarkably significant behavior.

The learning curves of each test subject were divided into five equal periods and using the paired t statistic, it was found that at the beginning of the assembly, the lobes and brain waves showed a significant activity. However, it was detected, that in the last periods the brainwaves remained unaltered and without much change compared to the start of the learning curve.

It is recommended that for higher accuracy, the learning curve should be divided into more periods. Also, it is recommended that an EEG device with more electrodes and higher resolution should be used.

Acknowledgements Felix Lira, gratefully acknowledges the graduate scholarship granted by CONACyT. Ana Garcia and Jorge de la Riva, acknowledge the financial support received by Technologic Nacional de Mexico through a Research Grant obtained in March 2019.

References

1. Ndaro, N.Z., Wang, S.-Y.: Effects of fatigue based on electroencephalography signal during laparoscopic surgical simulation. Shanghai 200093, China: Hindawi, **2018**(2389158), 1–6 (2018)
2. Benjamin, N., Freivalds, A.: Ingeniería industrial. Métodos, estándares y diseño del trabajo, Duodécima Mc Graw Hill, Mexico DF, Mexico (2009)
3. Michele, F., Luigi, D.G.: How much sleep do we need? Sleep Med Rev, Elsevier, Rome, Italy **5**, 155–179 (2001)
4. Nagel, A.G.: Manual de Electroencefalografia. MCGRAW HILL, España (2001)
5. Hall, J.E., Guyton, : Tratado de fisiología médica. Decimotercera edición, México, Elsevier (2016)
6. Cabrales, P.A.: Neuropsicología y la localización de las funciones cerebrales superiores en estudios de resonancia magnética funcional con tareas. Acta Neurológica Colomb. 92–100 (2015)
7. Peter, Y., Chen, T., Anderson, D., Yu, J., Nagata, S.: Legoworld: a multisensory environment for virtual prototyping. In: Proceedings of Spie, vol. 3295 (1998)
8. Ferreira, A., de Pinho, F.L.: Utilização de Bloquinhos de Montagem LEGO® para o Ensino dos Conceitos do Sistema Toyota de Produção. ENEGEP, 5525–5532. In: Proceedings of SPIE, pp. 313–321 (2005)

9. Emotiv Epoc.: EMOTIV EPOC+ 14 Channel Mobile EEG. View (2019). https://www.emotiv.com/product/emotiv-epoc-14-channel-mobile-eeg/
10. Emotiv.: EmotivPro software (2019). https://www.emotiv.com/emotivpro/
11. Lukodono, R.P., Ulfa, S.K.: Determination of standard time in packaging processig using stopwatch time study to find output standard. J. Eng. Manag. Ind. Syst. **2**, 87–94 (2017)
12. Matlab R2015a Copyright 1984–2005 The MathWorks, Inc
13. Delorme, A., Makeig, S.: EEGLAB: an open source toolbox for analysis of single-trial EEG dynamics including independent component analysis. J. Neurosci. Methods **134**(1), 1–14 (2004) [EE.UU]
14. Mullen, T., Kothe, C., Chi, Y., Ojeda, A., Kerth, T., Makeig, S., Cauwenberghs, G., Ping, T.: Real-time modeling and 3D visualization of source dynamics and connectivity using wearable EEG. Conf. Proc. IEEE Eng. Med. Biol. Soc. **2013**, 2184–2187 (2013)
15. Montgomery, D.: Design and analysis of experiments. 8th edition, Arizona State University, Tempe: Wiley (2012). ISBN 978-1-118-14692-7

Audio Features Extraction to Develop a Child Activity Recognition Model Using Support Vector Machine to Monitoring Security in a Smart City

Antonio García-Domínguez, Carlos E. Galván-Tejada, Laura A. Zanella-Calzada, Jorge I. Galván-Tejada, Alberto Ochoa-Zezzatti, and Javier Martínez

Abstract Children activity recognition and classification is a subject of novel interest on which different works have been presented, where the data source used to classify the activities is determinant to define the other components of the classification model. This work uses environmental sound as data source to perform the recognition and classification of activities. A model for classifying children's activities is presented based on the Support Vector Machine (SVM) algorithm, which is trained with features extracted from the audio samples of each of the activities defined for analysis: running, playing, crying and walking. As a result, a model able to predict, based on its contained information, the class to which a new sample of analyzed audio belongs is obtained.

1 Introduction

The human activity recognition and classification is a subject of recent interest, on which many works have been presented and proposed different applications, as works presented by Aggarwal [1] and Hamid [2], to facilitate the daily life of

A. García-Domínguez (✉) · C. E. Galván-Tejada · J. I. Galván-Tejada · J. Martínez
Unidad Académica de Ingeniería Eléctrica, Universidad Autónoma de Zacatecas, Jardín Juárez 147, Centro, 98000 Zacatecas, Zac, Mexico
e-mail: antonio.garcia@uaz.edu.mx

C. E. Galván-Tejada
e-mail: ericgalvan@uaz.edu.mx

J. I. Galván-Tejada
e-mail: gatejo@uaz.edu.mx

L. A. Zanella-Calzada
LORIA, Université de Lorraine, Campus Scientifique BP 239, 54506 Nancy, France
e-mail: laura.zanella-calzadad@univ-lorraine.fr

A. Ochoa-Zezzatti
Doctorado en Tecnologia, UACJ, Ciudad Juarez, Mexico

A. Ochoa-Zezzatti et al. (eds.), *Technological and Industrial Applications Associated With Industry 4.0*, Studies in Systems, Decision and Control 347,
https://doi.org/10.1007/978-3-030-68663-5_9

human beings by promoting an automated interaction with their environment. An important aspect to be considered in human activity recognition is the data source to be used since it depends on the processing of the data and the classification models used. As data source, the use of different types of sensors has been proposed, as in the work presented by Arnon [3]. In recent years, for the area of recognition and classification of children's activities, has been proposed the use of different data sources, such as video cameras, accelerometers, and radio-frequency devices, as in the work presented by Kurashima and Suzuki [4]. Most of the works described in this area collect information for analysis by embedding the sensor directly into a child's garment to record activity data, such as the works presented by Nam and Park [5]. This way of data capture has the disadvantage that the devices or sensors used, which are placed in the garments, can interfere directly with the natural action of the children, not allowing them to perform normally the activities to be analyzed.

One way to solve the problem mentioned above is to change the data source to one that does not interfere with the activities to be performed by the study subjects. Under this idea, the environmental sound has been used as data source to recognize and classify human activities, as in the work presented by Salomons et al. [6], since data capture passes inadvertently to the study group, thus not interfering with the activities to be analyzed. The use of environmental sound as data source for the recognition and classification of children's activities is an approach not yet addressed and on which the present work is centered.

Environmental sound as data source in child activity classification models is a major challenge due to the complexity of the audio signal analysis process as well as due to the different environmental factors that may interfere during the data's capture process, causing that the samples taken do not have the necessary characteristics for their analysis. For this reason, the adequate data processing (audio samples) and the choice of an appropriate model that optimizes the activity recognition process becomes of vital importance. For the correct audio signal processing, it is necessary to perform a feature extraction process on which the classification model will be based. Given these features, it is possible, with a set of training dara (samples), to label the classes (the type of sound to which the samples belong), to construct and train a model that predicts the class of a new sample. In the present work, Support Vector Machines (SVM) are used as the learning algorithm for the model.

2 Materials and Methods

This section describes the features of the dataset used for the creation and training of the classification model, as well as the methods that define how the model is created and its operation for the child activity recognition process.

2.1 *Dataset Description*

For this work, the dataset is composed of audio recordings of four activities performed by children in the age range of 12–36 months. This age range is chosen based on the pediatric ages defined by the World Health Organization (WHO), which define that from 12 months old they cease to be babies and from 36 months old they become in pre-school children, the intermediate period being a stage in which most children share very similar behaviors, allowing a more precise analysis of the activities they perform.

The activities to be analyzed are: walking, running, crying and playing (manipulating plastic blocks), and audios were captured in a controlled environment (inside a room). Table 1 shows a description of each of the activities included in the analysis.

2.1.1 Metadata

The recorded audio is in WAV (WaveForm Audio File) format, in order not to experience loss of quality that other formats present due to the compression process. The general features such as sampling rate, channels and resolution are shown in Table 2.

The features presented in Table 2 ensure an acceptable quality for recorded audio files, and they define the parameters required for future recordings in order to expand the dataset.

2.1.2 Devices

For the data acquisition process, the audios of the activities were recorded using the microphones present in two mobile devices. The main features of the devices used are shown in Table 3.

Table 1 General description of activities

Activity	Description
Crying	Emitting crying sound in reaction to some event
Playing	Handling plastic pieces
Running	Moving quickly from one place to another
Walking	Moving from one place to another at medium speed

Table 2 Metadata

Activity	Sampling rate (Hz)	Channels	Resolution
Crying	44,100	Stereo	16 bit
Playing	44,100	Stereo	16 bit
Running	44,100	Stereo	16 bit
Walking	44,100	Stereo	16 bit

Table 3 Recording devices

Device	Processor	OS
Lanix Ilium s620	MediaTek MT6582 quad-core	Android 4.2.2
Motorola Moto G4	Snapdragon 617	Android 6.0.1

Table 4 Commonly audio features used

Time-domain features	Frequency-domain features
RMS (Root Mean Square)	Fundamental frequency
Energy	Frequency components
ZCR (Zero Crossing Rate)	Spectral centroid
Maximum amplitude	Spectral flux
Minimum energy	Spectral density
	Spectral rollof

2.2 *Data Analysis*

Once the audio signals are captured, it is necessary to perform the features extraction process with which the classification model is generated. For this work, a feature extraction approach and a Support Vector Machine (SVM) algorithm to generate the classification model are proposed. This process is explained in detail below.

2.2.1 Feature Extraction

An audio feature is any qualitatively or quantitatively measurable aspect of a sound, these measures are a series of parameters that compactly represent the sound information and can be used to characterize it. It is possible to obtain many features that can be used to classify audio signals [7], these features can be separated into two types: time-domain and frequency-domain. Table 4 shows some of the audio features commonly used for analysis.

2.2.2 Classification Model

SVM is a supervised machine learning algorithm that can be used for both classification or regression challenges. However, it is mostly used in data classification problems as in the works presented by Kallas et al. [8], Qi et al. [9], and Misumi et al. [10], and in works of sounds classification as the work presented by Tran and Li [11]. As is explained by Yu and Kim [12], SVM is a type of linear classifier that can be used to solve binary classification problems. It works by correlating data to a large feature space so that data points can be categorized, even if the data cannot be separated linearly in another way. A separator is detected between the categories and the data are transformed so that the separator can be extracted as a hyperplane. Thereafter, the features of the new data can be used to predict the group to which the new record belongs.

2.3 Model Validation

The classification model is generated by an induction process, training it with the dataset (features extracted from the audios), which is called training dataset, to have a model that can predict the class to which belongs a new sound evaluated. The training of the classification model refers to the process of taking audio samples of which it is known to which class they belong and based on them to generate the classification model, this is called a supervised classification system. The validation process of the model is a process of deduction that consists of performing tests of classification of sounds that represent the activities analyzed, in order to determine the degree of certainty with which the model classifies the audio samples in the correct classes. Figure 1 shows the previously described process.

3 Experiments

3.1 Audio Recordings

As the first step in the experimentation process, it is necessary to obtain the audio recordings of the activities to be analyzed. Through the process of audio capture for the activities in the present work, a total of 31 recordings were obtained, the quantities of audio files for each activity are shown in Table 5.

3.2 Feature Extraction

For the generation and training of the classification model to be used it was worked on the Python programming language because it has features and specialized libraries that facilitate this type of analysis. In particular, it was used the library PyAudio-Analysis which contains a large number of features for audio analysis, included [13]:

- Extract audio features and representations
- Classify unknown sounds
- Detect audio events and exclude silence periods from long recordings
- Perform supervised segmentation
- Perform unsupervised segmentation

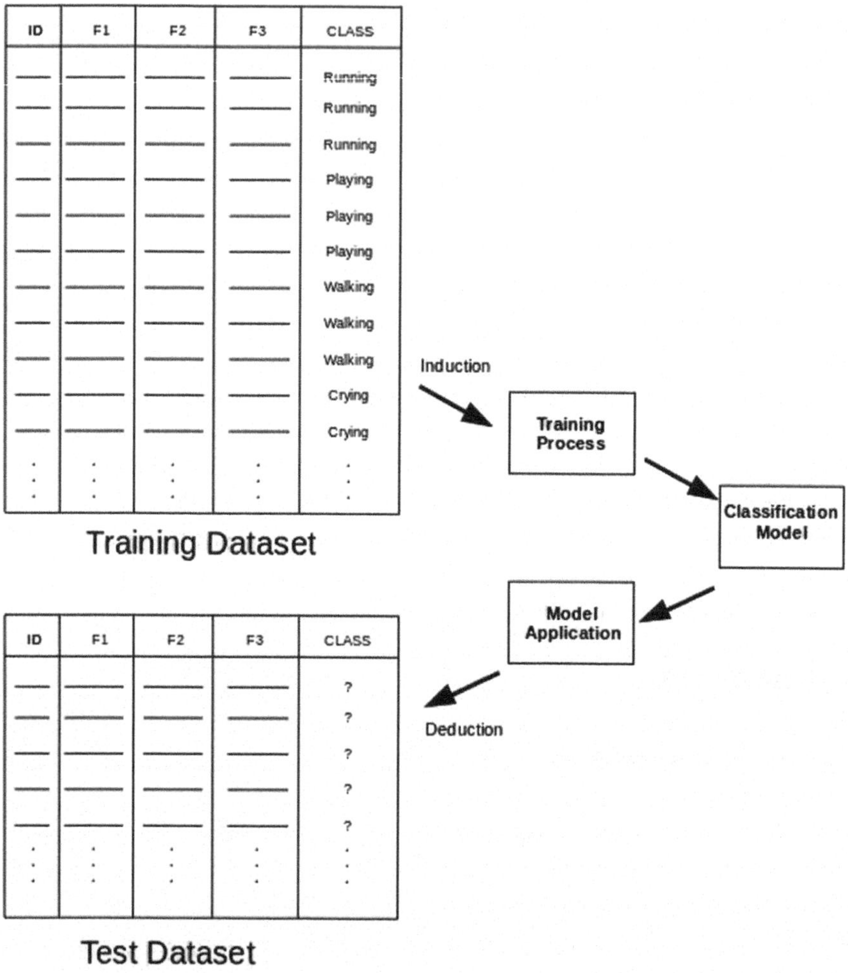

Fig. 1 Training process and model validation

Table 5 Audio file by activity

Activity	Recordings
Crying	8
Playing	8
Running	8
Walking	7

Table 6 Extracted audio features

Feature ID	Feature name	Description
1	Zero crossing rate	The rate of sign-changes of the signal during the duration of a particular frame
2	Energy	The sum of squares of the signal values, normalized by the respective frame length
3	Entropy of energy	The entropy of sub-frames normalized energies. It can be interpreted as a measure of abrupt changes
4	Spectral centroid	The center of the gravity of the spectrum
5	Spectral spread	The second central moment of the spectrum
6	Spectral entropy	Entropy of the normalized spectral energies for a set of sub-frames
7	Spectral flux	The squared difference between the normalized magnitudes of the spectra of the two succesive frames
8	Spectral rollof	The frequency below wich 90% of the magnitude distribution of the spectrum
9–21	MFCCs	Mel Frequency Cepstral Coefficients form a cepstral representation where the frequency bands are not linear but distributed according to the mel-scale
22–23	Chroma vector	A 12-element representation of the spectral energy where the bins represent the 12 equal-tempered pitch classes of western-type music (semitone spacing)
34	Chroma deviation	A The standard deviation of the 12 chroma coefficients

- Extract audio thumbnails
- Train and use audio regression models
- Apply dimensionality reduction to visualize audio data and content similarities.

Before starting with the generation and training of the classification model, it was necessary to extract the features of the audio signals that were to serve as input to the model. Table 6 shows the extracted features implemented in the library; these features later served as a basis for generating the classification model.

3.3 *Classification Model*

The next step is the generation of a classification model with the data arranged for that purpose, which in this sense refers to the features extracted from each of the audio files captured from the activities. As mentioned previously, for this work the SVM algorithm is used to generate the classification model. This is how the classification model is created and trained from the audio files previously captured.

Table 7 Confusion matrix

	Cry	Play	Run	Walk
Cry	21.75	0	3.25	0
Play	0	19.5	0.5	5
Run	3.5	0.75	14.5	6.25
Walk	0	5.25	1.25	18.5

3.4 Classification Process

Once the classification model has been generated and trained with the data extracted from the audio files, the process of classifying audio is done by performing the extraction of features, to load the classifier model and tries to classify the audio file into one of the classes with which the model was previously generated and trained. Table 7 shows the resulting confusion matrix in the creation and training of the classification model. It is common to use confusion matrices for problems of classification and selection of features, as in the work presented by [14], since in these it is possible to visualize the behavior that has the classification model when trying to classify the test data.

From the confusion matrix obtained it can be interpreted that the model classifies most of the sounds in the correct class, this can be observed by analyzing the percentages of the main diagonal of the matrix, which indicates in which classes (columns) are being classified the sounds (rows). It can also be observed that some of the sounds are not classified in the correct class, this is normal behavior since the audios may present variations or strange noises that do not allow the correct classification.

4 Conclusion and Future Research

From the experimentation presented above, it can be concluded that the model generated by the use of the SVM algorithm and trained with the extracted features of the captured audio, acceptably classifies audio recordings of the activities performed by infants from 12 to 36 months, achieving a 74.25% of correct classifications, which can be observed in the confusion matrix. As future work, it is proposed, first, to expand the size of the dataset so that the model created is more robust, in addition, to include more activities and thus have a more complete classification model. Another important aspect to be considered is the feature selection process, to optimize the model, including only features really useful in the process of generating the classification model and improve the percentage of correct classifications.

A complementary aspect of the Intelligent Activity Monitoring System is to cross-reference the information with IoT data in order to identify and prevent accesses in recreational parks, as can be seen in Fig. 2.

Fig. 2 Our intelligent system can monitor up to 47 children at the same time and identify possible risk activities in a child park

A future research goal is to adapt our Intelligent System to identify possible risks or imminent hazards in water parks where the noise of water can distort the monitoring of children's activities, as can be seen in Fig. 3.

Fig. 3 Identification of dangerous or high-risk activities in children in a water park

Considering that by 2050 there will be 27 million boys and girls aged between 1 and 15 years is an activity highly demanded by the generation Z for the care of their infants.

Acknowledgements The authors would like to thank the people who provided the audio recordings of children's activities, through which the classification model was created in this research.

References

1. Aggarwal, J.K.: Human activity recognition-a grand challenge. In: International Symposium on Signals, Circuits and Systems, vol. 2, pp. 10–13. IEEE (2007)
2. Hamid, R.: Classifier boosting for human activity recognition. In: Ensemble Machine Learning, pp. 251–272 (2012)
3. Arnon, P.: Classification model for multi-sensor data fusion apply for human activity recognition. In: 2014 International Conference on Computer, Communications, and Control Technology (I4CT) (2014)
4. Kurashima, S., Suzuki, S.: Improvement of activity recognition for child growth monitoring system at kindergarten. In: IECON 2015—41st Annual Conference of the IEEE Industrial Electronics Society, IEEE (2015)
5. Nam, Y., Park, J.W.: Child activity recognition based on cooperative fusion model of a triaxial accelerometer and a barometric pressure sensor. IEEE J. Biomed. Health Inf. **17**(2), 420–426. IEEE (2013)
6. Salomons, E., Havinga, P., Leeuwen, H.V.: Inferring human activity recognition with ambient sound on wireless sensor nodes. Sensors **16**(10), 1586 (2016)
7. Fahmeeda, S., Tabassum, A.: Audio data security and feature extraction over cloud. Int. J. Comput. Appl. **168**(10), 33–37 (2017)
8. Kallas, M., Francis, C., Kanaan, L., Merheb, D., Honeine, P., Amoud H.: Multi-class svm classification combined with kernel pca feature extraction of ECG signals. In: 19th International Conference on Telecommunications (ICT) (2012)
9. Qi, X., Silvestrov, S., Nazir, T.: Data and generalized support vector machine (2007)
10. Misumi, M., Orii, H., Sharmin, T., Mishima, K., Tsuruoka, T.: Image classification for the painting style with svm (2016)
11. Tran, H.D., Li, H.: Probabilistic distance svm with hellinger- exponential kernel for sound event classification. In: International Conference on Acoustics, Speech and Signal Processing (ICASSP). IEEE (2011)
12. Yu, H., Kim, S.: Svm tutorial—classification, regression and ranking. In: Handbook of Natural Computing, pp. 479–506 (2012)
13. Giannakopoulos, T.: Pyaudioanalysis: an open-source python library for audio signal analysis. Plos One **10**(12) (2015)
14. Visa, S., Ramsay, B., Ralescu, A.L., Van Der Knaap, E.: Confusion matrix-based feature selection. In: Proceedings of the Twenty second Midwest Artificial Intelligence and Cognitive Science Conference, vol. 710, pp. 120–127 (2011)

Sentiment Analysis Using Natural Language Processing Through a Speech Recognition System Using a Hybrid Mobile App

Alejandro Acosta, Alberto Ochoa-Zezzatti, Lina M. Aguilar-Lobo, and Gilberto Ochoa-Ruiz

Abstract Determining people's feelings when they speak is relatively easy due to the tone and language with which they express themselves. This happens if another person is listening and understanding what the first person wants to imply. However, for an algorithm that is capable of capturing the sound of a person's voice, it is difficult on its own to interpret the feeling or even analyze what the person wants to say. To achieve this, it is necessary to rely on some computer science and linguistics techniques such as Natural Language Processing (NLP). With sentiment analysis algorithms in combination with voice recognition and the general use of NLP, it is possible to create intelligent systems that allow the interpretation of people's feelings based on the audible message that they emit.

Keywords Natural language processing · Voice recognition · Sentiment analysis · AI algorithms

1 Introduction

According to Bonastre et al. [1], The voice is the most casual and important way for people to communicate between humans and as lately also between humans and machines. One thing that sometimes might be underrated but it is important, voice becomes part of the personality of the persons itself.

A. Acosta (✉)
Universidad Cuauhtémoc, Cuauhtémoc, Mexico
e-mail: alejandro.acosta@uabc.edu.mx

A. Ochoa-Zezzatti
Universidad Autónoma de Ciudad Juarez, Ciudad Juárez, Mexico

L. M. Aguilar-Lobo
Universidad Autonoma de Guadalajara, Zapopan, Mexico

G. Ochoa-Ruiz
ITESM Campus Guadalajara, Zapopan, Mexico

A. Ochoa-Zezzatti et al. (eds.), *Technological and Industrial Applications Associated With Industry 4.0*, Studies in Systems, Decision and Control 347,
https://doi.org/10.1007/978-3-030-68663-5_10

In the tech industry, is very known that Voice-based solutions exists in order to improve the process of businesses and the industry in general. These type of solutions or systems are limited. Systems are pre-programmed to support sort of type of voices that were configurated in order to work efficiently once installed. Let's draw the following scenario. A security system can recognize the voice of a person that was setup at the beginning when it was installed. After that, the user (person) can use the voice recognition system by using its voice as can be noticed. However, the limitation in this is that the user can only dictate the programmed commands that the actual system supports. The system will not understand or do anything else but what it was configurated by the engineers that designed it. Going back to Bonastre et al. [1], there are a few types of voice base systems such as: Automatic speaker verification (ASV). Text-to-text speech (TTS). Voice conversion (VC). Computer speech and language (CSL). Each of these have their very own characteristics, limitations, uses and applications.

In order to enhance the normal voice recognition systems, the current work contains 3 main sections to present. The first one it will be about how to recognize voice and convert into text by using an algorithm created in Go. The second part contains the explanation to analyze text and can tell what the sentiment of that message is by using a rating system developed in Go. And the last part is about to create a hybrid mobile app to record voice messages and display the sentiment of it using Google Flutter framework to develop the mobile app.

2 Voice Recognition Algorithm Using Go

Overview. In this section, voice recognition works in a way that it can be treated like a new topic, but it is not, voice recognition exists from 1952 when the system "Audrey" [2] was created by Balashek, Biddulph and Davis. Voice recognition also can be applied by many different ways in the nowadays. The way that now it will be presented is to create a Go algorithm that recognize some voice commands in order to be analyzed after.

There are other ways to consider but those are not going to be implemented at this time. The work done by Dhakal et al. [3] that is about a real-time speaker recognition user interface, stands to exploit the advantages of hybrid feature extraction techniques that contain the features of Gabor Filter (GF), Convolution Neural Networks (CNN), and statistical parameters as a single matrix set. This is a great option for the future of this work that will be explain further below.

Let's see how the architecture of this solution will look. See the next diagram (Fig. 1).

There is no doubt that using the offered power that current PaaS does such as Amazon with their AWS, will allow to create reliable, trustful, powerful and great services to provide an optimal solution like the one shown in the diagram above.

User having an App installed in their smartphones, will upload a voice messages up to the API where that message will be stored in the S3 (Storage) service, then

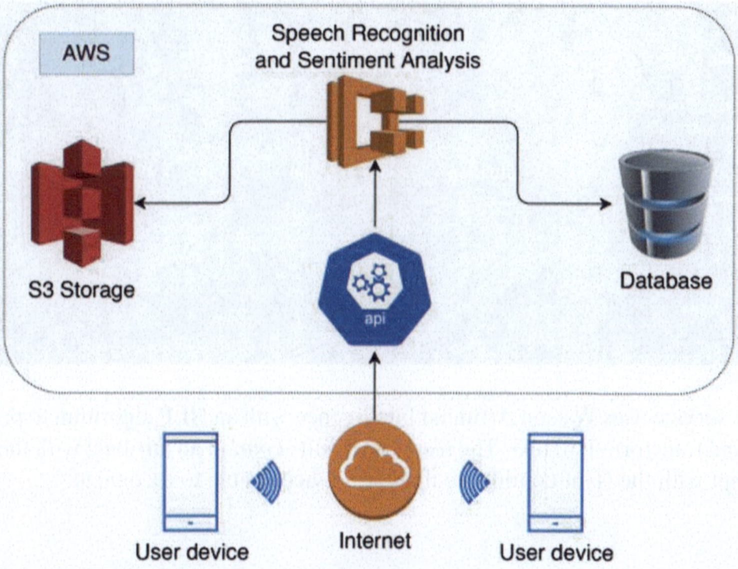

Fig. 1 Diagram of connectivity and voice processing architecture

processed and compared with proper datasets stored in the database to be able to recognize and analyze the sentiment of the message provided.

Now, let's see how the Speech Algorithm look like in Go. First of all, we will use IBM Cloud Watson API SpeechToText API service to synthetize voice in order to get the text from voice message.

```
var result, _ responseErr = speechtotext.Recognize(
 &speechtotextv1.RecognizeOptions{
  Audio: audioFile,
  ContentType: core.StringPtr("audio/flac"),
  Timestamps: core.BoolPtr(true),
  WordAlternativesThreshold: core.Float32Ptr(0.9),
  Keywords: []string{"colorado", "tornado", "tornadoes"},
  KeywordsThreshold: core.Float32Ptr(0.5),
} }, Model: core.StringPtr("en-US_BroadbandModel"),
```

The piece of code above uses Watson Speech to Text synthesize speech input process to retrieve the result text. Uses audio format "audio/flac" to recognize based on an English from United States.

Watson service return a response that look like this:

```
{
  "results": [
   {
     "final": true,
     "alternatives": [
       {
       "transcript": "the situation with the virus is really bad",
       "confidence": 0.98
       }
     ]
   }
 ],
 "result_index": 0
}
```


This service uses Watson Artificial Intelligence with an NLP algorithm to process audio and transform it to text. The response results contain an attribute with the final transcript with the % of confidence that the service is able to recognize.

3 Sentiment Analysis Using a Rating System with Go

Overview, in this section we are going to talk about what sentiment analysis is, how it is important and how it can be applied to text. In particular, text processed by the previous algorithm that transform voice into text.

Troussas and Virvou [4] said that Sentiment analysis refers to the use of expert methods (such as natural language processing, text analysis, computational linguistics) to systematically identify, extract, quantify, and study affective states and subjective information [4].

Habimana et al. [5] also said that the sentiment analysis appeared to be an important tool that allows the automation of getting insight from the user-generated data. In this occasion it is messages that people may record somewhere and by having its sentiment score, decisions of any kind can be made. There are many techniques that makes sentiment analysis possible, such as deep learning. With that, the need for sentiment analysis ranges from individuals to large organizations [5] and deep learning is used to make this possible worldwide and with an incredible power and scalability because it can be deployed over most (all) platforms around the cloud. This gives the proper solutions for problems that requires big data analysis or high-performance solutions.

According to Habimana et al. [5]: "Conducting sentiment analysis is more than classifying a document or a sentence into positive or negative classes. Indeed, finding the sentiment discussed in every aspect or feature of the entity is of prime importance. Therefore, depending upon the granularity required, the sentiment analysis task is performed at the document, sentence and aspect level" [5].

We are going to continue working with Go, at this point the next algorithm is receiving a text value that correspond to the voice text transformed.

```
model, err := sentiment.Restore()
if err != nil {
  panic(err)
}

var analysis *sentiment.Analysis
var text string

analysis = model.SentimentAnalysis(text sentiment.English)

if analysis.Score == 1 {
  fmt.Println("\"", text, "\"\nThe Sentiment is Positive\nScore = ", analysis.Score)
} else {
  fmt.Println("\"", text, "\"\nThe Sentiment is Negative\nScore =", analysis.Score)
}
```

This algorithm uses a library that already contains a trained model of NLP in English language that is able to apply a clustering method to identify written words, analyze them and detect a score to indicate if the text is interpreted either positive or negative.

Here is the response when the algorithm processes the text "the situation with the virus is really bad".

```
" the situation with the virus is really bad "
The Sentiment is Negative
Score = 0
```

By having this feedback, we can assume that a Negative sentiment the score is 0.

To perform these sentiment tasks, exists a diverse type of datasets that allow systems to integrate them. See the following table (Table 1).

4 Hybrid Mobile App with Google Flutter Framework

With the rise of smartphones and tablets, travelers now have remarkable connectivity that allows them to make travel arrangements and share information before, during, and after their trips. Companies recognize that mobile is increasing in importance, they do not yet understand its full reach and impact. Compared to PCs, mobile use patterns, behaviors, and expectations are different, and each phase of the travel cycle requires a distinctive approach for mobile [6]. Said by Linton et al. [6]. Islam et al. [7] states that mobile application uses, and development is a new and rapidly growing sector. The use of mobile apps is becoming facilitate and people, society of developing are upgrading themselves and making a new type of IT infrastructure. Mobile apps are running on a small hand hold mobile device which is moveable, easy to use and accessible from anywhere and any place. Many people are using mobile application to contact friends, browse internet, file content management, document creating and handling, entertainment etc. People can do many things of his daily life

Table 1 Details of the datasets commonly used in sentiment analysis

Dataset	#Samples	#'ftrain	#Dev	#Test	#Classes	Sentiment analysis task	Language
IMDB	50,000	25,000		25,000	2	Sentence sentiment classification	English
1MDB2	348,415	User choice	User choice	User choice	10	Document sentiment classification	English
SST-5	11,855	8544	2210	1101	5	Sentence sentiment classification	English
SST-2	9613	6920	872	1821	2	Sentence sentiment classification	English
Amazon	I10 K 143 M]	User choice	User choice	User choice	2, 5	Sentence and ABSA	English
SemEval2014-Dl	2931	2292		639	3	ABSA and ATE	English
ScmEval2014-D2	4712	3591	-	1121	3	ABSA and ATE	English
SemEval2017	61,873	13,601	5988	12,284	3	Sentence sentiment classification	English
STS	1.049 M	1.048 M	-	498	2	Sentence sentiment classification	English
STS-Gold	2034	User choice	User choice	User choice	2	Sentence sentiment classification	English
STS-Gold	58	User choice	User choice	U ek-i choice	2	Entity sentiment classification	English
Yelp	Above 1.2 M	User choice	User choice	User choice	2, 5	Sentence, document and ABSA	English
HR	24,348	User choice	User choice	User choice	2	Sentence sentiment classification	Chinese
MR	10,662	8655	961	1046	2	Sentence sentiment classification	English
Sanders	5513	User choice	User choice	User choice	4	Sentence sentiment classification	English
Deutsche Bahn	21,821	User choice	User choice	User choice	3	Multi-lingual sentiment	Deutsch
ASTD	10,006	User choice	User choice	User choice	5	Multi-lingual sentiment	Arabic
YouTutje	47 videos	User choice		L'Ber choice	3	Multimodal sentiment	English
CMU-MOSi	93 videos/	52 videos/	10 videos/	31 videos/	7	Multimodal sentiment	English
	2199 utterances	1284 utterances	229 utterances	686 utterances		classification	
CMU-MOSEI	3229 videos/	2250 videos/	300 videos/	679 videos/	7	Multimodal sentiment	English
	22,676 utterances	16,216 utterances	1835 utterances	4625 utterances		and emotion classification	

and business life. Not only the mobile application has an impact for user but also it plays an important role in business [7].

Having this in mind, we can conclude that smartphones and mobile apps are one of the most interesting inventions from the last years. Now with all of this explanation about mobile apps, let's talk about what this mobile app is going to be about.

First of all, the framework that is going to be used to build it. Flutter, is a framework from Google that allows to create productive and powerful apps that can be used on the 2 more mobile platforms these days, iOS and Android.

Flutter uses Dart programming language to build its components. This is the main variables that handles the messages and scores to display the proper sentiment visualization, happy or sad emotion.

```
bool _sentiment_score = true;
String _sentiment_text = "[Click play button to send message]";
String _sentiment_image_path = 'images/positive.png';
```

Then when click the play button when the message gets analyzed, the following code evaluate the score by switching its value.

```
void _requestSentiment() {
  setState(() {
    _sentiment_score = !_sentiment_score;

    if (_sentiment_score) {

      _sentiment_text = "\"the situation with the virus is getting better\"";
      _sentiment_image_path = 'images/positive.png';

    } else {

      _sentiment_text = "\"the situation with the virus is really bad\"";
      _sentiment_image_path = 'images/negative.png';

    }

  });
}
```

Finally, this piece of code build the UI (User Interface) on the mobile device simulator.

```
return Scaffold(
  appBar: AppBar(
    title: Text(widget.title),
  ),
  body: Center(
    child: Column(
      mainAxisAlignment: MainAxisAlignment.center,
      children: <Widget>[
        Text(
          '$_sentiment_text',
        ),
        Text(
          '$_sentiment_score',
          style: Theme.of(context).textTheme.display1,
        ),
        Image.asset('assets/$_sentiment_image_path'),
      ],
    ),
  ),
  floatingActionButton: FloatingActionButton(
    onPressed: _requestSentiment,
    tooltip: 'Switch score to show proper emoticon',
    child: Icon(Icons.play_arrow),
  ),
);
```

Also, emoticons and other icons need to be declared in the project configuration.

```
# The following section is specific to Flutter.
flutter:

  # The following line ensures that the Material Icons font is
  # included with your application, so that you can use the icons in
  # the material Icons class.
  uses-material-design: true

  # To add assets to your application, add an assets section, like this:
  assets:
    - assets/images/positive.png
    - assets/images/negative.png
    - assets/images/upload.png
    - assets/images/recording.png
    - assets/images/processing.png
```

Below you will find the screens that shows the idea of the entire process when the message is being recorded and then it is uploaded to be analyzed as positive or negative and the proper emoticon is shown (Fig. 2).

These screens above, allows the user to record a voice message by clicking in the button on the right corner at the bottom. For first versions, it is recommended to only record voice messages up to 6 s so the backend services processes can operate within an optimal time frame.

Also, the recording is going to be stored in the local storage of the mobile device. Once the recording is done, it will be ready to be uploaded (Fig. 3).

When is ready to be uploaded, it is enough to only push the button at the bottom and the mobile app will push it into the API gateway (Fig. 1). It is important to have a good internet connection on the mobile device to have a good experience during this process.

Fig. 2 Screen to recording voice message

While the voice is being uploaded, a new screen appears with a message. There is a button to cancel the processing to tell the web service to stop processing and the mobile app will go to the main screen again.

During the processing, the web services are connecting to Watson IBM speech to text service and running the sentiment analysis after that to cluster and rate the transformed message returning the response needed to the mobile app (Fig. 4).

Having the process finalized, now the mobile app has the response from the web service. If the score detected by the sentiment analysis is "1", the app will assume that it is a "Good" message, but if the score detected is 0, then the app will assume that it is a "Bad" message.

For both cases, the app will display the emoticon that represents the proper analyzed score.

Future Research

For the next steps on this work, we are looking to deploy all the projects into the proper platforms.

Go projects.

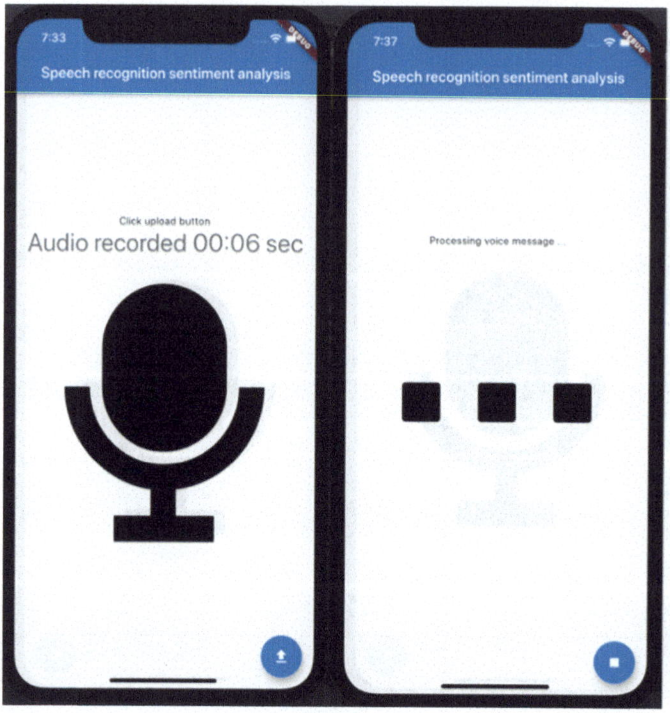

Fig. 3 Screen to upload and process the voice recording

- Speech recognition and Sentiment analysis into an API mounted on AWS ECS service.
- Create an API gateway to connect as third-party service.
- Audio file conversion from mp3 into flac.

Flutter project.

- Record audio from smartphone microphone.
- Save the recording into the local storage of the device.
- Send recording as mp3 to be converted by the API.
- Use API endpoints to analyze and display the right results given the response.
- Design the final UX and UI of the app to converge all the screens shown above.
- Publish mobile app in main stores, Apple Store and Google Play.

Other part that will be created in the near future is the support for S3 storage and the Databases.

S3 storage, it is going to be used to store the recordings so the web services can save it there and the speech to text process by Watson can read from. This service will store both original and converted files (mp3 and flac).

Fig. 4 Screens when analyzing text is being displayed with a proper emoticon

Databases are going to be used to keep track of the recordings processed like main data: IP address, Device, recording length. Also, some result metadata like analysis score, text transformed, accuracy, length of process and more (Fig. 5).

The API Gateway diagram will be modular and will have concrete functionality each. It needs an HTTP POST endpoint for the devices to access and upload voice recordings. Then the endpoint will convert the file into "flac" format and store them

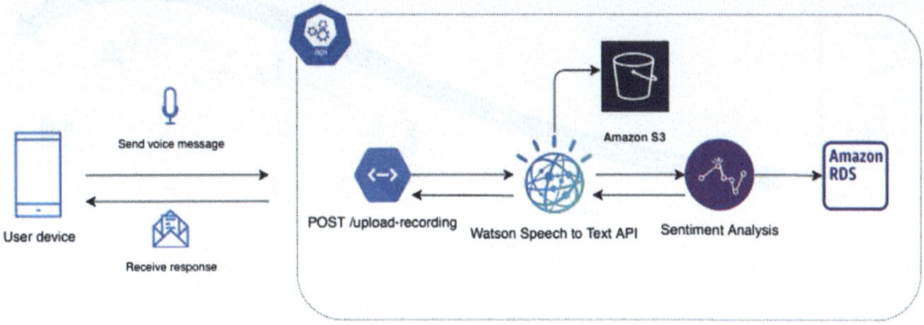

Fig. 5 API gateway internal functionality

in Amazon S3 service (storage). Once it has the converted file, it will be passed into Watson service API in order to get the response of that result. The response will contain the text needed to be analyzed by the sentiment analysis module. When the score is calculated and the other data is grabbed, all of it will be stored in a Amazon RDS database service based on MariaDB (open source) in order to have all the record of the process.

The approach that was made about the developed environment in which the situations described in this work are simulated, is easily extensible to environments with a greater number of agents without more than increasing or modifying the decision variables and even the criteria itself. Undoubtedly, guaranteeing its reliability by means of personal assistants for scenarios with high uncertainty is a way to counteract any malicious manipulation in the process of correct decision making.

It will be necessary to act, on the one hand, in the formative scope, preparing the new generations in technical knowledge able to create and to sustain the necessary technological infrastructures; but also, specially, in the promotion of creative abilities, of social relation, of decision making in uncertain surroundings, of facilitating leadership, among others; abilities necessary for the management of the complexity of the new productive surroundings. An important aspect is to describe a Business Model that contemplates vital aspects in the improvement of productivity as it can be observed in Fig. 6.

It should be considered that in the very near future Industry 4.0 will require planning with well structured strategies and stratagems for strategic and tactical plans, much more than for operational plans.

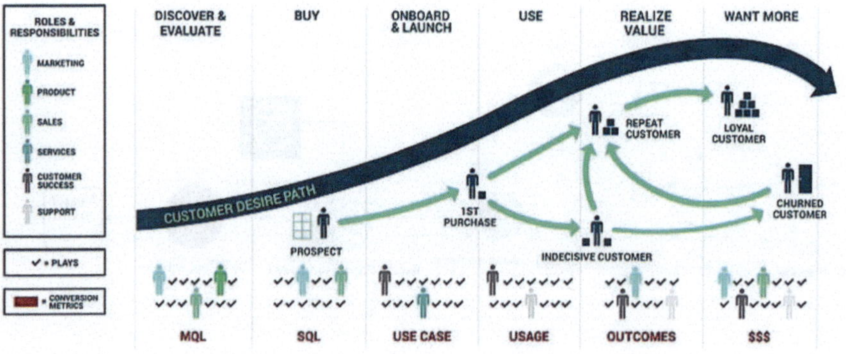

Fig. 6 A proposed model for the implementation of competitiveness improvement in Industry 4.0 considering the factor of generation Z

References

1. Bonastre, J.-F., Kinnunen, T., Larcher, A., Yamagishi, J.: Introduction to the special issue "speaker and language characterization and recognition: voice modeling, conversion, synthesis and ethical aspects." Comput. Speech Lang. **101021** (2019). https://doi.org/10.1016/j.csl.2019.101021
2. Pieraccini, R.: From AUDREY to Siri. https://www.icsi.berkeley.edu/pubs/speech/audreytosiri12.pdf
3. Dhakal, P., Damacharla, P., Javaid, A., Devabhaktuni, V.: A Near real-time automatic speaker recognition architecture for voice-based user interface. Mach. Learn. Knowl. Extr. **1**(1), 504–520 (2019). https://doi.org/10.3390/make1010031
4. Troussas, C., Virvou, M.: Advances in social networking-based learning. Intell. Syst. Ref. Libr. (2020). https://doi.org/10.1007/978-3-030-39130-0
5. Habimana, O., Li, Y., Li, R., Gu, X., Yu, G.: Sentiment analysis using deep learning approaches: an overview. Sci. China Inf. Sci. **63**(1) (2019). https://doi.org/10.1007/s11432-018-9941-6
6. Linton, H., Kwortnik, R.: The mobile revolution is here: are you ready? Cornell Hosp. Rep. **15**, 6–18 (2015)
7. Islam, Dr., Rashedul, M.D., Mazumder, T.: Mobile application and its global impact. Int. J. Eng. Technol. **10**, 72–78 (2010)

Logistics of Hospitalization Patients with COVID and Ambulances Required

Marco Del Moral, Alberto Ochoa, Alberto Lasserre, and Gastón Cedillo

Abstract The shortest path problem is a typical problem of optimization. This chapter presents an innovative model associated with the use of case-based reasoning to solve a problem of routing vehicles in a Hospital of El Paso, United States. In this research, diverse components are described to characterize this problem through the use of a Knowledge system. The algorithm was developed in Java, thus obtaining a tool which determines the best tracks to the vehicles associated with ambulances. An experiment was realized to probe the validations; the results were used to compare it with the Dijkstra algorithm and determine the quality of the results. The future research of this intelligent tool is to determine an innovative perspective related to episodic knowledge applied to resolution of diverse ambulances and as this topic is determinative to find and remember the best solutions quickly, additionally we compare it with a code from other postgraduate students trying to implement an algorithm similar to Logistics but using a Hybrid Intelligent system.

Keywords Distribution of ambulances in a hospital · Vehicle routing problem · Case-based reasoning · Ant colony algorithm · Shortest path

1 Introduction

COVID'2019 pandemics is a novel sickness where the city of El Paso, in Texas, is located and affects about one million people, in a database obtained by DSHS and associated with 47,500 records compiled in different events of eleven months. Each patient hospitalized in the main hospital of the city of El Paso and linked to at least

M. Del Moral (✉) · A. Lasserre
Instituto Tecnológico de Orizaba, Orizaba, Mexico
e-mail: marcojulioarg@gmail.com

A. Ochoa
Juarez City University, Ciudad Juarez, Mexico

G. Cedillo
IMT, Mexico, Mexico

A. Ochoa-Zezzatti et al. (eds.), *Technological and Industrial Applications Associated With Industry 4.0*, Studies in Systems, Decision and Control 347,
https://doi.org/10.1007/978-3-030-68663-5_11

27 symptoms whose involvement is associated with this type of pandemics events, it is crucial to know this for this research project, each of the scenarios that occur each time a COVID'2019 pandemic occurs in order to adequately link a requirement of a set of ambulances for the population at risk during this climatological event and that is recurrent during the minus seven months of the year. Nowadays, digital maps are increasingly common to greatly improve the optimization of evacuations performed by emergency vehicles such as ambulances or fire trucks [1]. With the progress that has been made in technology, these maps are becoming more sophisticated, in the way that they are able to find specific locations, draw routes and so forth [16]. Another thing that is noteworthy is that they show how the information has improved dramatically, as they changed from traditional maps to maps with real images taken from the air, satellite, or even a hybrid version of these two. The motivation of this project is specifically focused on the use of this increased interaction nowadays, in order to achieve an improvement in the logistics after a huge COVID'2019 pandemic which affects many people and determine what is the best way to organize the ambulances to move patients to diverse hospitals [18]. The objective of this work is to develop a system to help create routes on the basis of the emergencies given in El Paso, Texas, through the use of a system of neighborhoods of ants that allows them to create routes to take care of patients affected by a COVID'2019 pandemic or other types of emergencies in a quick way. This is important because the life of the people is at risk. In the United States, to minimize the time to arrive to the place of the accident, they do a comparison according to three possible emergencies at the same time and require other vehicles to respond to them. To provide assistance to citizens, paramedics in an ambulance need a route to arrive as quickly as possible to the place of the incident [17]. If there are many emergencies, they are classified in order of importance: Hospitalization related to a COVID'2019 pandemic, Rescue, and Prevention Action on public hazard. In all these activities, the time is vital because with a timely arrival the effect of the damage in a COVID'2019 pandemic can be decreased to prevent an explosion in the leak case and find alive persons among others. The bio-inspired algorithms are a technique of artificial intelligence focused on the solution of different problems, especially optimization problems. One of these algorithms is the swarm intelligence algorithm, where we can find the algorithm of the ant colony (ACO), particle swarm optimization (PSO), bees and so on [19]. The proposed algorithm to solve the routing problem in El Paso city is an Ant Colony System.

2 Descriptions of the Model Components

In this section, we offer details of each component related to the application domain that is involved in the problem, in our case we solve a Logistics problem related to the El Paso Health System's Hospitals by the use of a bioinspired algorithm to create routes of vehicles to attend emergencies.

2.1 The Shortest Route

The problem known as the shortest path or the shortest route, as its name suggests, tries to find the minimum or shortest route between two points. This minimum may be the distance between origin and destination points or the time to travel from one point to another. Mathematically, this system is described as a weighted graph $G = (V, A, d)$ where vertices are represented by $V = \{V0, V1, \ldots, Vn\}$, and arcs are represented by $A = \{(vi,\ vj)|i\ =\ j\}$ The distances associated with each arc are represented by the variable Cij measured by the Euclidean distance. The objective functions of the problem [7] are

$$\text{min}Z = \text{All the defined arcs } CijXij \tag{1}$$

Decision variables are as follows: Xij: action to move from node i to node j 0 indicates that there is no displacement and 1 indicates that yes there is movement. Cij: cost or time to get from node i to node j. Restrictions for the Total input flow = total output flow (external input into node j) + i All the defined arcs (i, j)

$$X_{ij} = (\text{external output from node } j) + k \text{ All the defined arcs}(j, k)\ Xjk \tag{2}$$

This type of optimization problems cannot be solved using exact methods. We cannot find its optimal solution with acceptable computational efforts. Since the early 50s, many algorithms have been developed to find a solution to this problem by finding good solutions but not necessarily optimal solutions. In the 80s, the solution techniques focused on the implementation of general-purpose meta heuristics including, among others, the ant colony, genetic algorithms, and taboo search.

2.2 The Shortest Path Algorithm

The shortest path algorithm, also called the Dijkstra algorithm, is an algorithm for determining the shortest path given in a source vertex to other vertices in a directed graph with weights on each edge. The shortest path algorithm belonging to the greedy algorithm [9] is an efficient algorithm of complexity O(n2), where n is the number of vertices, used to find the least cost path from a source node to all other nodes in the graph. It was designed by the Dutchman Edsger Wybe Dijkstra in 1959 [13]. The foundation on which this algorithm sits is the principle of optimality; the solution is built with the election of local optima in the hope of obtaining a global optimum.

3 Proposal Methodology

It is important to define that among a plethora of artificial intelligence techniques, the most appropriate one and the one that was selected for the present research was combined Case-based Reasoning and Ant Colony Optimization algorithm to improve a correct solution, which was validated with other similar investigations. The internal structure of the CBR systems is composed of two parts: the obtaining of the case and the reasoning of the case [20]. The first is responsible for finding the appropriate cases in the base of cases and the second is responsible for finding a solution to the problem given its description (see Fig. 1).

The CBR fulfills a cycle from the beginning of the problem until the solution is obtained, the cycle covers four parts: recover, reuse, review and retain (see figure two). This is also known as the 4 Rs (Retrieve, Reuse, Revise and Retain), according to [21]. The parts that comprise the CBR cycle depend on the case base or case library since in this case the previous cases that contain valuable information are stored for the CBR to be successful. We must remember that cases are problems that have a solution, for that it is necessary to obtain a representation of the cases so that they are stored in the base of cases since not all the information that is available about the problem is important to solve the problem. For this reason, we add one more part to CBR model that is the representation of cases, as is shown in Fig. 2.

Case representation: It is one of the most important parts of the CBR because the four parts of the CBR life cycle depend on it. Its importance lies in the fact that a case is a piece of knowledge that represents an experience and includes a problem, which is the description of the task to be solved and a solution, which corresponds to how the task was solved [21]. In turn, a set of cases is called case base or case library. Usually, a case is represented as an attribute-value pair; this represents the problem and the solution of the case. In some cases, the case contains a third element that is the

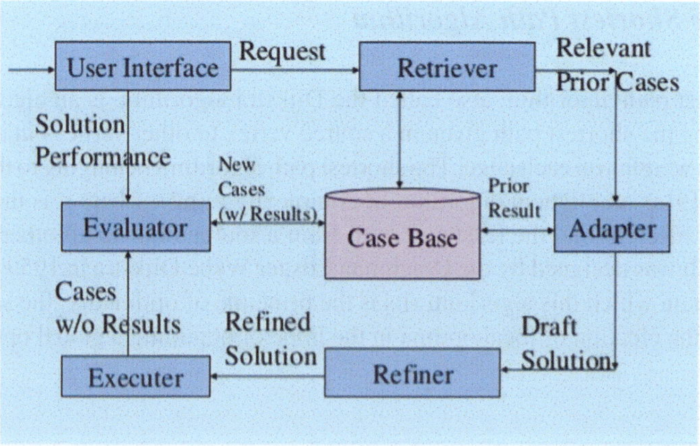

Fig. 1 Basic structure of an CBR system according to [21]

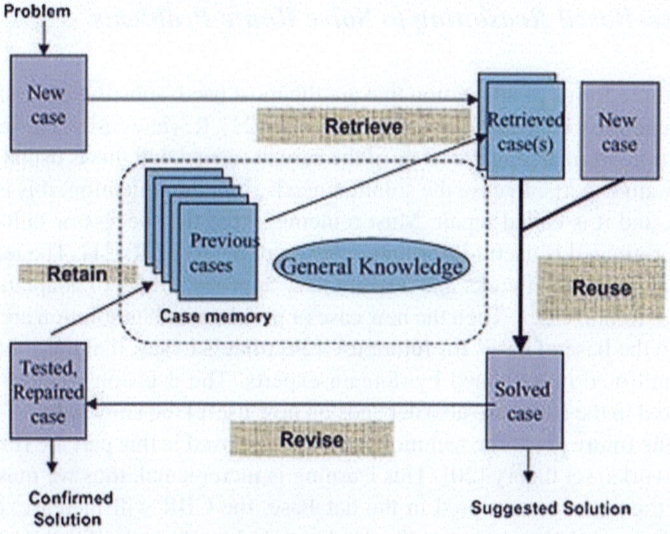

Fig. 2 Life cycle of the CBR according to [24]

result, that is, the state of the problem once the solution was applied. The experience can be represented in a different way; the classic one includes vector form, frame-based, object-oriented and textual, although there are already more sophisticated representations that are hierarchical cases, generalized cases, cases based on the based design, cases based on planning [22]. Recover cases: The quality in the result of the CBR systems depends on the similarity measures used for the recovery of similar cases. The soft computing techniques used in this part of the CBR are: diffuse indexing, diffuse grouping, case classification, probability, Bayesian models for the selection of cases [20], the nearest neighbor. In this phase, the current problem is checked against the problems stored in the base of cases. The comparison is a process of comparing two cases among themselves and determining the degree of similarity (DOS, for its acronym in English degree of similarity). Besides the measures of similarity, knowing the domain helps determine the similarity of the new case with a previous case and having the degree of similarity lead us to a degree of adequacy of the solution of the problem or current case [24]. Reuse cases: The reuse can be given by means of copying or integrating the solution of the cases that were recovered in the previous part. In reuse, interactive and conversational diffuse reasoning can be used, learning to reuse case knowledge and diffusional approaches [20]. This part is also known as adapting the solution since the solution that was obtained in some occasions is necessary to adapt it to be given a solution to the case, as explained by [24].

3.1 *Case-Based Reasoning to Solve Route Problems*

There are three forms of adaptation that are the most used: substitution, transformation, and generative adaptation, this is indicated in [21]. Review cases: The evaluation of the solution originated in the reuse of the case is carried out, this is usually carried out by domain experts. In case the solution needs some modification, this is done in this phase, and it is called repair. Must remember that the success or failure of the solutions originated is useful information to improve the CBR [21]. The techniques used here are neural networks and evolutionary approach, rules of adaptation using set theories. Retain cases: Then the new case or problem and its solution are retained or stored in the base of cases for future use [24], that is to say, that the solution was already confirmed or validated by domain experts. The decision whether the new case is stored in the case base also depends on how useful the knowledge of that case will be in the future [21]. The techniques that can be used in this part are fuzzy rules, neural networks, set theory [20]. This learning is incremental, thus we must bear in mind that the more cases stored in the database, the CBR will increase, as it will reach the time needed to maintain the database, and continues fulfilling its function [21]. Many authors described different applications to resolve this kind of problem. Lianxi Hong proposed an algorithm which permits to determine the time to reach a specific point in a city according to different possible scenarios [8]. On the other hand, in [24] is proposed a concept similar to ours, with relation to "emergencies," which may occur in any time and place [11], and organize the demand to the vehicles, in our case is to the rescue units, try to minimize the effort to attend the demands in a day with the estimation of arrive, and solve another emergence, for example bee swarm which can be damage to children. In Fig. 3 is proposed our model inner a Smart City.

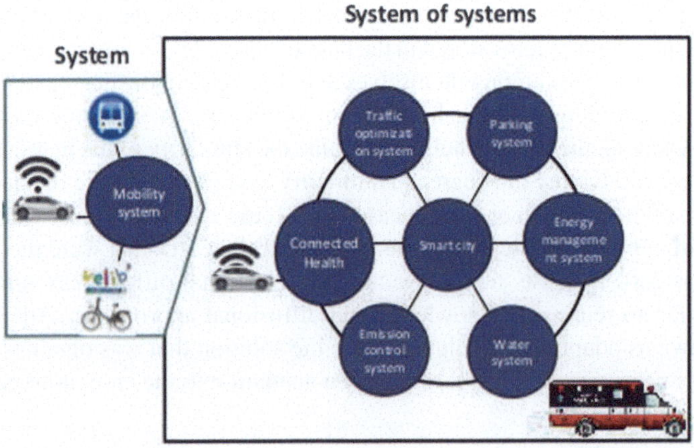

Fig. 3 An ambulance in a transport model in a Smart City

4 The Proposed Model of Our Hybrid Algorithm

An intelligent tool was developed using Case-based Algorithm based on Ant Colony Algorithm and the programming language Java (J2SE) and as the first step we begin with the creation of the graph for the central area covering the Main Hospital in El Paso, a total of 2451 streets, avenues and boulevards (edges) and 1710 nodes. Subsequently, an entity was designed called "object" to store information about each node, as the impact to neighboring nodes and their respective distance.

These objects were related to a data structure called a multidimensional array which saves computer resources because this structure does not cause the overflow of memory related with the cells which compound the grid and generate a square incidence matrix, it stores only the necessary track which is visualized in their analysis. The Ant Colony algorithm has been proved effective to solve NP-Hard problems [3] when they use multidimensional arrays [2]. The structure of the generic algorithm is as follows [5]: Algorithm:

1. Optimization based on Ant Colony
2. Initialize parameters ()
3. while not stop condition ()
4. for ant = 1 to n construct solution ()
5. evaluate solution ()
6. update pheromones ()
7. end for end while
8. The optimization quantity is the distance of the route. Thus, the truck movement cost between loading spots i and j is a function of all separate costs for each factor which affects the track route:

9. $dij = \alpha daij + \beta dbij + \gamma dcij + \cdots.$ (3)

10. Let *tij* (*t*) be the intensity of trail on edge (*i, j*) at time *t*. Each ant at time t chooses the next node, where it will be at time $t + 1$. Therefore, if we call an iteration of the ACO algorithm the *n* moves carried out by the n ants in the interval (*t*, *t* +1), then for every *n* iterations of the algorithm, each ant has completed a tour. At this point the trail intensity is updated according to the following formula:

11. $\tau ij\,(t + n)\rho \cdot \tau ij\,(t) + \Delta\tau ij\,,$ (4)

12. where ρ is a coefficient that represents the evaporation of trail between

13. $k = 1kij.$ (5)

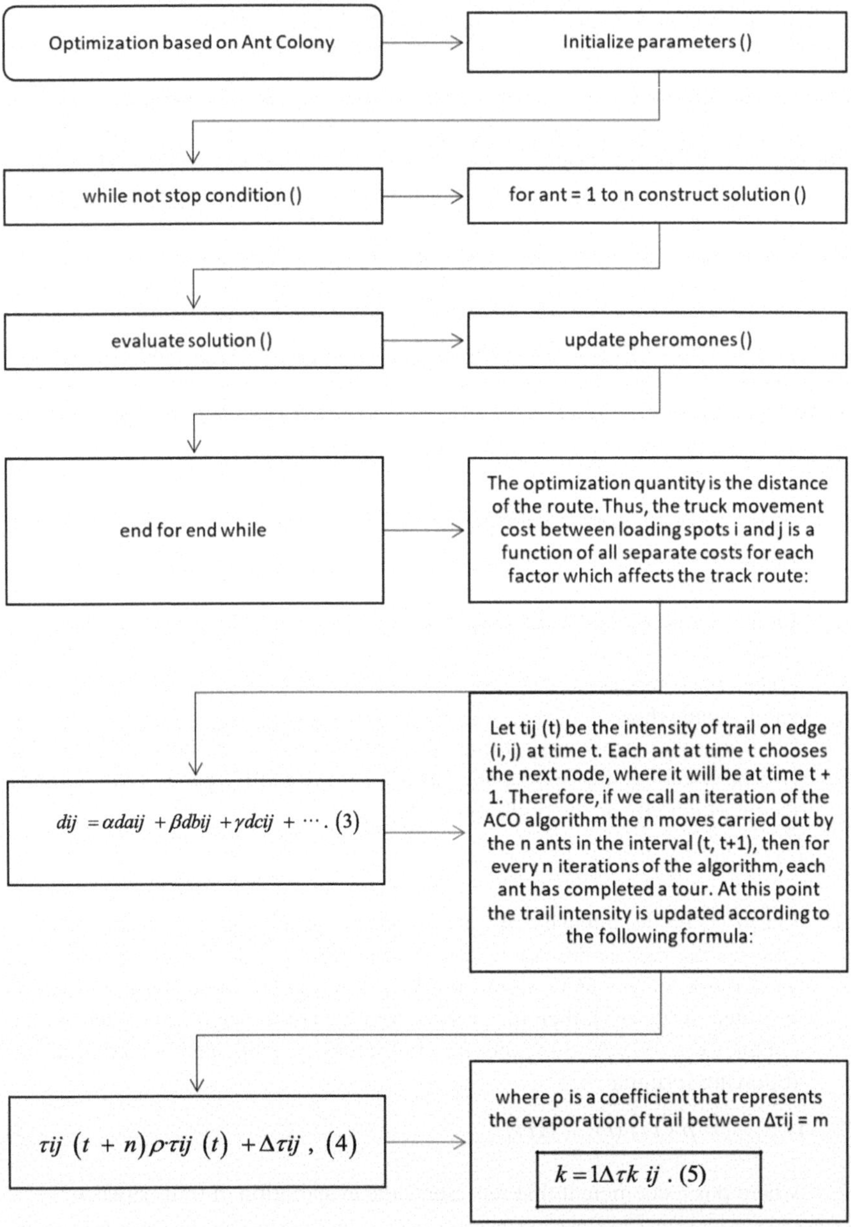

The coefficient ρ must be set to a value < 1 to avoid unlimited accumulation of trail (see note 1). In our experiments, we set the intensity of trail at time 0, *tij* (0), to a small positive constant c. In order to satisfy the constraint that an ant visits all the n different loading spots, we associate with each ant a data structure called the hlist, that saves loading spots already visited up to time t and forbids the ant to visit them again before n iterations (a tour) have been completed, as is proposed in [12]. When a tour is completed, the hlist is used to compute the ant's current solution (i.e., the movement cost of the path followed by the ant). The hlist is then emptied, and the ant is free to choose again,

$$\eta ij = 1/dij \tag{6}$$

We call visibility hij the quantity $1/dij$. This quantity is not modified during the run of the AS, as opposed to the trail, which instead changes according to the previous formula (4). We define the transition probability from loading spot i to loading spot j for the kth ant as

$$pk\,ij \;= \tau ij\ (t)\alpha \cdot \eta ij\beta k \in allowed\, k\tau ik(t)\alpha \cdot \eta ik\beta \tag{7}$$

The software implements the ability to block and alter the meaning of the streets, a fact that occurs in the central city of El Paso because of events, accidents, public works and so on. The method Initialize parameters enters the source node, the destination node, blocked streets and the number of ants involved in the search for the solution similar to the proposal in [14]. Construct solution takes place when ants move randomly with both probabilities using the Monte Carlo method if there is already a trail of pheromone as in [4, 6, 15]. Once an ant has found the evaluate solution, the destination node determines if the journey is of good quality, discarding those paths that do not decrease the distance obtained by other ants, and Updating the pheromone if you have found a shorter route.

The user interface displays the found routes to the destination, with the option of display all of them or one in particular in a map (Fig. 4), which has the options of adding landmarks (churches, schools, hospitals, parks, rivers), zoom, viewing the different layers, storing in the route file [10], exporting the map as an image and

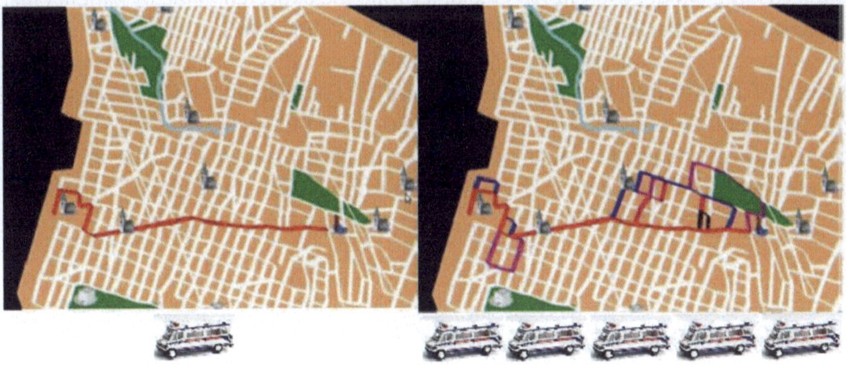

(a) (b)

Fig. 4 **a** Drawing of a single route. **b** Drawing of five routes

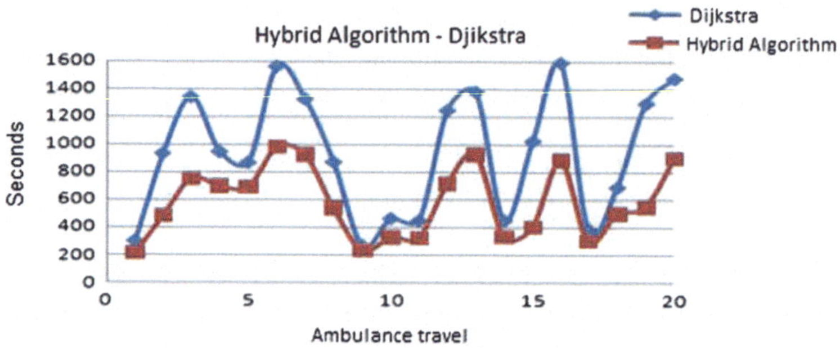

Fig. 5 ACO and Case-based Reasoning and Dijkstra comparison

sending it via Bluetooth to a mobile device. A comparative related with performance of our proposed model and Djikstra algorithm is presented in Fig. 5.

5 Experimental Results

The proposed algorithm was compared to the algorithm of Operations Research: The shortest path (Dijkstra). The comparison was carried out with the generation of 20 runs starting from the central zone associated with El Paso Hospital (node 759) to different nodes (Table 1).

The results were obtained with $\mu = 25.15$ seg and $\sigma = 15.65$ seg, in 35% of cases. While the Ant Colony gives better results than the shortest path algorithm (1, 4, 7, 10, 11, 14 and 18), in 20% the results were similar (8, 9, 12 and 17) and 45% was surpassed by the shortest paths as in Fig. 4. We select from a list of 1487 sickness related with the effects of a COVID'2019 pandemic, two incidences and with a Kriging Model used in ArcGis, determine the expected value of patients that required an ambulance to travel to Hospital, as it is shown in Fig. 6.

Another comparison was using the same instances and information obtained from Table 1 of twenty different ambulances from El Paso Health System's Hospitals, with the intention of building a robust design of experiments to try to understand the accumulative number of optimal solutions to reach the best track using the search space to three different codes of Hybrid Algorithm, PSO and Cultural Algorithm. The results will be observed in Fig. 7.

6 Conclusion and Future Research

The algorithm currently implemented gives good quality solutions to an NP-hard problem, improving by 35% of cases the routes provided by the shortest path algorithm. The 45% where the shortest path algorithm exceeds the ACO which

Table 1 Result for each ambulance travel from the houses of the patients to a Hospital in El Paso

#	Origin	Destination	Dijkstra	Hybrid Algorithm
1	945		222	212
2	614		464	507
3	903		755	841
4	941		732	698
5	1044		693	709
6	1202		984	1093
7	1094		953	927
8	1057		538	538
9	1418		231	231
10	170	759	338	328
11	526		347	324
12	462		718	718
13	846		859	1030
14	524		359	333
15	809		365	406
16	1107		886	1011
17	698		302	302
18	1062		517	499
19	1342		519	564
20	1199		885	984

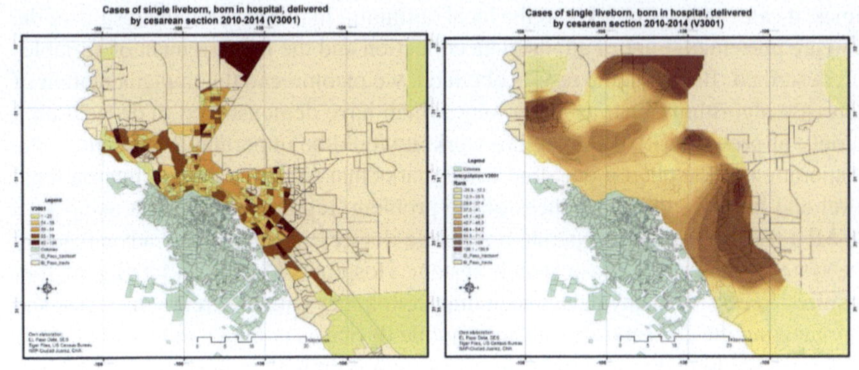

Fig. 6 Incidence of a specific sickness in Neighborhoods of El Paso, and a Kriging Model of incidence of this sickness in the future with the necessity of more ambulances to transport more patients during a COVID'2019 pandemic

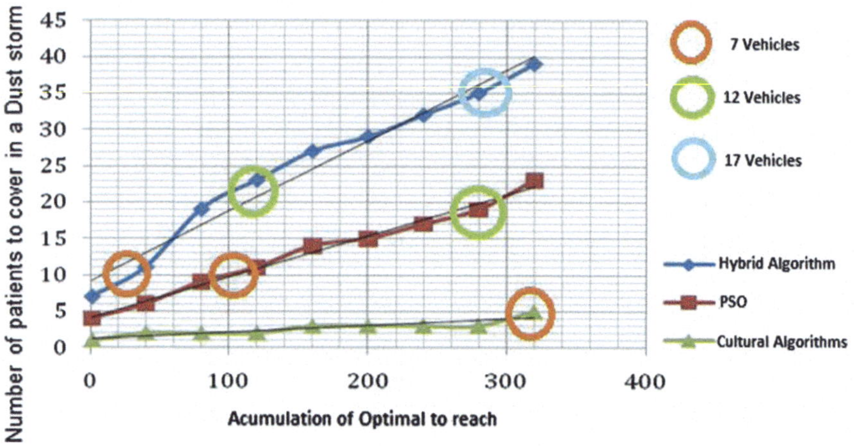

Fig. 7 Comparative analysis of a Hybrid Algorithm, a PSO and a Cultural Algorithm for an nstance of three emergencies dataset associated with a COVID'2019 pandemic at the same time, when we consider its performance on the basis of Table 1

is attributed to not yet implementing the evaporation of the pheromone [15], the pheromone amount in nature may remain a few hours to several months depending on different aspects, such as ant species or soil type [7], causing a minor influence on the effect of evaporation in the process of finding the shortest path. Due to the long persistence of the pheromone, it is difficult for the ants to "forget" a path that has a high level of pheromone but have found a path even shorter. Keep in mind that if this behavior is transferred to the computer to design a search algorithm sometimes, it can converge quickly to the local optimum. In this section, the results of the trial are presented. First of all, the data collection and the measurement of variables are described. Based on the results obtained, we recommend the implementation of heuristic algorithms such as ant colony, which have demonstrated to do well on a variety of problems [3, 4]. As future work, it would be important to implement the evaporation of the pheromone, find benchmarks that are being used at international level and prove to those instances of the problem, replicate the project using Java (J2ME) for the system to operate on mobile devices which provide advantages to the system in units of El Paso Health System' Hospitals. Making an Intelligent Tool requires Access from any device including cellphones with different screen size and reorganizing the correct decisions in a mobile device as is shown in Fig. 8.

We decide to make a comparison of our algorithm with relation to a PSO Algorithm and a Cultural Algorithm. We discovered the proposal of combining Case-based Reasoning, as is proposed in [23] and ACO Algorithm to obtain three different paths to a successful number of emergencies occurring at the same time, and that its performance improves by 22% the performance of PSO Algorithm and by 37% the performance of Cultural Algorithms. In the future research, it is important to describe the different times in other quadrants (the city is divided into four regions named Quadrants) of the city in the border zone which covers Hospitals and covers

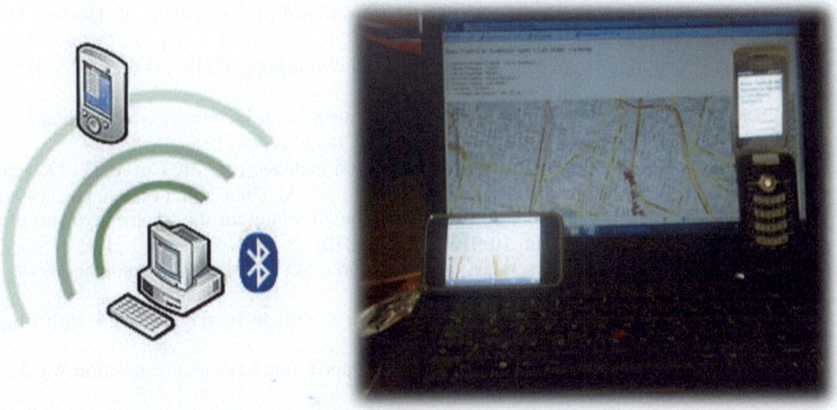

Fig. 8 Using Bluetooth, it is possible to use our proposed model of ambulances in a huge contingency related to a COVID'2019 pandemic

only the 48.7% of territorial space of the city. Additionally, there are several sub-problems associated with research that need to be improved, among them are ambulance drivers'fatigue as a groutier [25], the change of epidemiological traffic lights that can determine additional waiting times to normal ones, the accommodation of patients in the same ambulance to optimize space, this will be proposed with a Bin Packing model adapted to the context, and the management of respiratory equipment in an ambulatory way to be able to specify waiting times in a time window according to the health of the patients, a project of a framework for humanitarian logistics is the goal to achieve in a short term. However, the development of a holistic framework that considers each crucial aspect of humanitarian logistics in the midst of the Covid'2019 pandemic is a titanic task due to the complexity of determining what is the best technique to apply for each sub-problem and the appropriate response time for proper decision making including an output through a narrative script in an appropriate and specialized intelligent dashboard that allows to offer the greatest amount of detailed relative data to specify the appropriate approach for the proposed solution.

References

1. Barcos, L., Rodrıguez, V.M., Alvarez, M.J., Robuste, F.: Routing design for less-than-truckload motor carriers using ant colony techniques. Business Economics Working Papers, Series Universidad Carlos III de Madrid. Departamento de Econom´ıa de la Empresa, paper number wb043814 (2004)
2. Bell, J.E., McMullen, P.R.: Ant colony optimization techniques for the vehicle routing problem. Adv. Eng. Inform. **18**, 41–48 (2004)

3. Birattari, M., Pellegrini, P., Dorigo, M.: On the invariance of ant system. In: Dorigo, M., Gambardella, L.M., Birattari, M., Martinoli, A., Poli, R., Stutzle, T. (eds) Ant Colony Optimization and Swarm Intelligence, 5th International Workshop, ANTS 2006, LNCS 4150, pp. 215–223. Springer, Berlin (2006)
4. Dorigo, M., Birattari, M., Stutzle, T.: Ant colony optimization: artificial ants as a computational intelligence technique. IEEE Comput. Intell. Mag. **1**, 28–39 (2006)
5. Dorigo, M., Di Caro, G.: The ant colony optimization metaheuristic. In: Corne, D., Dorigo, M., Glover, F. (eds.) New Ideas in Optimization, pp. 11–32. McGraw Hill, London, UK (1999)
6. Ghoseiri, K., Nadjari, B.: An ant colony optimization algorithm for the bi-objective shortest path problem. Appl. Soft Comput. **10**, 1237–1246 (2010)
7. Grimaldi, R.P.: Matematicas Discretas y Combinatoria, 5th edn. Editorial Addison Wesley, Mexico (1995)
8. Hong, L.: An improved LNS algorithm for real-time vehicle routing problem with time windows. Comput. Oper. Res. **39**, 151–163 (2012)
9. Johnsonbaugh, R.: Matematicas Discretas, 6th edn. Grupo Editorial Pearson Education, Mexico (2005)
10. Kolavali, S.R., Bhatnagar, S.: Ant Colony Optimization Algorithms for Shortest Path Problems, in Network Control and Optimization. Lectures Notes in Computer Science, vol. 5425. Springer, Berlin (2009)
11. Lei, H., Laporte, G., Guo, B.: The capacitated vehicle routing problem with stochastic demands and time windows. Comput. Oper. Res. **38**, 1775–1783 (2011)
12. Ponce, J.C.: Diseno de un Modelo de Optimización de Colonias de Hormigas Aplicado al Problema del Clique Maximo. Tesis Doctoral. Universidad Autonoma de Aguascalientes, Doctorado en Ciencias de la Computacion, Aguascalientes, Mexico (2010)
13. Taha, H.A.: Investigacion de Operaciones, Editorial Alfaomega, Mexico, 5th ed. No. de paginas 960 2008
14. Xiong, W., Wang, L., Yan, C.: Binary ant colony evolutionary algorithm. Int. J. Inform. Technol. **12**, 10–20 (2006)
15. Yu, F., Li, Y., Wu, T.: A temporal ant colony optimization approach to the shortest path problem in dynamic scale-free networks. Physica A **389**, 629–636 (2010)
16. Mohammad, S.R., Mei, Y., Hamilton, M., Salim, F.D.: CAPRA: A contour-based accessible path routing algorithm. Inform. Sci. **385–386**, 157–173 (2017)
17. Talarico, L., Meisel, F., Sörensen, K.: Ambulance routing for disaster response with patient groups. Comput. Oper. Res. **56**, 120–133 (2015)
18. van Barneveld, T.C., Bhulai, S., van der Mei, R.D.: The effect of ambulance relocations on the performance of ambulance service providers. Eur. J. Oper. Res. **252**(1), 257–269 (2016)
19. Şimşek, A., Kara, R.: Using swarm intelligence algorithms to detect influential individuals for influence maximization in social networks. Expert Syst. Appl. **114**, 224–236 (2018)
20. Simon, C., Shiu, K., Li,Y., Zhang, F.: A fuzzy integral based query dispatching model in collaborative case-based reasoning. Appl. Intell. **21**(3), 301–310 (2004)
21. Garcia, I., Sebastia, L., Pajares, S.: Eva Onaindia: GRSK: a generalist recommender system. WEBIST **1**, 211–218 (2010)
22. Bergmann, R., Kolodner, J.L., Plaza, Enric: Representation in case-based reasoning. Knowl. Eng. Rev. **20**(3), 209–213 (2005)
23. Behbahani, M., Saghaei, A., Noorossana, R.: A case-based reasoning system development for statistical process control: case representation and retrieval. Comput. Ind. Eng. **63**(4), 1107–1117 (2012)
24. Begum, S., Ahmed, M.U., Funk, P., Xiong, N., Folke, M.: Case-based reasoning systems in the health sciences: a survey of recent trends and developments. IEEE Trans. Syst. Man Cybern. Part C **41**(4), 421–434 (2011)
25. Rahman, H., Ahmed, M.U., Barua, S., Begum, S.: Non-contact-based driver's cognitive load classification using physiological and vehicular parameters. Biomed. Signal Process. Control. **55** (2020)

A Heuristic Method for Oil Distribution Networks Applied to the Switching Behavior in the Oil Industry

Mario M. Monsreal-Barrera and Oliverio Cruz-Mejía

Abstract This work develops a novel method of designing distribution networks for petroleum distribution. The method takes advantage of techniques including outliers filtering as well as traditional methods combined into a single heuristic. One of the main contributions of this study is the fact that the method has the ability to group client nodes based on their reachability, which is a local density measure derived from outliers' detection, and refers to the typical distance at which a point can be reached from its neighbors. The novel algorithm is able to offer much better results reducing distribution costs significantly.

Keywords Distribution network · Petroleum distribution · Center of gravity · Reachability

1 Introduction

Due to the raise of competition stemming from industry's constant globalization process, organizations constantly need to develop new and better ways of delivering their products to their clients. A clear example of this is the transition many countries are undergoing in the energy sector. The opening of the oil industry in some regions represents an important opportunity to emerging or global companies to enter local markets, and a huge challenge for those already established. Considering the complexity of the energy sector and industry, the investment of large amounts of capital is necessary in order to increment its efficiency, thus assuring that the benefits will be higher than the investments and operations costs.

One of the main characteristics of the oil/hydrocarbon industry is the fact that the products of the competing companies are extremely similar, so companies can only

M. M. Monsreal-Barrera
Texas A&M Transportation Institute, College Station, USA
e-mail: m-monsreal@tti.tamu.edu

O. Cruz-Mejía (✉)
Universidad Autónoma del Estado de México, Toluca, Mexico
e-mail: ocruzm@uaemex.mx

A. Ochoa-Zezzatti et al. (eds.), *Technological and Industrial Applications Associated With Industry 4.0*, Studies in Systems, Decision and Control 347,
https://doi.org/10.1007/978-3-030-68663-5_12

be more competitive by finding, extracting, producing and distributing petroleum and its derivatives in a more efficient way than their competitors [11]. Consequently, the organization and the management of the supply chain are crucial for the competition between the different oil/hydrocarbon companies: They don't compete in products but in service and price. This type of competition is more critical because of the magnitude and volume of the operations, so any improvement in their processes may result in substantial benefits. Very few industries can benefit more from an efficient chain supply management than the oil and hydrocarbon industry [11].

Another key issue for the industry is its environmental impact. A better design of the distribution network and a more efficient use of the assets not only represent a cost reduction, but also a lower impact on the environment stemming from a better use of energy and lower greenhouse gas emissions. A country's efficient and sustainable distribution network of energy, conditions its competitiveness within a globalized context.

The present work introduces a new method of designing networks, applied to the analysis of a hydrocarbon distribution network. The method uses techniques of outliers filtering, and traditional methods combined into a single heuristic. One of the main contributions of this study is the fact that the method has the ability to group client nodes based on their reachability, which is a local density measure derived from outliers' detection, and refers to the typical distance at which a point can be reached from its neighbors. However, the difference is that in our case reachability is calculated based on vehicle autonomy, which makes it more realistic for application purposes. An additional advantage of this new method, besides its simplicity and that it can be easily applied to any industry, is that the method determines simultaneously the capacity and location of the distribution hubs (e.g. depots)—based on demand and client's location. Because hub location and capacity depend on demand, location, and on the number of the clients nodes that it will serve, traditional methods generally require first either the hub capacity to determine the hub location, or the location to determinate the capacity, which result in sub-efficiencies. The study uses data of actual and current distribution network of gasolines in Mexico. Results show substantial savings in distribution costs for this product, when using the new proposed method.

2 Literature Review

Facility location problems are normally strategic in nature and implicate long-term decisions and investments, binding organizations to many uncertainties during the operational lifetime of a facility [5]. Due to the importance and impact of the decision many models and approaches to facility location have been developed over the years.

Singh et al. [3], for example, construct their model considering a set of risk factors, their probability of occurring and their expected cost impacts, with the objective of reducing the total cost of the supply chain. Lim et al. [10] take the issue a step further, they investigate the impact of wrongly estimated disruption probabilities. With a continuous location model they conclude that the increase of the expected total costs of the disruptions is greater due to the underestimation than the overestimation of the disruptions' occurrence probability.

Baron et al. (2010) for their part center on the demand and its uncertainty, and apply robust optimization to the problem. They consider a multi-period fixed-charge network location problem and apply a robust optimization approach in which they formulate the problem to include alternative levels of uncertainty over the periods. Martinez-Gomez et al. [8] propose a mathematical programming formulation for the optimal location and reallocation accounting for future growth of the organization based on a multi-annual framework. Athawale and Chakraborty [6] employ preference ranking organization method for enrichment evaluation, which is an effective multi-criteria decision-making tool often used to solve complex manufacturing problems, for the selection of the location out of a set of alternatives.

Another approach is the one of a greener supply chain. Elhedhli and Merrick [9] focus on reducing the carbon emissions with their model, along with fixed and variable location and production costs. The authors employ a Lagrangian relaxation to decompose the general problem into a capacitated facility location problem with single sourcing and a concave knapsack problem. Zhang et al. [7] present a two-stage methodology to identify the best location for a biofuel production facility in which stage 1 uses a Geographic Information System to identify feasible locations and stage 2 employs a total transportation cost model to identify the preferred position.

Important also to location problems is the nature of the objective network. Zhang et al. [2] introduce a model for a supply chain with bidirectional flows, their model is a nonlinear mixed-integer program that optimizes the location of the distribution centers and the allocation of the retailers. The model considers the possibility for retailers to return the products directly to the supplier for reprocessing without going through the distribution centers. Lopez [1] develops a model based on a hybrid-mixed integer programming method to solve the Territory Design Problem, the problem which is focused on grouping small geographical areas into lager clusters. From a large group of basic areas with specific attributes and certain restrictions for the resulting territories, an optimal districting plan is obtained.

A key aspect of the solution of facility location problems is the impact of outliers, i.e. very distant clients. Charikar et al. [4] recognize the effect of outliers can have on the possible solutions for a facility location problem. In their work, the authors study various generalized facility location problems, K-center, K-median, non-capacitated, etc., where only a specified fraction of the customers are to be served, and provide various approximation algorithms that take that constraint into account.

3 Methodology

The methodology consists in developing three scenarios to compare the performance of the new method: Actual, Center of Gravity and New Method. The "Actual" scenario calculates incurred costs when supplying to the different client nodes (terminals) with the current network configuration considering actual demand for each terminal. The "Center of Gravity" scenario uses the well-known center of gravity method, which determines the best facility location (supplying refineries) based on the actual geographic location and demand of the client nodes (terminals). However, this method does not considers transportation costs for the allocation, and it uses the current grouping of client nodes because it requires that the group of client nodes be already established in order to calculate the location of the service node [12].

The "New Method" scenario uses the same demand data and location of the client nodes; however, this new method considers the reachability criteria to perform the terminal grouping, the assignment, and to define the location and capacity of the service nodes. This method uses also the Center of Gravity method for the last step. Figures 1 and 2 show the steps for the "Center of Gravity" and the "New Method", green shaded boxes in Fig. 2 show the different and additional steps of the new method.

At the heart of the new method lies the reachability grouping. This reachability is determined by considering the maximum vehicle autonomy as a threshold. Then clients are sorted by X and Y coordinates, calculating the distance between each of them per spherical Earth axis (the study employs spherical distances considering the Earth curvature for all scenarios). The algorithm determines the grouping based on these spherical distances and the threshold, by assessing the nearest neighbor and either accept it in current group or reject it. This grouping simultaneously determines the required capacity of the service node. Once the groups are created the Center of Gravity technique is applied to determine the service node location. We use Pseudo Code to describe the reachability grouping decision of the algorithm, (due to document size constraints we kept other parts of the algorithm unaddressed in this description, also for confidentiality we do not disclose the threshold or vehicles autonomy).

Fig. 1 Steps of the center of gravity method

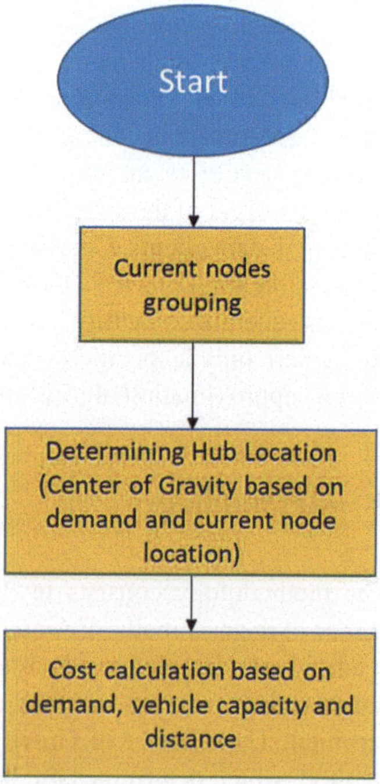

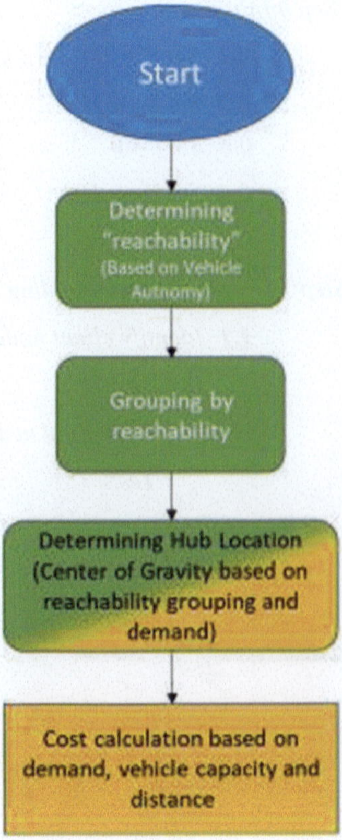

Fig. 2 Steps of the new method

We define:

K: Client nodes set to be evaluated

D_0*: Distance set in the generated group*

R_0*: Client nodes set to visit in the generated group*

$\mathfrak{R}$*: Generated group set*

{()}: Empty set

$d_{j,i}$ = *Transit distance in meters from client node j to client node i, (or from origin when j = 0 to client i, or from last visited client node to the origin when i = 0)*

k = Maximum values of indixes j and i in the set

i: Index of client nodes as destination I

j: Index of client nodes as origin J

I: 0, 1, 2, ... k

J: 0, 1, 2, ... k.

Step 0 *Initialisation*

0.1 Set $K = \{(1,2,3\ldots k)\}$

0.2 Set $K' = K$

0.3 Set $j = 0$

0.4 Set $D_0 = \{()\}$

0.5 Set $R_0 = \{()\}$

Step 1 *Group iterative building*

1.1 *Identify client nodes*

Is $K' \neq \{()\}$*?*

1.1.1 *Find nearest client node*

Yes:

Are there any $i \in K'$*, with*

$d_{j,i} + d_{i,o}$ *< Threshold?*

Yes:

Find min $(d_{j,i})$, $i \in K'$

Add $d_{j,i}$ *to* D_0, $i \in \min(d_{j,i})$

Add $d_{i,j}$ *to* D_0, $i \in \min(d_{j,i})$

Is there other $d_{j,i} = \min(d_{j,i})$, $i \in K'$*?*

Yes:

Go to 1.1.1

No:

Add i to R_0, $i \in \min(d_{j,i})$

No:

Print $\Re$

End

No:

Print $\Re$

End

4 Results

Results are in the form of savings, which in turn are based on the distribution costs. All distribution costs are calculated using the same and current unitary values as well as the same capacities and vehicle mixes.

Due to the Reachability criteria, and based on the costs, the new method can discriminate the customer nodes or terminals that can be cost-efficiently served, as opposed to those "out of the reach" or "too expensive". In this case, the method suggests that 41% of the terminals should not be attended.

The following table presents the annual savings in distribution costs for both methods: Center of Gravity and New Method, when compared to the current network. The amounts are shown in US Dollars for the terminals identified with cost-efficient distribution (Table 1).

The service discrimination for profitability is useful in industries where the strategy is focused on operative performance. Gasoline distribution, however, has a social and competitive function. Therefore, the priority is demand fulfilment. For this reason, Table 2 shows results (annual savings) when performing the adjustment for 100% of demand (all terminals).

It is important to point out that the present study focuses on the analysis of network designs and their impact on distribution costs (exclusively), but does not analyze the cost–benefit. The previous is relevant since, for decision making on network design, the savings offered by the new method should be assessed in light of the implementation costs.

The present analysis offers a new perspective for distribution network design by using real data of the gasoline commerce in Mexico. The application to this industry is not random since currently Mexico finds itself in important changing period regarding the hydrocarbon industry, where the operations and resources optimization is a key issue.

The study presents a considerable improvement in operational costs of the analyzed distribution network, which points to the existence of important areas opportunity needed to be evaluated to continue the logistic development of the industry. More sophisticated algorithms could be explored to increase accuracy and saving; however the tradeoff may be the applicability of these new tools.

Table 1 Annual distribution costs savings—cost-efficient distribution terminals

Annual distribution costs savings (USD millions)	
Center of gravity	New method
\$173.3	\$456.2

Table 2 Distribution costs annual savings—all terminals

Annual distribution costs savings (USD millions)	
Center of gravity	New method
\$363.1	\$863.3

References

1. Lopez, F.: Multi-objective territory design with capacity & geographic constraints. Int. J. Comb. Optim. Probl. Inform. **1**(1), 3–12 (2010)
2. Zhang, Z.-H., Berenguer, G., Shen, Z.-J.: A capacitated facility location model with bidirectional flows. Transp. Sci. **49**(1), 114–129 (2014)
3. Singh, A., Mishra, P.K., Jain, R.: Design of global supply chain network with operational risks. Int. J. Adv. Manuf. Technol. **60**(1), 273–290 (2012)
4. Charikar, M., Khuller, S., Mount, D., Narasimhan, G.: Algorithms for facility location problems with outliers. In: Proceedings of the Twelfth Annual ACM-SIAM Symposium on Discrete Algorithms, pp. 642–651. Society for Industrial and Applied Mathematics (2001)
5. Baron, O., Milner, J., Naseraldin, H.: Facility location: a robust optimization approach. Prod. Oper. Manag. **20**(5), 772–785 (2011). https://doi.org/10.1111/j.1937-5956.2010.01194.x
6. Athawale, V., Chakraborty, S.: Facility location selection using PROMETHEE II method. In: Proceedings of the 2010 International Conference on Industrial Engineering and Operations Management, pp. 9–10 (2010)
7. Zhang, F., Johnson, D., Sutherland, J.: A GIS-based method for identifying the optimal location for a facility to convert forest biomass to bio-fuel. Biomass Bioenerg. **35**(9), 3951–3961 (2011)
8. Martinez-Gomez, J., Nápoles-Rivera, F., Ponce-Ortega, J., Serna-González, M., El-Halwagi, M.: Optimization of facility location and reallocation in an industrial plant through a multi-annual framework accounting for economic and safety issues. J. Loss Prev. Process. Ind. **33**, 129–139 (2015)
9. Elhedhli, S., Merrick, R.: Green supply chain network design to reduce carbon emissions. Transp. Res. Part D Transp. Environ. **17**(5), 370–379 (2012)
10. Lim, M., Bassamboo, A., Chopra, S., Daskin, M.: Facility location decisions with random disruptions and imperfect estimation. Manuf. Serv. Oper. Manag. **15**(2), 239–249 (2013)
11. Chima, C.: Supply chain management issues in the oil and gas industry. J. Bus. Econ. Res. **5**(6), 27–36 (2007)
12. Carro Paz, R., González Gómez, D.: Localización de instalaciones. Administración de Las Operaciones **13**, 1–23 (2010)

Metaheuristics for Order Picking Optimisation: A Comparison Among Three Swarm-Intelligence Algorithms

Jared Olmos, Rogelio Florencia, Vicente García, Martha Victoria González, Gilberto Rivera, and Patricia Sánchez-Solís

Abstract Nowadays, the Order Picking Problem (OPP) represents the most costly and time-consuming operation of warehouse management, with an average ranging from 50 to 75% of the total warehouse management cost. So, OPP is being analysed to improve logistics operations in companies. The OPP consists of dispatching a set of products, allocated in specific places in a warehouse, based in a group of customer orders. In most traditional warehouses, the optimisation methods of order picking operations are associated with time, whose model is based on the Traveling Salesperson Problem (TSP). The TSP is considered as an NP-Hard problem; thus, the development of metaheuristics approaches is justified. This chapter presents a comparison among three different optimisation metaheuristic approaches that solve the OPP. An analysis is used to evaluate and compare ant colony optimisation, elephant herding optimisation, and the bat algorithm. This study considers the number of picking aisles, the number of extra cross aisles, the number of items in the order, and the standard deviation in both the x and y axis of the product distribution in the warehouse.

Keywords Order picking problem · Ant colony optimisation · Bat algorithm · Elephant herding optimisation · Traveling salesperson problem · Swarm intelligence

1 Introduction

Nowadays, production and distribution companies are continually searching for innovative ways to generate the highest possible profit. So, the right management of a wide variety of processes throughout the product's lifecycle is essential. This study is focused on the Order Picking Problem (OPP), a phase of the product's lifecycle process, and its resolution using different optimisation algorithms. Order Picking is defined as the recollection of products stored in a warehouse, satisfying a set of

J. Olmos · R. Florencia · V. García · M. V. González · G. Rivera (✉) · P. Sánchez-Solís
Universidad Autónoma de Ciudad Juárez, 32310 Juárez, Chihuahua, Mexico
e-mail: gilberto.rivera@uacj.mx

A. Ochoa-Zezzatti et al. (eds.), *Technological and Industrial Applications Associated With Industry 4.0*, Studies in Systems, Decision and Control 347,
https://doi.org/10.1007/978-3-030-68663-5_13

customer orders, and it is considered as the most costly and time-consuming operation of warehouse management [1, 2].

Consequently, the OPP is a high priority area in warehouse logistics to improve. Optimisation techniques play an essential role when company managers search for minimising cost or maximising profit. Some of the most used optimisation techniques are the ones based on swarm intelligence (SI-based metaheuristics) [3]. These algorithms exploit the collective intelligence and behaviour of self-organised systems such as foraging of social insects or other animals.

The objective of this study is to run, evaluate and compare three SI-based algorithms: Ant Colony Optimisation (ACO), Elephant Herding Optimisation (EHO) and the Bat Algorithm (BA), facing an set of instances.

The remainder of this chapter is organised as follows: In Sect. 2, we present a literature review. Section 3 describes the instances. Section 4 presents the computational experiments and results. Finally, In Sect. 5, we discuss some conclusions and directions for future work.

2 Literature Review

This chapter compares three optimisation algorithms to solve the OPP. Hereafter, in this section, it is presented a literature review of order picking and the three SI-based metaheuristics evaluated.

2.1 Order Picking

In the OPP, there is a set of orders, consisting of a subset of items stored in the warehouse; each order must be supplied [4]. The "picker" is the element that collects the articles in the order, starting from the depot and finishing in the same spot. Beroule in [5] stated that optimisation methods of order picking operations are mostly associated with time and are based on the Traveling Salesperson Problem (TSP). The literature specialised in OPP assumes that the travelled distance is the element that mainly affects the performance; thus, optimising lengths is the main objective [5].

The TSP is defined by a salesman who must find an optimal route that passes through a set of costumers; the vendor must visit each of them only once, and (s)he starts and finishes at the same place. The aim is to find an optimal route in terms of cost, usually focusing on the distance, satisfying the restrictions described above. In our context of order picking, each location of a product is like a customer to visit. The TSP is classified as an NP-hard problem; so, metaheuristics are usually used to face it in a reasonable time [6].

The TSP is formalised as, given a positive integer n—number of cities—and a bidimensional distance matrix c, there exists a tour defined as a sequence of integers where each subsequent integer, taken from the set $[1, 2, 3, \ldots, n]$ appears at least once (the initial and final integers are identical). The tour can be represented as follows:

$$t = (i_1, i_2, i_3, \ldots i_{n-1}, i_n, i_1), \quad (1)$$

where t represents the tour and i_j represents the jth costumer; t has a sequence of traversed edges x; so, the optimal solution is sought by minimising the objective function described in Eq. 2.

$$z(x) = \sum_{(i,j)\in x} c_{i,j} \quad (2)$$

Here, x is composed of the ordered pairs of t as follows:

$$x = \langle (i_1, i_2), (i_2, i_3), \ldots, (i_{n-1}, i_n), (i_n, i_1) \rangle. \quad (3)$$

2.2 *Ant Colony Optimisation*

ACO is a well-known approach for solving combinatorial problems such as the OPP. Dorigo et al. [7] proposed this algorithm, and several variants have emerged from it [8–10]. This algorithm is inspired by the behaviour of ant colonies when searching for food supplies. Ants have the ability to produce and release a pheromone to communicate with each other. Ants start exploring the surface at random, emitting traces of pheromones while finding routes to the food supply [11]. Once the pheromone trail is on the surface, subsequent ants follow this trace. However, the pheromone trail tends to disappear over time, which causes shorter routes to be selected by the ants. The chosen pathways receive reinforcement of pheromones from each ant that walks on.

For this reason, more and more ants choose these shorter routes. The probability of selecting a path increases proportionally with the number of ants that walks by that route [12]. Figure 1 represents this behaviour.

In summary, the ACO algorithm contains two rules or phases [14]. The first one is the update of local pheromone while constructing the solutions. The second one occurs when the pheromone is globally updated once all ants have built the solution [15].

The following equation describes the probability of choosing a path by an ant.

$$P_{ij}^k = \frac{(\tau_{ij}^\alpha)\left(n_{ij}^\beta\right)}{\sum_{l\in(N_i^k)}\left(\tau_{ij}^\alpha\right)\left(n_{ij}^\beta\right)}, \text{ if } j \in N_i^k \quad (4)$$

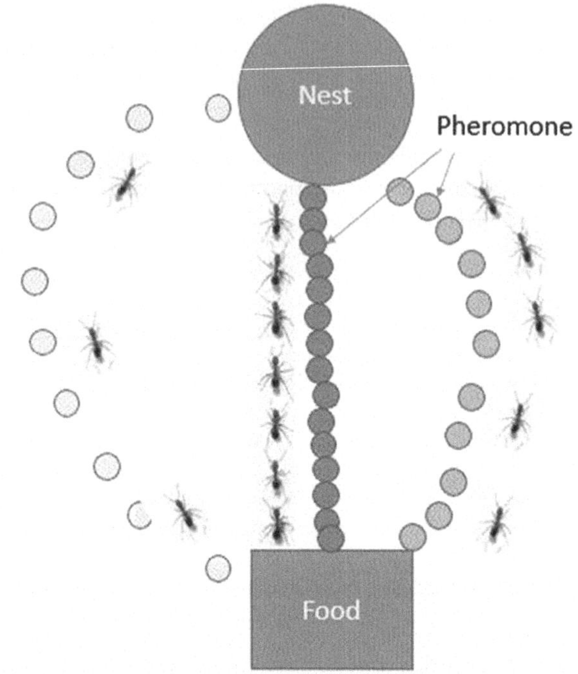

Fig. 1 Representation of the release of pheromone. *Source* [13]

Equation (4) calculates the probability of success of an ant k reaching a position j from a position i. In the first iteration, ants have the same likelihood of getting any node j because of the lack of pheromone trail on the routes. The term τ_{ij} represents the value of the amount of pheromone that exists between i and j, reflecting the attraction of moving from node i to node j [7]. The term n_{ij} is a heuristic value between nodes i and j, and it is inversely proportional to the distance between the two nodes. It is calculated as $n_{ij} = \frac{1}{d_{ij}}$. The relative importance of the pheromone trail is represented by the parameter α; and the parameter β is the relative importance of the heuristic information. Here, N_i^k is the set of nodes that the kth ant has not visited (while standing in position i); so, the end of the iterative process of visiting the nodes occurs when $N_i^k = \emptyset$. The pheromone update on the routes is affected at the end of the process [14].

The evaporation rate is represented by Eq. 3 and is bounded, so its value must be between 0 and 1. The term ρ is a parameter that avoids the unlimited accumulation of pheromone. This process causes the reduction of pheromones, where the ant traffic is lower.

$$\tau_{ij} = (1 - \rho)\tau_{ij}, \forall(i, j) \in L, \quad 0 < \rho < 1 \tag{5}$$

Equation 6 and Eq. 7 are used to update the pheromone trace of the routes given ants' traffic.

$$\tau_{ij} = \tau_{ij} + \sum_{k=1}^{m} \Delta\tau_{ij}^{k}, \forall(i, j) \in L \tag{6}$$

$$\Delta\tau_{ij}^{k} = \begin{cases} 1/L^{k} & \text{if } (i, j) \in T^{k}, \\ 0 & \text{otherwise.} \end{cases} \tag{7}$$

The pheromone value of the selected routes is increased, favouring the probability of choosing that route. The ants will prefer ways with intense pheromone, which positively reinforces the pheromone trails on the best routes [16]. The term L^k is used to represent the sum of the length of the edges that belongs to a route T^k. According to Eq. 4, the increase of pheromone is inversely proportional to the length of the path, using the $\lim_{L^k \to +\infty} \frac{1}{L^k} = 0$ to represent it. Finally, when this process is repeated—ants travel the graph, the evaporation is performed, and the pheromone is updated—it results in near-optimal solutions [10, 12, 17].

2.3 *Elephant Herding Optimisation*

Elephant Herds have the following behaviour when searching for food: they divide into two groups, one consisting of male elephants that forage for food in large distances, and the other one composed of female elephants that form groups, performing local searches near their matriarch [18–22]. Once the first group of male elephants finds a food source, the matriarch and the whole group moves toward it.

Wang in [18] lists four main assumptions used to model the elephant's biological behaviour; these assumptions are described below.

1. Each elephant has a particular visual range, which helps them to look out for other animals that might threaten their personal space. Male elephants have a wider visual range due to the instinct of protection they should have, allowing them to move randomly while searching for food. This range, in the algorithm, is calculated by the Euclidean distance.
2. When two elephants see each other, a contest begins to demonstrate who is the strongest one. This action is represented in the algorithm by comparing the fitness value of both elephants. The one with the highest value is considered the strongest one, and the loser must leave the area. Figure 2 illustrates that behaviour.
3. A clan is made up of only one female group, and they are always together.
4. There exists a maximum lifespan in each elephant. When an elephant dies, a new baby must be born. An impressive characteristic is that the gender of this new baby is inherited from the elephant that just died; this keeps balanced the clan's gender.
5. EHO splits elephants into k clans, considering that each jth member in the ith clan follows the movement of the clan's matriarch. The matriarch is the elephant

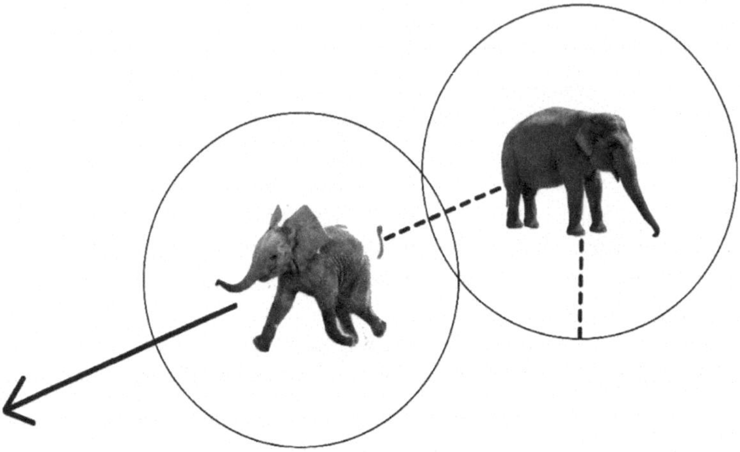

Fig. 2 Elephant with the lowest fitness flees from another elephant with better fitness. *Source* [23]

with the best value in each generation. The EHO algorithm identifies four types of elephants: the baby, the male, the female, and the matriarch.

Matriarch

The oldest living female elephant is considered the leader; this elephant—a.k.a. the matriarch—has the best fitness value, which enables her to dictate the direction that the whole clan must follow. Equation 8 describes the matriarch elephant movement [24, 25].

$$x_{i,j}^{new} = (\beta)x_i^{centre} \tag{8}$$

In Eq. 8, i is the index of the clan, x_i^{centre} represents the matriarch movement. The parameter β is a binary value that controls the influence of $x_i^{centre} = \{x_{i,1}^{centre}, x_{i,2}^{centre}, \ldots, x_{i,D}^{centre}\}$, defined by [24] as:

$$x_{i,d}^{centre} = \frac{1}{n_i}\sum_{j=1}^{n_i} x_{i,j,d}, \tag{9}$$

where $1 \leq d \leq D$ indicates the dth dimension, n_i represents the number of elephants in the clan i, $x_{i,d}^{centre}$ is the centre, calculated by the average of the whole current solutions in the clan, and $x_{i,j,d}$ represents the dth dimension of the point of elephant j belonging to the ith clan.

Female Elephants

The female elephants always follow the matriarch staying close to the clan. They perform a local search under the direction of the matriarch of the i clan. Their moving behaviour is represented by [24]:

$$x_{i,j}^{new} = x_{i,j} + \alpha\left(x_i^{best} - x_{i,j}\right)r, \tag{10}$$

where $x_{i,j}^{new}$ represents the new position for elephant j of the ith clan, $x_{i,j}$ represents the last position, x_i^{best} represents the best solution in the ith clan, the parameter $\alpha \in [0, 1]$ indicates the matriarch influence over the clan, and $r \in [0, 1]$ is a random number, used to diversification.

Male Elephants

Male elephants are responsible for searching for food in a more profound sense (randomly). They expand the search scope by exploring the area. The algorithm represents the elephants with the worst fitness movement, m_i, as follows [24]:

$$x_i^{worst} = x_{\min} \cdot (x_{\max} - x_{\min}) \cdot random() \tag{11}$$

The terms $x_{\max}$ and $x_{\min}$ describe the upper and lower bound, respectively, and a random-number function is used to represent randomness.

Baby Elephants

These baby elephants are born with the same gender of the most recently dead elephant, inheriting its fitness value. Usually, they stay with female elephants until reaching adulthood; then, they must leave the clan if they are male.

2.4 Bat Algorithm

Yang in [26] proposed the BA to solve combinatorial optimisation problems. This algorithm is based on the echolocation behaviour of bats while searching for prey. The author establishes three premises described below:

1. Bats use echolocation to trace their preys and hunt them.
2. There exists a fly randomly made by bats. This fly has a velocity v_i at a certain position x_i and a frequency $f_{\min}$ with a variation of the wavelength λ and loudness A_0 to search for their prey.
3. The parameter of the loudness varies reaching an $A_{\min}$.

With these premises, the author proposes the following equations to model the behaviour of bats

$$v_i^t = v_i^{t-1} + \left(x_i^t - x^*\right) f_i \tag{12}$$

$$f_i = f_{\min} + (f_{\max} - f_{\min})\beta \tag{13}$$

$$x_i^t = x_i^{t-1} + v_i^t \tag{14}$$

$$x_{new} = x_{old} + \in A^t \quad (15)$$

Equation 12 helps to describe the random velocity at a specific time.

The term x_i represents a position at a fixed frequency $f_{\min}$. The wavelength varies and the volume A_0 is taken to find the prey. The loudness frequency can vary randomly in many ways. The position x_i, velocity v_i and the frequency f_i are initialised, and Eq. 10 is used to update and find the best solutions. The term β represents a random vector, and x^* represents the current best solution. This process is iterated, updating the loudness and the frequency of the impulse emission as follows [27]

$$A_i^{t+1} = \propto A_i^t \quad r_i^{t+1} = r_i^0[1 - \exp(-\gamma t)], \quad (16)$$

$$A_i^t = 0, r_i^t \to r_i^0, \text{ as } t \to \infty, \quad (17)$$

where $\propto$ and γ are constant.

3 Data Description

The set of instances used in this work has been adapted from [28]. There are four different files per test instance, which represent different features. The first one is the "list" file, which is a list of the products and their description; it is illustrated in Fig. 3. The second one is the "order" file, which is a set of customer orders specifying the number of products to be ordered, the identifier of the product, and the quantity needed; it is illustrated in Fig. 4. The third file, named "productloc", describes the location of the products in the warehouse, each product is allocated in a single location; it is illustrated in Fig. 5. The final file, named "warehouse", describes the warehouse configuration, and it's the most important one because it includes eight parameters to be set and five data variables that will guide our tests; it is illustrated in Fig. 6. The parameters and their ranges are the following:

1. The number of aisles: Represents the total of corridors in the warehouse.
2. The number of extra cross aisles: Represents the total of additional cross aisles in the warehouse (not considering both the front and the rear corridors).
3. The number of shelves: Represents the total shelves per rack position, the value doesn't vary.
4. Minimum of products required: Represents the number of products needed in the list archive to fill the warehouse.
5. Aisle width: This parameter doesn't vary and is set in three units.
6. Rack depth: This parameter doesn't vary and is set in one unit.
7. Location width: This parameter doesn't vary and is set in one unit.
8. Cross aisle width: This set doesn't vary and is set in one unit.

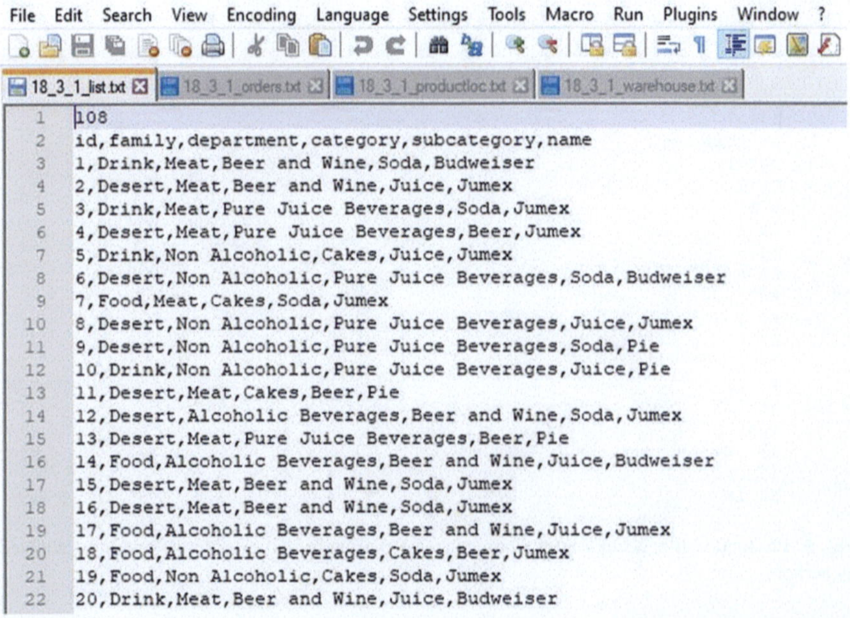

```
108
id,family,department,category,subcategory,name
1,Drink,Meat,Beer and Wine,Soda,Budweiser
2,Desert,Meat,Beer and Wine,Juice,Jumex
3,Drink,Meat,Pure Juice Beverages,Soda,Jumex
4,Desert,Meat,Pure Juice Beverages,Beer,Jumex
5,Drink,Non Alcoholic,Cakes,Juice,Jumex
6,Desert,Non Alcoholic,Pure Juice Beverages,Soda,Budweiser
7,Food,Meat,Cakes,Soda,Jumex
8,Desert,Non Alcoholic,Pure Juice Beverages,Juice,Jumex
9,Desert,Non Alcoholic,Pure Juice Beverages,Soda,Pie
10,Drink,Non Alcoholic,Pure Juice Beverages,Juice,Pie
11,Desert,Meat,Cakes,Beer,Pie
12,Desert,Alcoholic Beverages,Beer and Wine,Soda,Jumex
13,Desert,Meat,Pure Juice Beverages,Beer,Pie
14,Food,Alcoholic Beverages,Beer and Wine,Juice,Budweiser
15,Desert,Meat,Beer and Wine,Soda,Jumex
16,Desert,Meat,Beer and Wine,Soda,Jumex
17,Food,Alcoholic Beverages,Beer and Wine,Juice,Jumex
18,Food,Alcoholic Beverages,Cakes,Beer,Jumex
19,Food,Non Alcoholic,Cakes,Soda,Jumex
20,Drink,Meat,Beer and Wine,Juice,Budweiser
```

Fig. 3 List file. *Source* https://homepages.dcc.ufmg.br/~arbex/orderpicking.html adapted by authors

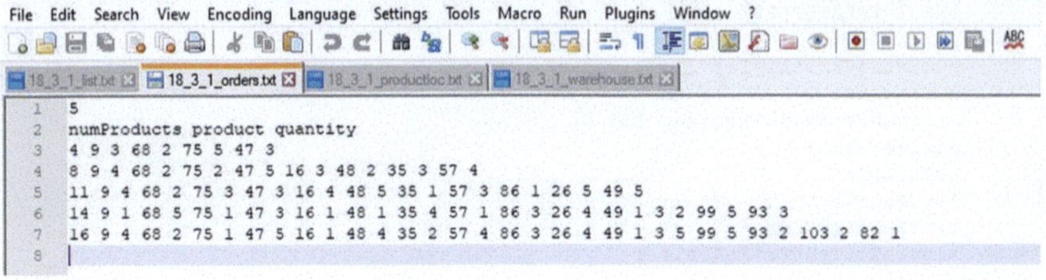

```
5
numProducts product quantity
4 9 3 68 2 75 5 47 3
8 9 4 68 2 75 2 47 5 16 3 48 2 35 3 57 4
11 9 4 68 2 75 3 47 3 16 4 48 5 35 1 57 3 86 1 26 5 49 5
14 9 1 68 5 75 1 47 3 16 1 48 1 35 4 57 1 86 3 26 4 49 1 3 2 99 5 93 3
16 9 4 68 2 75 1 47 5 16 1 48 4 35 2 57 4 86 3 26 4 49 1 3 5 99 5 93 2 103 2 82 1
```

Fig. 4 Orders file. *Source* https://homepages.dcc.ufmg.br/~arbex/orderpicking.html adapted by authors

The data variables are the following:

1. The number of locations per aisle: Represents the total location per corridor in the warehouse.
2. Total number of locations: It is calculated by multiplying the number of locations per aisle side times the number of shelves times the double of the number of aisles.
3. The number of vertices: Represents the sum of product vertices—where the picker can get a product—and artificial vertices—the ones used to help build the graph.

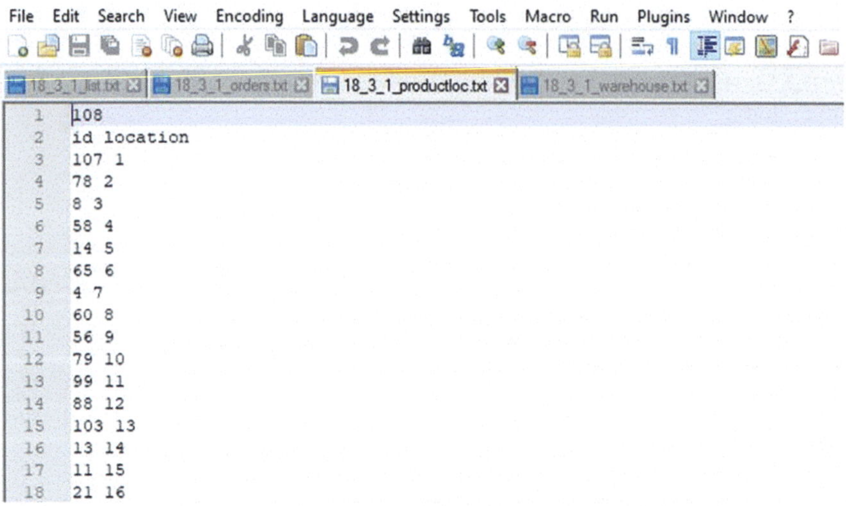


```
108
id location
107 1
78 2
8 3
58 4
14 5
65 6
4 7
60 8
56 9
79 10
99 11
88 12
103 13
13 14
11 15
21 16
```

Fig. 5 Productloc file. *Source* https://homepages.dcc.ufmg.br/~arbex/orderpicking.html adapted by authors

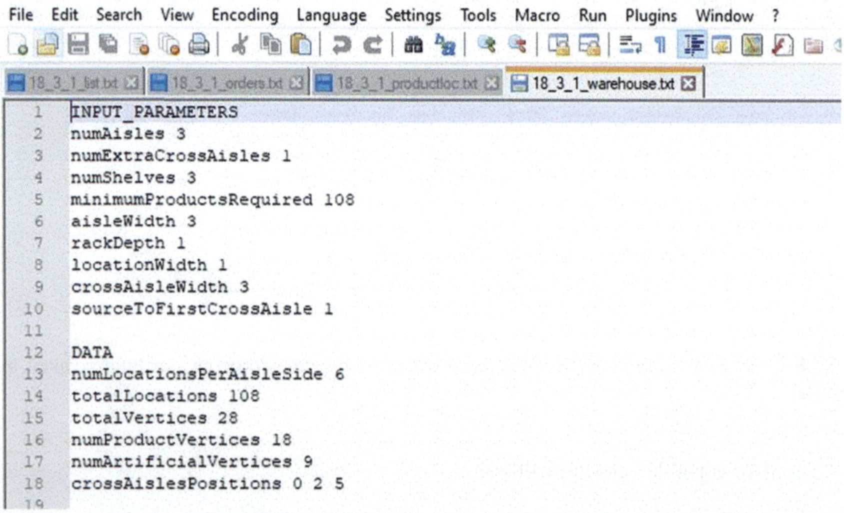


```
INPUT_PARAMETERS
numAisles 3
numExtraCrossAisles 1
numShelves 3
minimumProductsRequired 108
aisleWidth 3
rackDepth 1
locationWidth 1
crossAisleWidth 3
sourceToFirstCrossAisle 1

DATA
numLocationsPerAisleSide 6
totalLocations 108
totalVertices 28
numProductVertices 18
numArtificialVertices 9
crossAislesPositions 0 2 5
```

Fig. 6 Warehouse file. *Source* https://homepages.dcc.ufmg.br/~arbex/orderpicking.html adapted by authors

4. The number of product vertices: Represents the vertices where the picker can pick a product.
5. The number of artificial vertices: Represents the new vertices added.

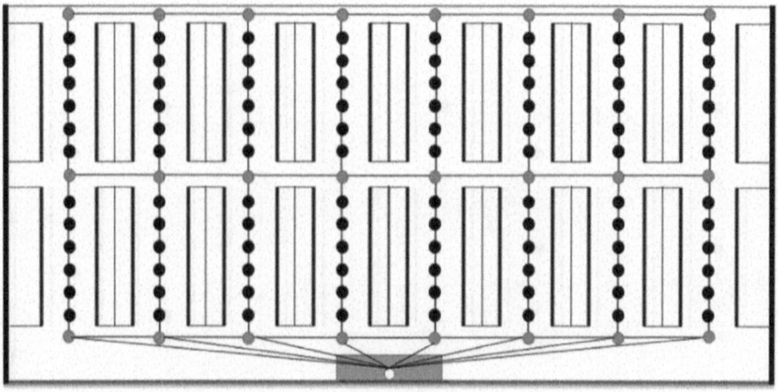

Fig. 7 Graph-oriented model configuration. *Source* https://homepages.dcc.ufmg.br/~arbex/orderpicking.html adapted by authors

6. Cross aisles positions: Represents the vertex number or position of the cross aisles.

The configuration described above is used and transformed into a graph-oriented model representing the pick points based on the locations of the warehouse. This transformation from an OPP to a TSP simplifies the problem-solution process. A Java program is used to transform the configuration. The graph-oriented model considers the possibility of picking up a product of either side of the aisle at every vertex. Figure 7 illustrates an example of this model.

In Fig. 7, it is described as a warehouse with eight aisles, one cross-aisle, and a centred depot. The black nodes represent the product picking vertices, the grey nodes represent the artificial vertices, and the white node represents the depot. With this model, we can quickly compute the distance travelled by a picker given a solution route. Once we have this graph model, a reduction process begins. This reduced graph is built based on the products of each order, eliminating the vertices that have no item in that specific order, and leaving the ones that have a product associated with them. An example of a reduced graph of Fig. 7, with eleven items in the order, is illustrated in Fig. 8. This reduced graph only indicates the product vertices needed to pick the items in the order.

This reduced graph simplifies the solution process of the problem; however, an extra reduction is needed to remove the artificial nodes that we do not need to use to solve the TSP. The elimination process uses a computation of the distance between the two types of nodes, the product, and the artificial ones. The length is evaluated, and then the shortest distance among every possible path from a node i to a node j is used. This resultant graph is used then to represent the instance as a TSP using the depot node as the starting and closing node of the solution. An adjacency matrix, as described in Fig. 9, is used to represent this graph in the different programs that run the algorithms.

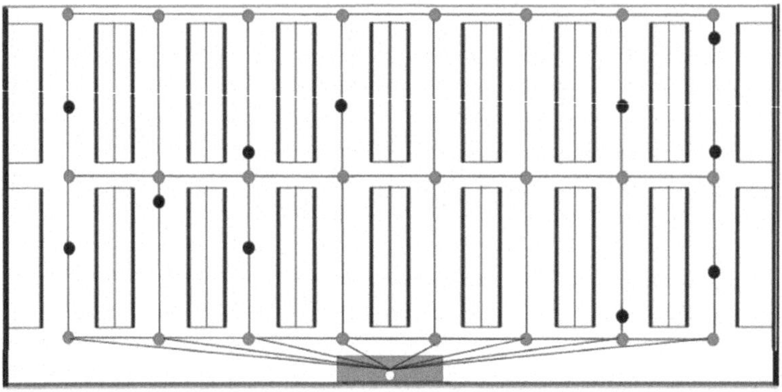

Fig. 8 Example of a simplified graph. *Source* https://homepages.dcc.ufmg.br/~arbex/orderpicking.html adapted by authors

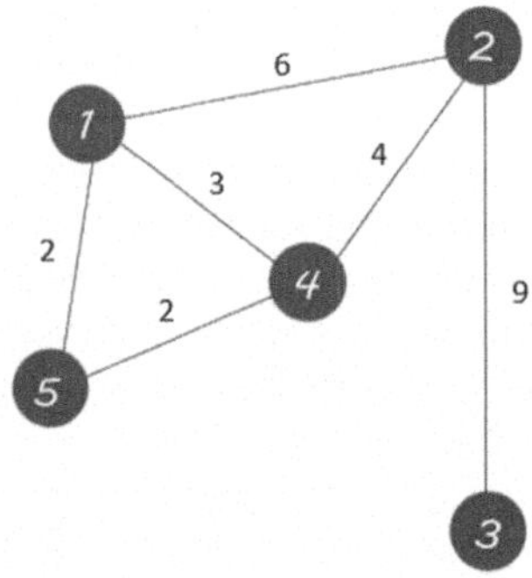

Adjacency Matrix

	1	2	3	4	5
1	0	1	0	1	1
2	1	0	1	1	0
3	0	1	0	0	0
4	1	1	0	0	1
5	1	0	0	1	0

Distance Matrix

	1	2	3	4	5
1	0	6	0	3	2
2	6	0	9	4	0
3	0	9	0	0	0
4	3	4	0	0	2
5	2	0	0	2	0

Fig. 9 Example of the adjacency matrix and distance matrix. *Source* https://math.stackexchange.com/questions/1890620/finding-path-lengths-by-the-power-of-adjacency-matrix-of-an-undirected-graph adapted by authors

4 Computational Tests

A computational test evaluates the performance and the runtime of the algorithms to determine which one is the best given specific scenarios. Five parameters vary to generate the test instances. These features are:

1. The number of aisles, from 3 to 15.
2. The number of extra cross aisles, from 1 to 9.
3. The number of products in the order, from 16 to 405.
4. The standard deviation in the horizontal axis of the product's distribution, from 1.12 to 25.19.
5. The standard deviation in the vertical axis of the product's distribution, from 1.13 to 8.82.

The parameter's configuration of each algorithm remained static, using the best setting for each one reported in [6]. For the ACO algorithm, it was established a parameter setting with five ants, 500 iterations, and the evaporation rate of 0.5, $\alpha = 1$, and $\beta = 5$. For the EHO algorithm, it was established 100 elephants, 800 generations, and ten clans. For BA, it was designated a loudness of 0.8, a rate of 0.3, 50 bats, and 100 generations. Table 1 describes the results of the algorithms giving the best result after 30 runs.

The ACO and EHO algorithms were implemented in NetBeans IDE 8.2 running JAVA, while the BA was implemented in Matlab R2013b. It was used a platform with the following specifications: Intel Core i3 7th generation at 2.4 GHz, with 8 Gb of RAM and Windows 10 Home version.

Table 1 summarises the results obtained by each algorithm; Column 1 contains the instance ID, Columns 2–6 present the values of the five features for each instance, and Columns 7–12 includes two data on the performance for each algorithm: distance traversed by the picker in the proposed route and run time. ACO had the best results in every instance run, excluding the first one with 16 products, on which EHO performed better. The BA was only executed on three instances (1, 25, 31) because of the delay in run time (the program was stopped after running three days).

5 Conclusions and Further Research

In this work, it is presented a comparison between three SI metaheuristic for solving an OPP. The OPP was transformed into a TSP to simplify its resolution. An analysis of the performance and the runtime of these three algorithms were made using the best parameter configuration of each one of them. Five features were varied in the instances: the number of aisles, the number of extra cross aisles, the number of items in the order, and the standard deviation in both the x and y axis of the product distribution in the warehouse to evaluate and analyse the algorithms. Based on the results, we conclude that the ACO algorithm is the best in performance and execution time, given the proposed parameter variations. The contribution of this work is to suggest an SI-based metaheuristic to OPP, and this choice is backed by simulation and based on the results of three SI metaheuristics: ACO, BA, and EHO.

As future research, more metaheuristics used in the literature could be compared to the ACO algorithm to identify which implementation is better for this kind of warehouse layout and orders.

Table 1 Results of the different optimisation algorithms

						ACO		EHO		BA	
	Aisles	Cross aisles	Items in the order	Standard deviation on x	Standard deviation on y	Distance	Time of execution	Distance	Time of execution	Distance	Time of execution
1	3	1	16	1.71269768	1.58640054	11	2 s	4	5 s	5	16 m 30 s
2	3	3	32	3.13779638	1.60361235	28	3 s	47	8 s	n. e.	n. e.
3	3	5	49	5.38453315	1.69106294	67	2 s	106	22 s	n. e.	n. e.
4	3	7	65	6.86802093	1.61319678	65	5 s	122	26 s	n. e.	n. e.
5	3	9	81	8.94432566	1.62137169	104	9 s	121	34 s	n. e.	n. e.
6	6	1	32	1.76776695	3.73073006	53	1 s	66	7 s	n. e.	n. e.
7	6	3	65	3.68514429	3.24948221	67	4 s	139	21 s	n. e.	n. e.
8	6	5	97	5.04702968	3.23821548	125	13 s	176	39 s	n. e.	n. e.
9	6	7	130	7.30662453	3.47736526	140	21 s	233	1 m 7 s	n. e.	n. e.
10	6	9	162	8.67408019	3.36805018	165	45 s	421	1 m 52 s	n. e.	n. e.
11	9	1	49	1.74549956	5.14773248	55	2 s	84	13 s	n. e.	n. e.
12	9	3	97	3.53030571	5.32964058	96	11 s	214	47 s	n. e.	n. e.
13	9	5	146	4.74337168	5.09793089	169	29 s	285	1 m 26 s	n. e.	n. e.
14	9	7	194	7.11998551	5.4072342	250	1 m 5 s	550	2 m 24 s	n. e.	n. e.
15	9	9	243	8.85892178	5.13937965	253	2 m 1 s	707	4 m 49 s	n. e.	n. e.
16	12	1	65	1.76327383	6.65376556	103	4 s	140	21 s	n. e.	n. e.
17	12	3	130	3.40668255	6.95535212	135	19 s	290	1 m 9 s	n. e.	n. e.
18	12	5	194	5.14701343	6.86365572	262	1 m 3 s	570	2 m 21 s	n. e.	n. e.
19	12	7	259	7.09508402	7.00288767	321	2 m 24 s	775	6 m 20 s	n. e.	n. e.
20	12	9	324	8.71463059	6.84328023	418	5 m 15 s	1050	8 m 31 s	n. e.	n. e.

(continued)

Table 1 (continued)

						ACO		EHO		BA	
	Aisles	Cross aisles	Items in the order	Standard deviation on x	Standard deviation on y	Distance	Time of execution	Distance	Time of execution	Distance	Time of execution
21	15	1	81	1.72078793	8.12001034	121	9 s	162	30 s	n. e.	n. e.
22	15	3	162	3.29392974	8.69677884	226	43 s	413	1 m 44 s	n. e.	n. e.
23	15	5	243	5.37344979	8.6689457	304	1 m 58 s	701	5 m 17 s	n. e.	n. e.
24	15	7	324	6.77700729	8.81784443	403	4 m 44 s	954	7 m 31 s	n. e.	n. e.
25	15	9	405	8.35836248	8.66205639	532	9 m 42 s	1307	16 m 32 s	no results	8 h 30 m
26	3	1	7	3.35232684	1.13389342	34	1 s	25	2 s	n. e.	n. e.
27	6	3	29	4.96837288	3.51912439	139	1 s	163	7 s	n. e.	n. e.
28	9	5	65	8.38562759	4.95280612	328	9 s	590	24 s	n. e.	n. e.
29	12	3	58	15.4832228	2.85736755	370	5 s	416	19 s	n. e.	n. e.
30	3	5	30	12.9217361	1.59056124	155	2 s	202	7 s	n. e.	n. e.
31	6	1	20	3.99308613	2.55208893	88	1 s	94	5 s	89	33 m 2 s
32	9	7	121	15.5615502	5.06025675	643	28 s	1088	1 m 12 s	n. e.	n. e.
33	12	1	40	3.26441576	7.42759926	208	3 s	206	10 s	n. e.	n. e.
34	15	5	151	11.9861252	8.56440604	718	41 s	1310	1 m 39 s	n. e.	n. e.
35	3	3	26	10.1166276	1.90424627	152	1 s	153	6 s	n. e.	n. e.
36	6	9	130	27.0007884	3.63229403	635	32 s	1110	1 m 11 s	n. e.	n. e.
37	9	7	156	21.4880533	5.35944934	646	48 s	1236	1 m 40 s	n. e.	n. e.
38	12	3	104	10.6889931	7.13164382	492	19 s	806	48 s	n. e.	n. e.
39	15	7	259	20.8340185	8.33078229	1043	3 m 30 s	2230	7 m	n. e.	n. e.

(continued)

Table 1 (continued)

						ACO		EHO		BA	
	Aisles	Cross aisles	Items in the order	Standard deviation on x	Standard deviation on y	Distance	Time of execution	Distance	Time of execution	Distance	Time of execution
40	3	9	72	25.1932748	1.68371562	271	8 s	406	26 s	n. e.	n. e.
41	6	7	115	23.6088549	3.35121823	527	24 s	932	58 s	n. e.	n. e.
42	9	5	130	18.5746491	5.34552126	563	33 s	1034	1 m 17 s	n. e.	n. e.
43	12	3	115	11.7680805	7.04396134	478	24 s	854	58 s	n. e.	n. e.
44	15	1	72	5.26970184	8.51095906	320	9 s	398	26 s	n. e.	n. e.

References

1. Hadi, M.Z., Djatna, T.: Implementation of an ant colony approach to solve multi-objective order picking problem in beverage warehousing with drive-in rack system. In: International Conference on Advanced Computer Science and Information Systems, pp. 137–142. Balic (2017). https://doi.org/10.1109/icacsis.2017.8355024m
2. de Koster, R., Le-Duc, T., Jan-Roodbergen, K.: Design and control of warehouse order picking: a literature review. Eur. J. Oper. Res. **182**(2), 481–501 (2007). https://doi.org/10.1016/j.ejor.2006.07.009
3. Ab-Wahab, M.N., Nefti-Meziani, S., Atyabi, A.: A comprehensive review of swarm optimization algorithms. PLoS ONE **10**(5), 1–36 (2015). https://doi.org/10.1371/journal.pone.0122827
4. Rosenthal, H., Ratliff, D., Arnon, S.: Order-picking in a rectangular warehouse: a solvable case of the traveling salesman problem. Oper. Res. **31**(3), 507–521 (1983). https://doi.org/10.1287/opre.31.3.507
5. Beroule, B., Grunder, O., Barakat, O., Aujoulat, O.: Order picking problem in a warehouse hospital pharmacy. Sci. Direct IFAC Papers OnLine **50**(1), 5017–5022 (2017). https://doi.org/10.1016/j.ifacol.2017.08.910
6. Ortiz-Zezzatti, A.O., Rivera, G., Gómez-Santillán, C., Sánchez, B.: Handbook of Research on Metaheuristics for Order Picking Optimization in Warehouses to Smart Cities. IGI Global, Hershey (2019). https://doi.org/10.4018/978-1-5225-8131-4
7. Dorigo, M., Maniezzo, V., Colorni, A.: Ant system: optimization by a colony of cooperating agents. IEEE Trans. Syst. Man Cybern. Part B Cybern. **26**(1), 29–41 (1996). https://doi.org/10.1109/3477.484436
8. Bastiani, S., Cruz-Reyes, L., Fernandez, E., Gómez, C., Rivera, G.: An ant colony algorithm for solving the selection portfolio problem, using a quality-assessment model for portfolios of projects expressed by a priority ranking. In: Design of Intelligent Systems Based on Fuzzy Logic, Neural Networks and Nature-Inspired Optimization, vol. 601, pp. 357–373. Springer, Cham (2015). https://doi.org/10.1007/978-3-319-17747-2_28
9. Gómez, C., Cruz, L., Schaeffer, E., Meza, E., Rivera, G.: Adaptive ant-colony algorithm for semantic query routing. J. Autom. Mob. Robot. Intell. Syst. **5**(1), 85–94 (2011)
10. Rivera, G., Gómez, C.G., Fernández, E.R., Cruz, L., Castillo, O., Bastiani, S.S.: Handling of synergy into an algorithm for project portfolio selection. In: Castillo, O., Melin, P., Kacprzyk, J. (eds). Recent Advances on Hybrid Intelligent Systems, vol. 451, pp. 417–430. Springer, Heidelberg (2013). https://doi.org/10.1007/978-3-642-33021-6_33
11. Gómez, C., Cruz, L., Schaeffer, E., Meza, E., Rivera, G.: Local Survival Rule for Steer an Adaptive Ant-Colony Algorithm in Complex Systems. In: Melin, P., Kacprzyk, J., Pedrycz, W. (eds) Soft Computing for Recognition Based on Biometrics, vol. 312, pp. 245–265. Springer, Heidelberg (2010). https://doi.org/10.1007/978-3-642-15111-8_15
12. Fernandez, E., Gomez, C., Rivera, G., Cruz-Reyes, L.: Hybrid metaheuristic approach for handling many objectives and decisions on partial support in project portfolio opti-misation. Inf. Sci. **315**, 102–122 (2015). https://doi.org/10.1016/j.ins.2015.03.064
13. Olmos, J., Florencia, R., López-Ramos, F., Olmos-Sánchez, K.: Improvement of the optimization of an order picking model associated with the components of a classic volkswagen beetle using an ant colony approach. In: Handbook of Research on Metaheuristics for Order Picking Optimization in Warehouses to Smart Cities, pp. 189–210. IGI Global, Hershey (2019). https://doi.org/10.4018/978-1-5225-8131-4.ch010
14. Gomez, C.G., Cruz-Reyes, L., Rivera, G., Rangel-Valdez, N., Morales-Rodriguez, M.L., Perez-Villafuerte, M.: Interdependent Projects selection with preference incorporation. In: García-Alcaraz, J., Alor-Hernández, G., Maldonado-Macías, A., Sánchez-Ramírez, C. (eds) In New Perspectives on Applied Industrial tools and techniques, pp. 253–271. Springer, Cham (2018). https://doi.org/10.1007/978-3-319-56871-3_13

15. De Santis, R., Montanari, R., Vignali, G., Bottani, E.: An adapted ant colony optimization algorithm for the minimization of the travel distance of pickers in manual ware-houses. Eur. J. Oper. Res. **267**(1), 120–137 (2018). https://doi.org/10.1016/j.ejor.2017.11.017
16. Cruz, L., Fernandez, E., Gomez, C., Rivera, G., Perez, F.: Many-objective portfolio optimization of interdependent projects with 'a priori' incorporation of decision-maker preferences. Appl. Math. Inf. Sci. **8**(4), 1517–1531 (2014). https://doi.org/10.12785/amis/080405
17. Nemhauger, M., Bellmore, G.: The traveling salesman problem: a survey. Oper. Res. **16**(3), 538–558 (1968). https://doi.org/10.1287/opre.16.3.538
18. Wang, G.-G., Deb, S., Coelho, L.D.S.: Elephant herding optimization. In: International Symposium on Computational and Business Intelligence, pp. 1–5. Bali (2015). https://doi.org/10.1109/iscbi.2015.8
19. Alihodzic, A., Tuba, E., Capor-Hrosik, R., Dolicanin, E., Tuba, M.: Unmanned aerial vehicle path planning problem by adjusted elephant herding optimization. In: 25th Telecommunication Forum (TELFOR), pp. 1–4. Belgrade (2017). https://doi.org/10.1109/telfor.2017.8249468
20. Bukhsh, R., Javaid, N., Iqbal, Z., Ahmed, U., Ahmad, Z., Nadeem-Iqbal, M.: Appliances scheduling using hybrid scheme of genetic algorithm and elephant herd optimization for residential demand response. In: International Conference on Advanced Information Networking and Applications Workshops, pp. 210–217. Cracow (2018). https://doi.org/10.1109/waina.2018.00089
21. Tuba, E., Alihodzic, A., Tuba, M.: Multilevel image thresholding using elephant herding optimization algorithm. In: International Conference on Engineering of Modern Electric Systems, pp. 240–243. Oradea (2017). https://doi.org/10.1109/emes.2017.7980424
22. Tuba, E., Capor-Hrosik, R., Alihodzic, A., Javanovic, R., Tuba, M.: Chaotic elephant herding optimization algorithm. In: World Symposium on Applied Machine Intelligence and Informatics, pp. 000213–000216. Košice (2018). https://doi.org/10.1109/sami.2018.8324842
23. Jimenez, R., Florencia, R., García, V., Lopez, A.: Use of elephant search algorithm to solve an order picking problem in a mobile atelier. In: Handbook of Research on Metaheuristics for Order Picking Optimization in Warehouses to Smart Cities, pp. 161–172. IGI Global, Hershey (2019). https://doi.org/10.4018/978-1-5225-8131-4.ch008
24. Wang, G.-G., Deb, S., Coelho, L., Gao, X.-Z.: A new metaheuristic optimisation algorithm motivated by elephant herding behaviour. Int. J. Bio-Inspired Comput. **8**(6), 394–408 (2016). https://doi.org/10.1504/IJBIC.2016.10002274
25. Bentouati, B., Chettih, S., El-Sehiemy, R., Wang, G.-G.: Elephant herding optimization for solving non-convex optimal power flow problem. J. Electr. Electron. Eng. **10**(1), 31–36 (2017)
26. Yang, X.S.: A new metaheuristic bat-inspired algorithm. In: González, J.R., Pelta, D.A., Cruz, C., Terrazas, G., Krasnogor, N. (eds) In Nature Inspired Cooperative Strategies for Optimization (NICSO 2010). Studies in Computational Intelligence, vol 284. Springer, Heidelberg (2010). https://doi.org/10.1007/978-3-642-12538-6_6
27. Florencia, R., Sanchez-Solis, J., Carvajal, I., Garcia, V.: Design of an Order Picking Reduce Module Using Bat Algorithm. In Handbook of Research on Metaheuristics for Order Picking Optimization in Warehouses to Smart Cities, pp. 211–225. IGI Global, Hershey (2019). https://doi.org/10.4018/978-1-5225-8131-4.ch011
28. Arbex-Valle, C., Beasley, J.E., da Cunha, A.S.: Modelling and solving the joint order batching and picker routing problem in inventories. In: Cerulli, R., Fujishige, S., Mahjoub, A. (eds) Combinatorial Optimization. ISCO 2016. Lecture Notes in Computer Science, vol 9849. Springer, Cham (2016). https://doi.org/10.1007/978-3-319-45587-7_8

Implementation of an Intelligent Framework for the Analysis of Body Movements Through an Avatar Adapted to the Context of Industry 4.0 for the Recruitment of Personnel

Javier Andres Esquivias Varela, Alberto Ochoa-Zezzatti, and Humberto Garcia Castellanos

Abstract In the Industry 4.0 is decisive the implementation of technology of great value that can determine correctly incorrect positions in workers of industrial processes that require of specialized attention constantly, in the present investigation we developed an application for mobile device that identifies possible future accidents when determining incorrect actions of a set of steps to assemble a harness, our proposal is an intelligent system that can monitor inadequate and incorrect movement and specify of a better form preventive actions to make.

Keywords Mobile device application · Intelligent ergonomics · Smart manufacturing

1 Introduction

The analysis of body movements is a key to success for companies in terms of cost savings due to adjustments to workstations that in theory need ergonomic modifications, in some cases the workstations are in full functionality, but the worker who operates them is the one who has some problem in performing their natural movements, this could be due to an accident he has had or if some muscle wear untreated. If it is determined first the analysis that the worker can make, the company will be able to redirect him to an area where it does not cause some conflict to the worker, therefore its labor efficiency in a labor day will not be affected and thus its efficiency

J. A. E. Varela (✉) · H. G. Castellanos
Engineering Sciences, Tecnológico Nacional de México IT Ciudad Juárez, Juarez City, Chihuahua, Mexico
e-mail: aesquivias6@gmail.com

H. G. Castellanos
e-mail: humber.gc@cdjuarez.tecnm.mx

A. Ochoa-Zezzatti
Research Technology, Autonomous University of Ciudad Juarez, Juárez City, Chihuahua, Mexico
e-mail: alberto.ochoa@uacj.mx

A. Ochoa-Zezzatti et al. (eds.), *Technological and Industrial Applications Associated With Industry 4.0*, Studies in Systems, Decision and Control 347,
https://doi.org/10.1007/978-3-030-68663-5_14

of production will be full and consistent. It is necessary to analyze the body movements of workers in their work stations, this way we can determine if the work station needs an aesthetic improvement or if it is necessary to relocate the worker to another area where it does not cause any problem in their movements.

Today the usefulness of technology within the ergonomic issues has marked a decisive change in data capture issues, as well as the agility of data processing, with these implementations we can use its advantages to help workers in their daily problems, as well as companies to improve their quality standards. It is of utmost importance to know the body movements that workers can make in order to redirect them to work areas where they can fulfill a full work efficiency without the need to be suffering from body pains. In most cases, when recruited, workers do not say that they suffer from fractures or muscular problems for fear of not being hired, but this omission of information leads to two problems in the long run: one is the discomfort that the worker will present during his or her workday, and the other is the inefficiency of the work that will cause the pain, which will significantly lower the worker's productivity.

In order to guarantee a better analysis of the data we will use recent technology in conjunction with computer equipment capable of processing the virtualizations and handling the data in the most efficient way. Using an avatar represented on a screen we will be able to make the worker replicate his movements that were previously predisposed for the analysis, this way we can determine if the worker fulfills the necessary requirements for the workstation. All this can be done with a set of deep cameras (Kinect Azure) that will be in charge of capturing the body movements through a system development software SDK that is provided by Microsoft, so the follow up of the worker's body movements can be analyzed with greater certainty.

These equipments would work in such a way that the data is managed through the database libraries together with the Python software, which can comply with the requests of the movements to be analyzed according to the task to be performed; that is, the system can be modified in order to be able to perform the necessary basic movements that will be evaluated. Modern optical technologies, such as 3D laser scanning by means of depth cameras, make it possible to capture measurements of the human body independently of physical contact [1].

The laser light projected on the human body is reflected and captured by the cameras to be mapped as a large set of points (point cloud) in a 3D coordinate system. These coordinates can be used to identify the body's reference points to extract linear, surface and volume measurements with greater accuracy and consistency than is possible with direct manual measurements [2, 3].

A set of software development tools (software development kit or SDK), allow the application developer to create a software package for a specific specialized system to visualize the 3D digital model and manually extract the anthropometric measurements, in conjunction with the analysis of the data captured and stored in individual worker libraries.

The Microsoft Company provides an SDK created for the implementation of the analysis in the movements of the human body and will work together with the depth cameras of the same company (KINECT Azure) to ensure good levels of quality and productivity, standard procedures are needed for the extraction of anthropometric data. Obtaining anthropometric measurements for even a small population based on manual extraction poses a great challenge which is necessary for the improvement of the worker's work as well as cost savings due to ergonomic modifications to the workstations.

The application seeks to be optimal and easy to use for both the operator and the people who will be in charge of making the analysis of the workers, so we can describe the following criteria to achieve this goal:

- Simple: Easy to use so that the user can replicate the basic movements that will be simulated through an avatar.
- Dynamic: Dynamism is necessary to be able to instantly evaluate the results so that the operator can obtain the necessary measures applicable to the worker being analyzed.
- Flexible: The simulations can be modified in their virtualization according to the necessary scenario to be analyzed, in an easy way and with the specific requirements according to the case of the analysis. This way, we can have more flexibility in the variety of tasks to be carried out within the simulation.
- Clarity: Storage of all data obtained in the analysis for verification and reproduction of results at any time.
- Integration: The connectivity of the scanner, computer and virtual reality equipment that must work in symbiosis since it will work with products from the Microsoft company, which have very high standards and quality levels both in the software supplied and in the hardware (Kinect azure depth cameras). For the acquisition of the 3D model, a digital file must be provided, which will be unique for each test to be evaluated.
- Persistence: The simulation can be repeated several times in case the person to be analysed does not understand the movements of the simulations.

2 Theory

Presence and realism are important concepts in relation to the experience of Virtual Reality [4, 5]. Today, with the evolution of screen resolutions, miniaturization and computing power, we are able to (re)produce more realistic Virtual Worlds. A combination of four technological dimensions (sensory, interaction, control and location) is expected to increase the sense of realism or presence and enhance the experience within Virtual Reality [6]. Yet realism is a construct with many competent parameters. On the one hand, it refers to resemblance, in which realism operates as the reproduction of something that is known and familiar to the observer. The object, environment or event in that case also exists in the "real" world, not mediated.

We can define realism as something that is perceived as real without any knowledge or reference to an object or event that is known to the observer. Even to the extent that the object or environment does not exist in the 'real' world [6]. In this study we focus on the latter, that is, users "subjective sense of being in the place represented by the Virtual Environment" [5] without knowing the (similarity with the) real situation, but reproducing a sensation in which the experience of Virtual Reality is like the real world [4].

The recommended methodology to obtain a virtual reality application must be committed to the basic parameters in technological issues and integration of humanitarian issues in order to make them work in harmony, below are some important points that will be implemented in the application:

- Audio technology
- Camera technology
- Display technology
- Ergonomics and user interfaces
- Geometric and trigonometric mathematics
- Image processing techniques and processes
- Optics and optometry
- Positional, tracking, and location methodology
- Processors (CPUs. GPUs, DSPs, FPGAs, and special purpose) and memory
- Projected augmented reality
- Semiconductor technology
- Software engineering, operating systems, APIs, drivers, computer graphics, game engines.

Projected augmented reality is a technique where the augmented reality overlay of information and/or graphics, is projected from the headset or head-mounted display (HMD) out onto the real world and objects within it resulting in projective displays. The three techniques may be applied at varying distance from the viewer: head-mounted, hand-held and spatial. Visual perception is the key to understanding, information transfer, and memory. Edgar Dale (1900–1985) was an American educator who developed the Cone of Experience. He postulated that we remember 10% of what we read, and 50% of what we see and hear (Fig. 1).

On the part of design tools, it should be specified that they refer to the technical specifications of the hardware necessary to carry out the prototype, which must work in complete harmony with the programs to be implemented, as well as the databases of the different software's that will be integrated into the prototype.

Within the design methodology, they refer to the basic needs that were raised to be able to search for the project's objective, thus giving a clear vision of the direction in which, we should focus its analysis and specifications, wherein we can include the scalability of the project.

The prototype which represents the increase in improvements can be implemented, to design more simulation models to be followed by the personal then analyzed, therefore we will have a selection of different changes without losing the essence of the project, which is the measurement of Basic movements of the human body to carry out operations without complications of free movement due to corporal problem that the personnel presents.

Validation and verification of data are of utmost importance within the project, the lines of the codes can be collected inside the parameters of each software used in

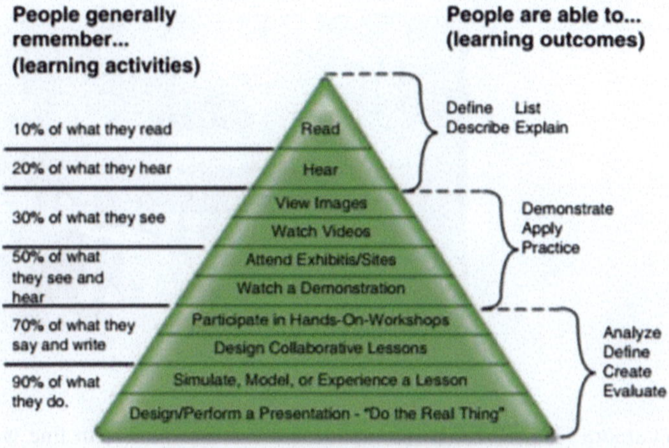

Fig. 1 Edgar Dale's Cone of Learning It relates to abstraction versus concrete and the greater use of sense (Credit: Jeffrey Anderson)

the databases, it is assumed in the same way, they can be integrated between them, and at the same time self-evaluate the movements captured by the depth cameras (Kinect azure) in conjunction with the SDK provided by the Microsoft company for the interpretation of data in the capture of images within the processed videos the same program.

The research to be carried out requires information collected through data and anthropometric surveys necessary for its veracity. According to Groves et al. [5], the survey is "A systematic method to collect information from a sample of entities to construct quantitative descriptors of the population of the quality of the entities that are members". In this research, it is proposed to use a comparative method where it will be evaluated using the software that contains the databases with the previous and established information. All anthropometric studies will need some kind of appraisals to compare and evaluate the results and see if the differences between individuals, sub-groups, and samples are detected.

3 Equipment

To collect the anthropometric measurements data, we are going to use a Microsoft Kinect Azure DK 2020 as seen in Fig. 2 the following components can be categorized by:

1. Depth sensor (1 MP) with wide and narrow field of view options that can be improved for an application.
2. Array of 7 microphones to capture far field sounds and voice.

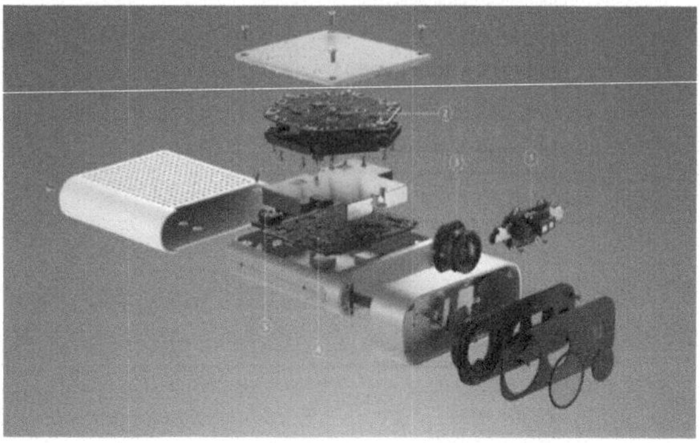

Fig. 2 Components Azure Kinect DK

3. Video camera (12 MP RGB) for additional color sequence in line with depth sequence.
4. Accelerometer and gyroscope IMU (specific type of sensor that measures angular rate, force and sometimes magnetic field. IMUs are composed of a 3-axis accelerometer and a 3-axis gyroscope) for sensor orientation and spatial tracking.
5. Cable sync connections to easily sync detection sensor sequences from multiple Kinect devices simultaneously.
6. Azure Kinect DK hardware with minimum capabilities for operation.

Azure Kinect DK has no processing capabilities. It can be connected to the computer of your choice, from a non-display device to a desktop workstation. The system requirements are:

PC with Windows® 10 or Ubuntu 18.04 LTS with the 7th generation Intel® Core ™ i3 processor (Dual Core 2.4 GHz with HD620 or faster GPU), USB 3.0 port and 4 GB of RAM. Body monitoring and other experiences may require more advanced PC hardware [6]. Azure Kinect hardware specifies integrating the latest Microsoft sensor technology into a single USB-connected accessory.

The body tracking PC host requirement is more stringent than the general PC host requirement. Recommended minimum Body Tracking SDK configuration for Windows is:

Seventh Gen Intel® Core™ i5 Processor (Quad Core 2.4 GHz or faster), 4 GB Memory, NVIDIA GEFORCE GTX 1070 or better, Dedicated USB3 port [6].

3.1 Display

A digital screen with a movement speed of 144 Hz is necessary, so we selected a model from the company LG 32GK850G-B 32 "Class QHD Gaming Monitor with G-SYNC shown in Fig. 3 [4]". That has the HDMI and USB necessary for the necessary connections for Kinect Azure and the PC to be used. The use of the screen

Fig. 3 LG display

facilitates the reproduction of the virtualization of the design made on the computer, as well as the visualization of the avatar that the person represents at the time of evaluation.

3.2 Virtual Reality Simulation Lens

A key part of the project hinges on a virtual reality glasses, which we will be using Oculus Go Standalone Virtual Reality Headset Fig. 4. This can be connected via Wi-Fi to the PC without the need for connection cables, this allows greater mobility to the person under evaluation. The following describes the functions and specifications of the lenses [7]:

- Personal Viewing: The littlest, big screen. Crystal clear optics 3D graphics.
- Portable & Easy to use: Experience portable, all-in-one VR. That means no PC, phone, wires or hassles.
- Controller: Effortless control. Whether it's navigating to your favorite shows or to a far-off land, the intuitive controller makes getting there a cinch.

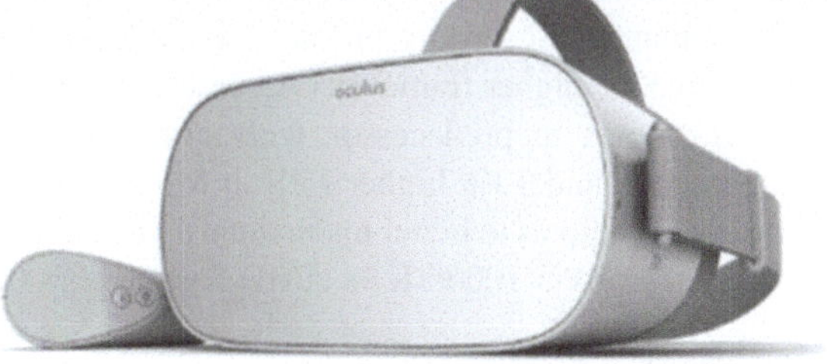

Fig. 4 Oculus Go VR lens

- Built-In Audio: Surrounded by sound. Spatial audio drivers are built right into the headset, providing dramatic, immersive sound without the need for bulky or tangled headphones. Oculus Go also features a 3.5 mm audio Jack.
- Set up with the Oculus mobile app and you're free to explore, watch, and play in VR right out of the box. No wires, no PC, and no extra hardware needed.
- Three Degrees of Freedom: 3DoF means this headset tracks your head movements and translates them into VR for 360 viewing. Sitting recommended. No external sensors needed.

3.3 The Mid-Range Cpu Intel Core I5-10600 K Versus Amd Ryzen 5 3600

With the official launch of Intel's 10th-generation Comet Lake-S processors, the Intel Core i5 10600K vs AMD Ryzen 5 3600 debate was inevitable. In the mid-range processor class, it's undeniable that the Ryzen 5 3600 isn't as "fast" as the competing Intel Core i5-9600K in terms of clock speed. But with hyper-threading technology, higher IPC (instructions per clock) performance and a lower price point, the Ryzen 5 has become a popular choice among non-OEM and budget-conscious consumers.

Price and availability: Intel Core i5-10600K launch price is set at $362, which is pretty much in line with the pricing for Intel's previous generation Core i5 chips. The Ryzen 5 3600, meanwhile, costs significantly less with a starting price of $205.

Specifications: The biggest change for the Core i5 10600 K over its predecessor is hyper-threading. The 10600K is a six-core, 12-thread processor with a listed base clock speed of up to 4.1 GHz, boostable to a maximum single-core turbo frequency of 4.8 GHz.

It comes unlocked, so it can be overclocked, and includes an integrated Intel UHD 630 GPU. Its Thermal Design Power (TDP) comes in at 125 W, 30 Ws higher than its predecessor. It supports 16 PCIe 3.0 lanes, up to 128 GB of dual-channel, non-ECC DDR4-2933 RAM, and has an L3 cache size of 12 MB. The AMD Ryzen 5 3600, mean-while, also has 6 cores and 12 threads, but with a lower base clock speed of 3.6 GHz. Its max turbo speed is 4.2 GHz. It can also be overclocked but does not come with an integrated GPU. Its TDP is much lower than the Intel Core i5-10600K's, coming in at 65 W, supports 20 PCIe 4.0 lanes, and up to 128 GB of dual-channel ECC DDR4-3200 RAM.

Performance: According to Intel, the desktop Comet Lake processors will provide up to 33% higher frames per second (FPS) while playing Mount and Blade II: Bannerlord over its predecessor, with up to 10% higher FPS in Player Unknown: Battlegrounds and 13% higher FPS in Monster Hunter World. These video game comparisons help us to better understand a virtualization of the kinekt body tracking SDK. Intel Core i5 10600 K vs AMD Ryzen 5 3600: which is going to be the best for the prototype?

The AMD Ryzen 5 3600 was able to match the gaming performance of Intel's 9th generation Core i5 chips at a significantly lower price point, so how likely is it that

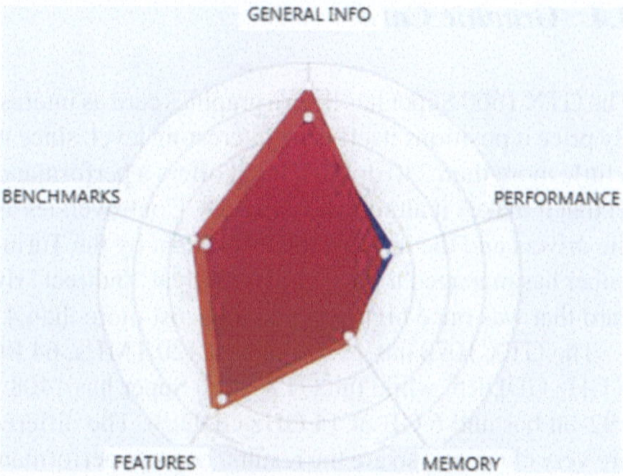

Fig. 5 Comparison chart Intel Core i5-10600K better than AMD Ryzen 5 3600

Intel's latest mid-range Comet Lake processor can bring gamers back to the Intel fold? While the Core i5-10600K features 4K-capable integrated graphics, expecting high-performance gaming out of an integrated GPU is a fool's errand. Most gaming consumers are going to be looking to utilize a more powerful discrete GPU for their rigs, so one of Intel's major advantages over the AMD Ryzen 5 3600 is someone moot.

What's more, the Ryzen 5 3600 matched the Core i9 9900K's performance nearly frame-for-frame so the 10–33% FPS improvements that Intel is promising for Comet Lake processors, while important, are likely to be tinkering on the margins when the Ryzen 5 3600 is already pumping out nearly 120 FPS. Assuming that the Intel Core i5 10600K performs as well as the Intel Core i9, is the difference between 100 FPS and 120 FPS. Comparison chart on Fig. 5.

On the high end, these kinds of performance differences might make sense, but for a gamer on a budget, those extra 20 FPS look awfully expensive, especially when your game already looks amazing—and that's assuming Intel can deliver that kind of performance bump at the Core i5 level, which still isn't certain. At this point, it still looks very much like the AMD Ryzen 5 3600 will remain the preferred CPU for mid-range gaming rigs [8].

- 13.89% faster CPU speed, 6 × 4.1 GHz versus 6 × 3.6 GHz
- 0.6 GHz higher turbo clock speed, 4.8 GHz versus 4.2 GHz
- 16.99% higher Pass Mark result (single), 3016 versus 2578
- 0.5 s faster Blender (bmw27) result, 234.5 s versus 235 s
- 6.14% higher Cinebench R20 (single) result, 501 versus 472
- 5 °C higher maximum operating temperature, 100 °C versus 95 °C
- Has NX bit
- Has integrated graphics.

3.4 Graphic Card

The GTX 1660 Super has been a graphics card as interesting as it has been questioned. By price it positions itself in an interesting level, since it is possible to find models for a little more than 230 dollars, but it offers a performance so similar to the GTX 1660 Ti that it makes it almost meaningless. Controversies aside, thanks to the maturity of the drivers and the improvements brought by the Turing architecture, the GTX 1660 Super has managed to face one of its great "indirect" rivals, the GTX 1070, a graphic card that was once high-end and that cost more than 450 dollars.

The GTX 1070 has 1920 shaders, 120 TMUs, 64 ROPs, 256-bit bus and 8 GB of 8 GHz GDDR5, while the GTX 1660 Super has 1408 shaders, 88 TMUs, 48 ROPs, 192-bit bus and 6 GB of 14 GHz GDDR6. The differences at the specification level are very clear, and so are the results. In gross performance the lightweight GTX 1070 has a very small advantage, and the GTX 1660 Super is almost at the same level as that if we make a medium balance. Some games perform better in the first one and others, the most current ones, work better in the second one. This confirms that driver level optimization has played a key role in pushing the GTX 16 series (and also the RTX 20) over the GTX 10. Comparison chart on Fig. 6.

This doesn't mean that it's worth upgrading from a GTX 1070 to a GTX 1660 Super, on the contrary, but it does confirm that with the price both have nowadays it's better to opt for the second one, even though it comes with 2 GB less graphics memory. The Nvidia GeForce GTX 1070 is the second fastest Pascal based graphics card in mid May 2016. It is a high-end desktop graphics card. The mobile pendant is also called GeForce GTX 1070, but features more shaders and a lower core clock. Both cards are based on a cut down GP104 chip as the GTX 1080 and manufactured in 16 nm FinFET at TSMC. Compared to the GTX1080, the GTX1070 features less CUDA cores (1920 versus 2560) that are clocked a bit lower (maximum Boost 1683 versus 1733). Instead of the new GDDR5X, the GTX 1070 uses slower GDDR5 graphics memory—but still 8 GB. The TDP is also reduced from 180 to 150 W. Therefore, gaming in 4 k at high details is possible with the card and also VR gaming is possible [8].

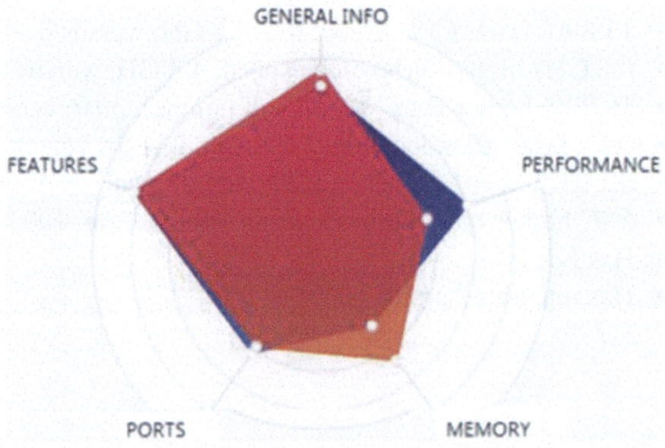

Fig. 6 Nvidia Geforce GTX 1660 Super better than Nvidia GeForce GTX 1070

- 24 MHz faster GPU clock speed, 1530 MHz versus 1506 MHz
- 6000 MHz higher effective memory clock speed, 14000 MHz versus 8000 MHz
- 80 GB/s more memory bandwidth, 336 GB/s versus 256 GB/s
- 25 W lower TDP, 125 W versus 150 W
- 102 MHz faster GPU turbo speed, 1785 MHz versus 1683 MHz
- 0.1 newer version of OpenGL, 4.6 versus 4.5
- 4 nm smaller semiconductor size, 12 nm versus 16 nm
- 1 higher version of GDDR memory. 6 versus 5.

4 Softwares

4.1 Body Tracking SDK Azure Kinect Integration

Use a body-tracking object to process Azure Kinect DK traps and generate body tracking results. It also maintains the global status of the crawler, renders queues, and output queue. There are three steps to use body tracker [9]:

1. Create a tracker: The first step to using body tracking is to create a tracker and requires passing the sensor calibration structure k4a_calibration_t. Sensor calibration can be queried with the k4a_device_get_calibration function of the Azure Kinect SDK.
2. Capture of depth and IR images from the device: The images are captured from the device in a correlated way. Each captured image contains a depth image, an IR image, a color image, or a combination of images. By default, the API will only return a capture once it has received all the images requested for the streaming mode. It can configure the API to return partial captures with only depth or color images as soon as they become available by deleting the "synchronized_images_only" parameter from the k4a device configuration.
3. Queue the capture and pop the results: The tracker internally maintains an in-put queue and an output queue to process Azure Kinect DK captures asynchronously more efficiently. Use the "k4abt_tracker_enqueue_capture ()" function to add a new capture to the input queue. Use the "k4abt_tracker_pop_result ()" function to extract a result from the output queue. The use of the timeout value depends on the application and controls the queue timeout.

4.2 Procedure of Conducting the SDK Kinect Software

Traditional Anthropometric Data. All the measurements taken have to be capture taking the following considerations:

1. The basic movements captures have to be taken according to the definitions of the selected database in SQL attending the tacit knowledge and combining the

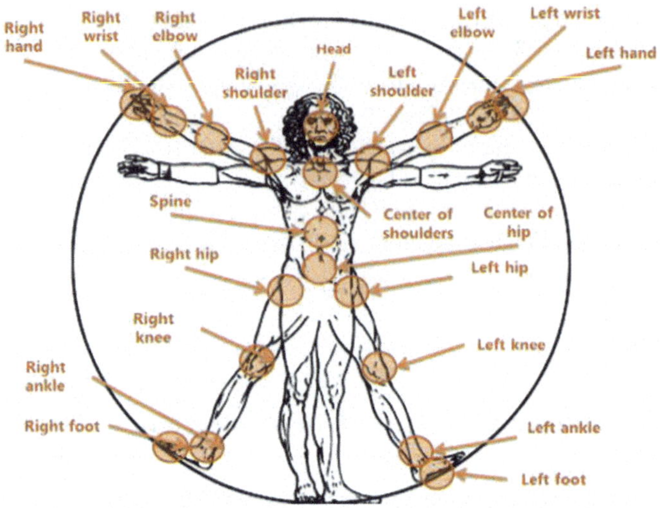

Fig. 7 The twenty (20) anatomical landmarks incorporated in Kinect SDK

Interactive ontologies of the dimensions of the selected movement as given in Pheasant [10] and adapting the human points of reference to the Kinect SDK (see Fig. 7). _get calibration function of the Azure Kinect SDK.

2. When being measured, the personal has to wear light clothes with bodyweight evenly distributed on both legs.
3. All Anthropometric measurements are based on protocols as outlined primarily in [10–12].

4.3 SQL Environment

Currently the SQL language is accepted by most databases, it provides a simple and intuitive way to read, enter or modify the information contained in the tables. Its knowledge is essential since practically all applications have a data repository that must be accessed at one time or another, using an SQL query. It could be said that the SQL language provides a means to formally represent instructions. The objective of the SQL language is to synthesize this natural way of expressing actions that refer to tables in a database.

The SQL language uses an intuitive syntax to faithfully collect the meaning of the colloquial phrases to which they correspond. Databases need to provide a means for users to insert, modify, or view information. SQL statements allow you to relate all the tables that make up a database, either to obtain a specific data set, ordered in a specific way, or, for example, to determine the collection of elements that should be removed.

5 Implemented Intelligent Application

Chatbot VRs are programmed under the concepts of artificial intelligence and machine learning. Therefore, beyond responding according to data stored in their memory, they interpret what the person says or does, give coherent responses, and learn from each of their interactions.

Virtual reality today is a technique that allows an improvement in the visual processing of movements, as well as its structure, which can be modified for the use of different scenarios. The key to the functionality of our prototype lies in the use of a VR chatbot, in which the person who is making the movements to be analyzed is in a friendlier environment, where the program will be integrated with the databases necessary to process movements in real-time. The combination of the depth cameras (Kinect Azure) will be those that capture the movements and provide us with the necessary data to be able to be processed within the chatbot.

The prototype in question on Fig. 8, integrates the use of a friendly avatar with features similar to that of human beings, where it's likely to find that the person who is carrying out the movement tests has an environment of harmony without falling into an environment full of camera cables and that can feel uncomfortable, achieving this environment of harmony and working with WIFI technology it's possible to interconnect the Kinect Azure, the screen with the person and the virtual reality glasses, already described in the article, then it will be represented by an avatar. Being of utmost importance the classification of the databases needed to create by means of program-ming in SQL, where one database will be with tacit movements and

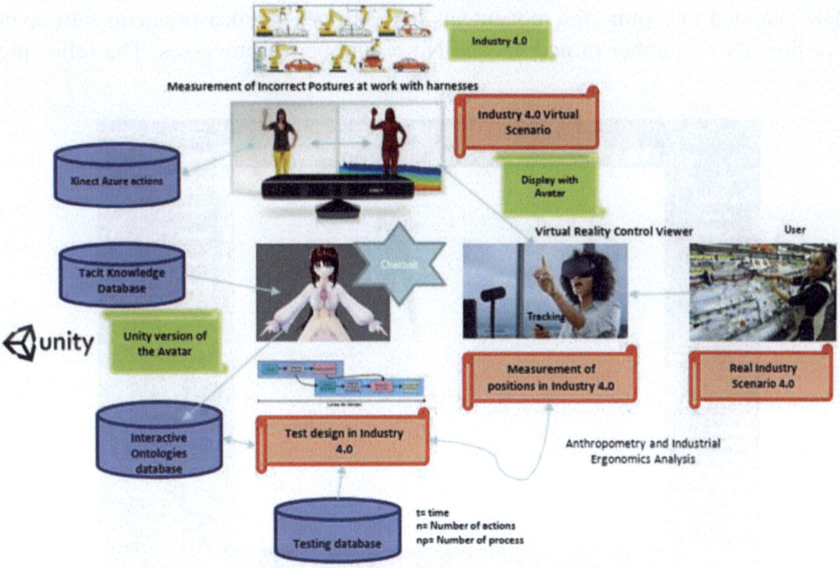

Fig. 8 An industrial process model related with Industry 4.0

another with inter-active ontologies that will be collected through the SDK software for tracking body movements. The company Microsoft, which works with the Kinect at the same time, as described in the hardware section, requires a PC with minimum specifications already clarified in order to make the prototype work smoothly and without complications.

Including VR devices such as Oculus lenses that allow the collection of even more external data can be very helpful in helping VR chatbots collect data on real-life human actions and interactions. There will need to be a strong and consistent backbone for that data to flow to ensure and improve the effectiveness of what these VR chatbots can do.

To harness the power of new computing devices and cloud computing to create virtual reality chatbots that have a "human" feel. A bot that looks, talks, gives hand gestures, and facial expressions like a human. Furthermore, the virtual reality robot must also fit into the application you are creating.

Today "AWS Sumerian" provides a program for the virtual chatbot, it has a variety of functions that help us explore and build in the shortest time possible. The service helps you with direct connections to "Amazon Polly" and "Amazon Lex bot" [13]. The service is implemented in the cloud, so it will not need a high-configuration computer either. It also helps with a bunch of assets like hosts and everyday objects like a table ETC as seen on Fig. 9. And of course, you can always import external assets into your Sumerian environment. It has over 20 different voices in different ac-cents, to choose the correct voice based on where the bot would launch.

The virtualization of the prototype will be designed in an environment using Unity in conjunction with the avatar designed to be able to generate the necessary scenario that will be followed by users, where together with the Kinect Azure and the Microsoft SDK intended to capture the movements in real time, the measurements will be in T = time, N = number of actions and Np = number of processes. The following

Fig. 9 Avatar on Amazon AWS Sumerian

Fig. 10 An industrial process model related with Industry 4.0

case study shows how the information collected by the prototype can be adapted and, results can be generated instantly from the evaluation. The forklift operator in question, Fig. 9, will have to be able to perform basic body movements to perform the necessary functions within the job. Likewise, it can make clear with the following case virtualized in Unity, obtaining results in various capture modalities and with these it was possible to solve the problem in question (Fig. 10).

5.1 Experiment and Implementation

For this research three factors (turns, units and days) to determine the load are ana-lyzed. Turns for three moments (M, T and N) where M = 8 a.m., T = 12 p.m., and N = 6 p.m. This because it is desired to identify which route is the must demanded. Three units (r5, r12 and r23) are analyzed. Finally discusses days a week from Monday to Friday (d1, d2, d3, d4, d5) where d1 = Monday, d2 = Tuesday, d3 = Wednesday, d4 = Thursdays and d5 = Fridays. The image is displayed with the analysis (Fig. 11).

The next step is to prepare the matrix in Excel with data prior to analysis in Stat graphics Centurion XVI. The result is the preparation shown below in the Table 1.

turn
M=8 am
T= 12 pm
N=6 pm

unit
r5
r12
r23

day
d1=Monday
d2=Tuesday
d3=Wednesday
d4=Thursday
d5=Friday

Fig. 11 Factors to analyze

Table 1 Excel matrix

	u5				
	d1	d2	d3	d4	d5
M	335	288	286	307	285
T	205	294	291	284	149
N	202	128	101	156	178

	c12				
	d1	d2	d3	d4	d5
M	207	32	246	299	292
T	84	168	135	113	113
N	252	341	246	0	130

	u23				
	d1	d2	d3	d4	d5
M	292	332	286	228	164
T	255	295	258	261	205
N	207	189	183	164	0

The data within each table represent the sum of the charges that are defined by the times defined in M, T and N. When the data is prepared, it's sent to Statgraphics; for the present work it has the version Centurion. The information of how the matrix is created in such software is displayed in Table 2.

There is a total of 45 records identifying time, unit, day and capacity. The aim of the experiment is to analyze the capacity variable and identify where there is more demand turn. The results are shown in the following images.

5.2 *Summary Statistics for Determine Industrial Process to Model*

In the Fig. 12 statistical capacity variables are displayed. Capacity distribution is shown in Fig. 13 it is possible to appreciate indicating statistics.

The following Fig. 14, shows in what turn exist more loads. Means and 95.0% LSD Intervals

The graph shows that the turn where more loads is in the M, approximately 290, followed by T, in the turn of 12 p.m., to 200 and less charge in N, that is, that at 8 p.m., people do not demand equally. The following graph (Fig. 15) shows a greater capacity per turn, identifying that M turn more capacity for r5 unit while for the same turn the unit r12 is the lower demand has.

Table 2 Excel matrix

Time	Unit	Day	Capacity
T	r23	3	258
M	r12	5	292
T	r12	1	84
T	r23	2	295
T	r23	1	255
M	r5	2	288
M	r23	3	286
T	r5	2	294
N	r5	2	128
T	r12	3	135
N	r12	4	0
N	r12	3	246
M	r5	4	307
M	r12	3	246
T	r23	5	205
N	r5	4	156
N	r5	3	101
T	r5	1	205
T	r12	5	113
N	r23	2	189
T	r23	4	261
N	r23	5	0
N	r23	1	207
M	r5	3	286
T	r5	5	149
N	r5	5	178
T	r5	3	291
N	r5	1	202
T	r12	4	113
M	r12	1	207
N	r12	5	130
M	r12	4	299
T	r5	4	284
N	r12	1	252
M	r12	2	32
M	r23	2	332
M	r5	1	335

(continued)

Table 2 (continued)

Time	Unit	Day	Capacity
M	r23	1	292
N	r23	3	183
M	r23	5	164
N	r23	4	164
N	r12	2	341
T	r12	2	168
M	r5	5	285
M	r23	4	228

Fig. 12 Factors to analyzed

Count	45
Average	210.356
Median	207.0
Variance	7722.69
Standard deviation	87.8788
Coeff. of variation	41.7763%
Minimum	0
Maximum	341.0
Range	341.0

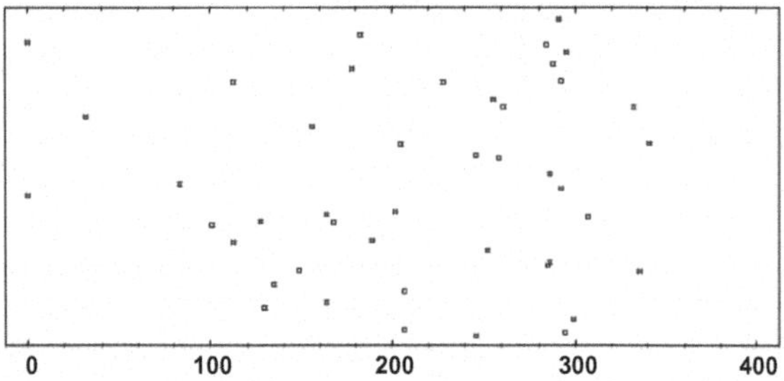

Fig. 13 Capacity distribution

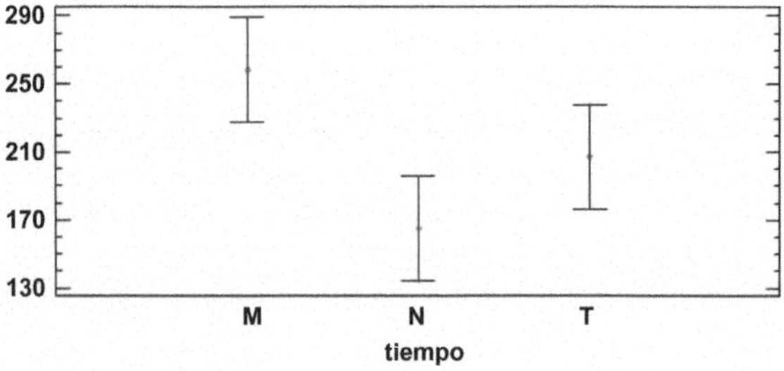

Fig. 14 Exist turns

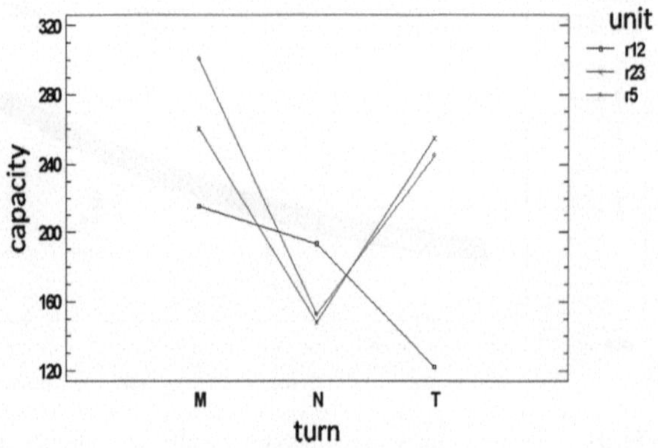

Fig. 15 Capacity per turn

6 Conclusions

The evaluation of human body movements as a common practice is not yet adapted to industry 4.0, observing within the natural parameters of corporate movements, to determinate the final location of people to be hired, therefore, minimize Engineering costs at workstations for an ergonomic fit. Being able to guarantee a job position in the quality of the controlled person and without problems that prevents from having free body movements that will allow to have better job performance.

The final standards for IoT have yet to be defined, in fact the protocols are in full development and there are several projects that intend to define standards [2]. Today a lot of business sectors are facing several communication challenges. Some applications are using Internet of Things with cellular modems just because it provides a positive ROI even when users are paying few dollars per month per sensor for the connectivity. But there are many applications with lower ROI requirements, and it is still being unconnected because of traditional communication technologies do not provide a positive return on investments with current monthly fees. Addressing the abovementioned problem LPWA networks offer an effective solution. With this technology, the initial cost of a radio module and amount of monthly service fee is much lower (cents/month as opposed to dollars/month). Whereas a cellular modem would cost over $30, and a WAVIoT IoT module can be built for a fraction of that. As is possible see in Fig. 16.

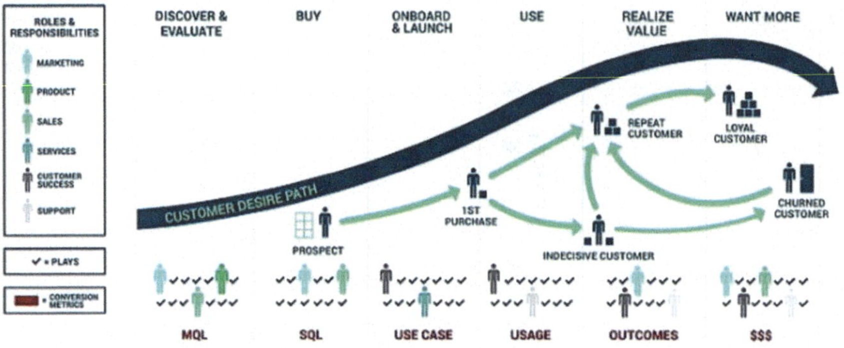

Fig. 16 Innovation model of a process involve an industrial process and industrial Internet of Things

References

1. Wang, M.J., Wu, W., Lin, K., Yang, S., Lu, J.: Automated anthropometric data collection from three-dimensional digital human models. Int. J. Adv. Manuf. Technol. **32**, 109–115 (2007). https://doi.org/10.1007/s00170-005-0307-3
2. Schranz, N., Tomkinson, G., Olds, T., Daniell, N.: Three-dimensional anthropometric analysis: differences between elite Australian rowers and the general population. J. Sports Sci. **28**(5), 459–469 (2010). https://doi.org/10.1080/02640411003663284
3. Stewart, A.D.: Kinanthropometry and body composition: a natural home for threedimensional photonic scanning. J. Sports Sci. **28**(5), 455–457 (2010). https://doi.org/10.1080/02640411003661304
4. Rettie, R.: Using Goffman's frameworks to explain presence and reality (2004)
5. Yu, I., Mortensen, J., Khanna, P., Spanlang, B., Slater, M.: Visual real-ism enhances realistic response in an immersive virtual environment. IEEE Comput. Graph. Appl. **32**(6), 36–45 (2012)
6. Van Gisbergen, M.S.: Contextual connected media: how rearranging a media puzzle, brings virtual reality into being. Breda: NHTV (2016)
7. Virtual Reality Lens Ocolus Specs, https://www.oculus.com/go/features/
8. CPU Comparison AMD Ryzen 5 3600 vs Intel Core i5-10600K, https://versus.com/en/amd-ryzen-5-3600-vs-intel-core-i5-10600k (November 15th 2020)
9. Body Tracking SDK Azure Kinect, https://docs.microsoft.com/en-us/azure/Kinect-dk/get-body-tracking-results
10. Pheasant, S.: Bodyspace, 2nd edn. Taylor & Francis, London (1996)
11. Hertzberg, H.T.: The conference on standardization of anthropometric techniques and terminology. Am. J. Phys. Anthropol. **28**(1), 1–16 (1968). https://doi.org/10.1002/ajpa.1330280110
12. Roebuck, J.A.: Anthropometric Methods: Designing to Fit the Human Body, Jr. Santa Monica, CA: Human Factors and Ergonomics Society, ISBN 0-945289-01-4 (1995)
13. Traditional Chatbot vs. Virtual Reality Bot I AWS Sumerian I by Vaibhav Malpani I Vaibhav Malpani's Blog I Medium; https://medium.com/vaibhav-malpanis-blog/traditional-chatbot-vs-virtual-reality-bot-awssumerian-4866e4f51693

Industry 4.0 Optimization and Its Future Effects on Z Generation Focused on the Paradigm Shift of an Innovation Ecosystem

Selection of Factors Influencing for Reliable Electrical Power Transmission Design in Industry 4.0

Rubén Jaramillo-Vacio, Javier Cruz-Salgado, and Alberto Ochoa-Zezzatti

Abstract At present, electrical energy is a very important factor in the productive chain of Industry 4.0, which is why guaranteeing the electrical supply, either by conventional energy or by clean energy, is mandatory for the utilities. The ampacity or current conduction capacity in underground cables can be mathematically modeled based on the thermal behavior of its elements, however, the availability of dedicated software that already has the programmed algorithm, helps make this task more efficient. Therefore, it is possible to determine with great accuracy the ampacity of lines that are currently in service. To minimize calculation errors with respect to those measured, parameters such as: characteristics of the power cable, configuration of the layout and grounding of the line, construction conditions and the surrounding environment, among others, must be recorded. In this chapter, an analysis of the sensitivity of ampacity with respect to the variability of the input parameters will be presented, in order to establish the statistical significance of the accuracy of the information.

Keywords Design of experiments · Industry 4.0 · Energy · Ampacity

1 Introduction

Overhead lines are inexpensive to build, but not reliable in the continuity of service, they are affected by environmental conditions, earthquakes, saline pollution, by local fauna, among others.

R. Jaramillo-Vacio (✉)
CFE-LAPEM, Irapuato, Guanajuato, Mexico
e-mail: ruben.jaramillo@cfe.mx

J. Cruz-Salgado
UDLAP, Cholula, Puebla, Mexico

A. Ochoa-Zezzatti
UACJ, Ciudad Juarez, Chihuahua, Mexico

A. Ochoa-Zezzatti et al. (eds.), *Technological and Industrial Applications Associated With Industry 4.0*, Studies in Systems, Decision and Control 347,
https://doi.org/10.1007/978-3-030-68663-5_15

The choice of whether to use overhead line or underground cable must be consistent with safety, reliability and operational constraints to ensure that the capacity of the transmission grid efficiently matches the supply and demand of electrical energy. The choice between overhead lines and underground cable is driven by technical, environmental and economic considerations.

The evolution and enhanced performance of underground cables with extruded insulations such as cross-linked polyethylene (XLPE) has refocused attention on the installation of underground extra-high-voltage, high-voltage and medium-voltage transmission circuits. The liberalization of the energy market and the need to connect new power plants to the grid has stimulated growing requirements to extend existing transmission systems.

The objective of determining the ampacity of an underground line is to know what the temperature rise within the conductor is due to the heat generated during nominal cable operation. The temperature of the cable is limited by the insulation material, so that if the limit is exceeded, the residual life of the insulation decreases exponentially. For example, according to the Arrhenius equation, the reactions that occur at room temperature increase twice when the temperature rises 10 °C [1]. Due to this change in reaction rate, the power cable ages or deteriorates more rapidly under thermal stress. By controlling this parameter of influence of the residual life expectancy, the chargeability can be projected to always keep it within a safe operating range. In [2] the operational thermal limits are established as a function of the duration of the current, so that there are three standardized ranges: continuous operation, in emergency condition and in short-circuit condition.

Something very important to point out is that the ampacity is not only defined by the gauge or cross-sectional area, but also by the conductor material, the arrangement of conductors, the interaction of nearby lines, the grounding, the depth of the installation, among others. Few studies of the variability of parameters have been carried out, however, the consideration of how they interact between them is not verified [3].

1.1 Thermal Behavior of the Power Cable

To develop the thermal model of the power cable, it is necessary to divide it into two generic groups: the heat generated by the conductor and the heat generated by the insulation. At present, models as complex as the one presented in [3] have been developed. Figure 1 shows the construction elements of a 115 kV aluminum power cable, where the screens that constitute it are schematized. Losses in metallic elements such as conductor and metallic screens are the most significant due to the Joule effect.

The losses in the metallic elements of the power cable are a function of the operating frequency (f) and temperature (t), and are proportional to the square of the current. The expression that correlates the frequency and the temperature in an equivalent model of resistance in alternating current (R_{ac}) is the following [4]:

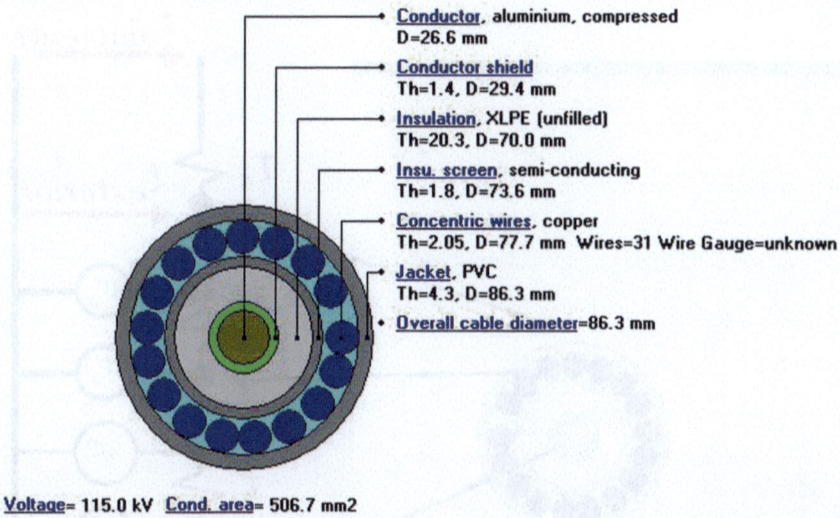

Fig. 1 Power cable construction

$$W = R_{ac}(f, t)I^2 \tag{1}$$

It is important to mention that insulating materials also produce heat and it is of great importance when operating at high voltage. That is why main insulation, screens, semiconductor covers, trusses and covers are modeled based on dielectric losses [4]:

$$W_d = \omega C V^2 tan\delta \tag{2}$$

Considering these concepts, it is possible to define the thermal model of the power cable, subdividing the conductor into areas. Heat sources are replaced by current sources, thermal resistances by electrical resistances and thermal capacitances by electrical capacitances. Figure 2 shows the correspondence between the power cable and the electrical elements for the calculation of the ampacity in steady state.

As seen in Fig. 3, the potential difference at the nodes is analogous to the temperature at the power cable elements. The potential difference of the terminals with respect to W_c is the temperature rise of the conductor with respect to the environment (Δt), this is represented by:

$$\Delta t = \left(W_c + \frac{1}{2}W_d\right)T_1 + (W_c + W_d + W_s)T_2 + (W_c + W_d + W_s + W_a)(T_3 + T_4) \tag{3}$$

With the previous expression, an expression of the ampacity can be defined, since the dielectric losses can be referred to a loss factor λ in proportion to the losses in

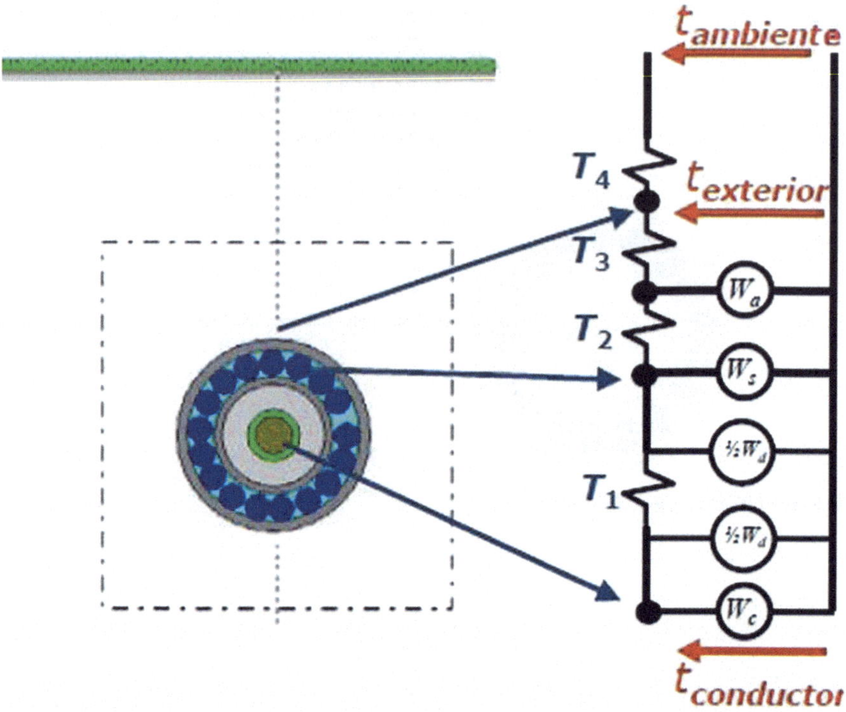

Fig. 2 Electro-thermal model

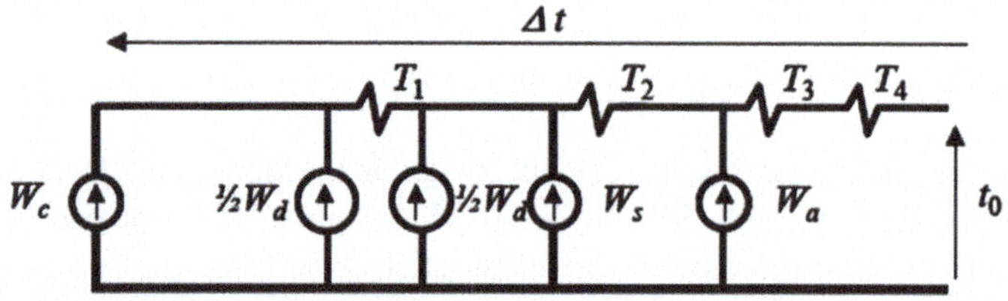

Fig. 3 Electrical analogy of power cable

the conductor, so that:

$$W_s = \lambda_1 W_c W_a = \lambda_2 W_c W_c = R_{ac} I^2 \tag{4}$$

So re-grouping we have:

$$I = \sqrt{\frac{\Delta t - W_d\left(\frac{1}{2}T_1 + T_2 + T_3 + T_4\right)}{R_{ac}T_1 + R_{ac}(1 + \lambda_1 + \lambda_2)(T_3 + T_4)}} \tag{5}$$

2 Underground Cable System in Industry 4.0

In the development of Industry 4.0, energy management and energy efficiency overall has taken center stage. As a matter of fact, knowing the regulations, specific power needs and increasing importance of energy to drive the smart factory and smart industry, there is more attention than ever for energy efficiency in industrial markets.

It's not exactly a secret that factories and many other industrial facilities are not just among the main "consumers" of power, but that power is equally critical in ensuring continuity and quality of the end-to-end manufacturing process. There is a reason why in Industry 4.0 there is so much attention for topics such as maintenance and asset management. There is also a reason why suppliers of industrial robots such as ABB Robotics are among the first to have built IoT-enabled connected services to enhance maintenance and minimize disruptions with regards to industrial robots in a market where every second counts and downtimes need to be avoided.

Energy efficiency in Industry 4.0 is equally a matter of the integration of energy sources and the connection of smart power networks in but also beyond the company.

Just as an unexpected failure in a critical component in the manufacturing process of, for instance a large car manufacturer, results in enormous costs, so do issues with regards to electricity and power.

In this scenario, underground systems become an energy security option. However, from the design and selection, all influencing factors must be considered to determine the safest and most reliable installation.

There are a few factors that influence the proper performance of the underground system, from its configuration, grounding, whether it is directly buried, whether it is in concrete, among others.

Considering all the combination of factors that ensure an adequate performance of the electrical supply, the techniques of design of experiments support to determine the most optimal conditions in which our system will operate (Fig. 4).

Fig. 4 Schematic diagram of the need for a reliable power supply and Industry 4.0

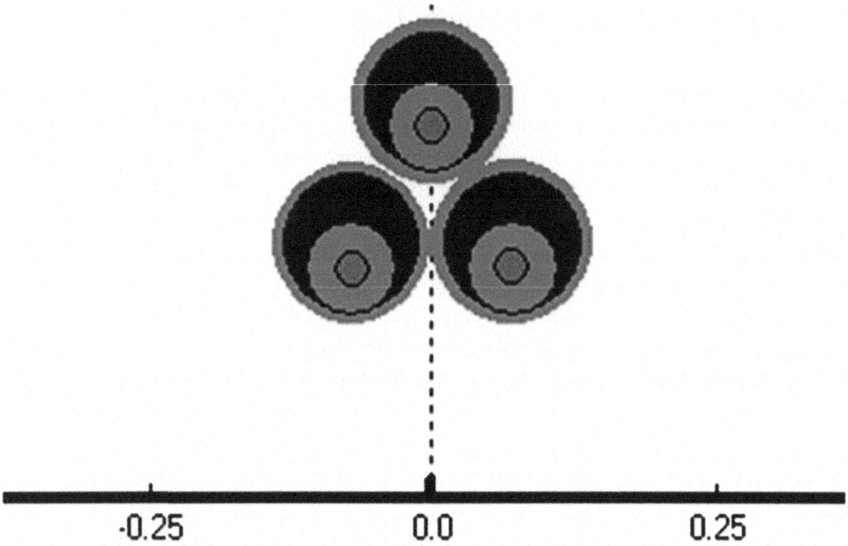

Fig. 5 Trefoil configuration

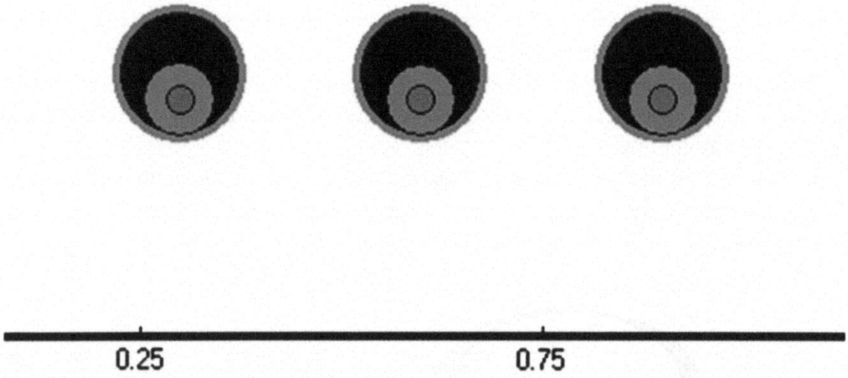

Fig. 6 Flat configuration

Below are some examples of basic configurations that should be considered when designing an optimal system (Figs. 5, 6, 7, 8 and 9).

3 Design of Experiments

Manufacturing organizations today are required to understand their manufacturing processes. To achieve this, it is common to use designs of experiments in order to increase their knowledge about the process behavior. A series of tests are conducted

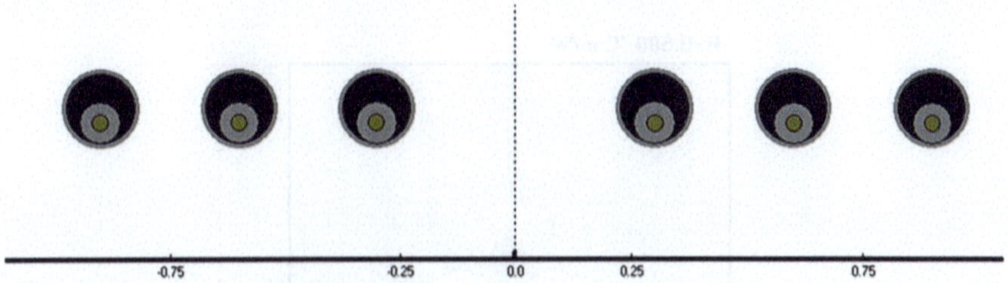

Fig. 7 Flat—buried—adjacent combined

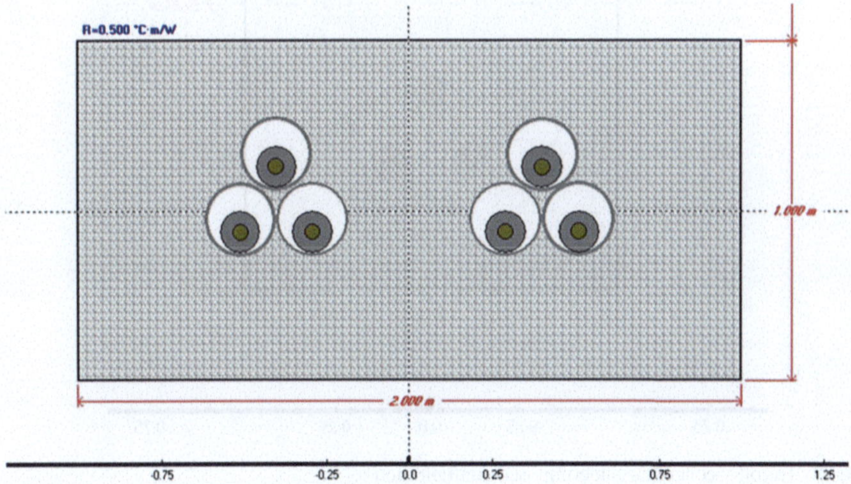

Fig. 8 Trefoil – concrete – adjacent configuration

to produce outcomes, as an experimental strategy. The main point is to establish continuous improvement of process quality. It is mandatory to understand the amount of variability an its impact on processes [5].

In an engineering environment, experiments are often conducted to explore, estimate or confirm. Exploration refers to understanding the data from the process. Estimation refers to determining the effects of process variables or factors on the output performance characteristic. Confirmation implies verifying the predicted results obtained from the experiment [6].

The most used experimental designs in most manufacturing companies, are full and fractional factorial designs at 2-levels. These kinds of designs allow an experimenter to study the effect of the studied factors over a response.

A full factorial designed experiment consists of all possible combinations of levels for all factors. The total number of experiments for studying k factors at 2-levels is 2k. The 2k full factorial design is particularly useful in the early stages of experimental work, especially when the number of process parameters or design parameters (or

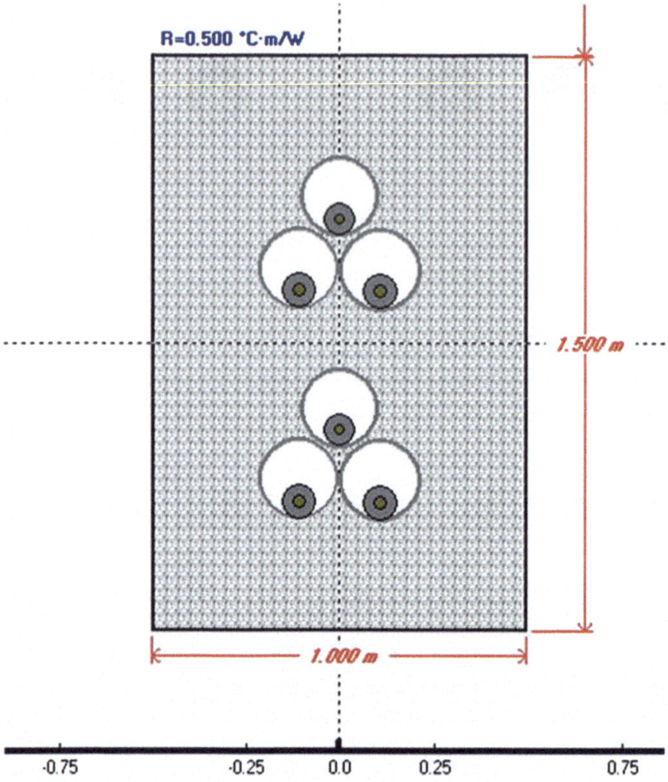

Fig. 9 Trefoil—concrete—superimposed configuration

factors) is less than or equal to 4. One of the assumptions we make for factors at 2-levels is that the response is approximately linear over the range of the factor settings chosen [6].

Experimenters can choose a full factorial design or fractional factorial design. This chapter is primarily focused on fractional factorial designs at 2-levels only. However, if readers wish to learn about experimental design for full factorial design, the authors would suggest them to refer to [7].

When an experiment is carried out for the first time, is normal that the experimenter wants to include many factors in order to guarantee that the all possible source of variability is going to be analyzed. Nevertheless, for 5 or more factors, the number of necessary trials tend to be too large. Manufacturing organizations do not have adequate time, resources or budget to carry out full factorial experiments. It can be reasonably assumed that the main effects and two-order interactions can be obtained by running only a fraction of the full factorial experiment, given that certain higher-order interactions (third order and higher) are not important.

A type of orthogonal array design which allows experimenters to study main effects and desired interaction effects in a minimum number of trials or experimental runs is called a fractional factorial design. These fractional factorial designs are the most widely and commonly used types of design in industry. These designs are generally represented in the form $2^{(k-p)}$, where k is the number of factors and $\left(\frac{1}{2}\right)^p$ represents the fraction of the full factorial 2^k. For example, 2^{5-2} is a 1/4th fraction of a 2^5 5 full factorial experiment. This means that one may be able to study 5 factors at 2-levels in just 8 experimental trials instead of 32 trials [6].

4 Experimentation and Results

The following section describes an example of a fractional factorial design with resolution III. This example involves an experiment to study the effect of eleven factors at 2-levels using thirty-two trials. The response of interest for the experiment is AMPACITY. Table 1 illustrates the list of factors and their levels used for the experiment (Table 2).

The runs were performed in random order on eight successive days. This is a 2^{11-7} fractional factorial design (1/128 fraction) resolution III, with one replica (i.e. main effects are confounded with two-factor interactions). Statgraphics software is used for statistical analysis of data. The first step in the analysis is to identify the most important factors which influence the AMPACITY. A Pareto plot is constructed to identify the key factors (Fig. 4). The graph shows that Ambient Temp (B), Thermal Resistivity (C), Conduct Gauge (D), Material conductor (E), Grounding(F), Duct Distance (H) and Pipeline Support (J) have a significant effect on the AMPACITY.

Figure 10 presents the main effects plot for the experiment. Main effects B, C and E appear to produce higher conduction capacity on their lower levels, while effects D, F, H and J appear to produce higher conduction capacity on their lower levels.

In Table 3 the analysis of variance (ANOVA) is presented. The ANOVA divides the error variability into separate pieces for each effect, then, test the statistical significance for each effect by comparing its mean square against an estimate of experimental error. The ANOVA shows that the selected linel model adequately represented the data obtained. In this case, 7 effects have a P-value less than 0.05, indicating that are significantly different from zero with a confidence level of 95.0%. The effect of a factor is defined to be the change in response produced by a change in the level of the factor. This is frequently called a main effect because it refers to the primary factors of interest in the experiment. Table 3 show that the seven main effect have a statistically significant effect over the response (Capacity conduction). On the other hand, the main effects Grounding(F), Conduct Gauge (D) and Thermal Resistivity (C) has the strongest effects (Fig. 11).

The R-Squared in this case was 91.5%, and indicates that the model adjusted explains 91.5% of the variability. The standard error of the estimate shows that the standard deviation of the residuals is 66.19. The absolute mean error (MAE), that is the average value of the residuals, is 44.11. The Durbin-Watson statistic (P = 0.5251) tests the residuals if there is any significant correlation based on the order in which the data is presented in the file. Since the P-value is greater than 5.0%, there

Table 1 Experimental design layout of the experiment

Run	Block	Duct type	Ambient temp	Thermal resistivity	Conduct gauge	Material conductor	Screen	Ground level	Duct distance	Location	Pipeline support	Disp	Ampacity
1	1	8	20	1.5	1600	Aluminum	Cross B	1.5	0.3	Superimposed	Buried	Trefoil	795
2	1	8	20	1.5	800	Aluminum	Multi	2	0.5	Adjacent	Buried	Trefoil	382
3	1	8	40	3	800	Aluminum	Multi	1.5	0.3	Adjacent	Concrete	Flat	273
4	1	6	20	3	1600	Aluminum	Multi	1.5	0.5	Superimposed	Concrete	Trefoil	485
5	1	8	40	3	1600	Aluminum	Cross B	2	0.5	Superimposed	Concrete	Flat	540
6	1	6	40	1.5	1600	Aluminum	Multi	2	0.3	Superimposed	Buried	Flat	367
7	1	8	40	1.5	1600	Copper	Multi	1.5	0.5	Adjacent	Concrete	Trefoil	607
8	1	6	40	3	800	Copper	Multi	2	0.5	Superimposed	Buried	Trefoil	251
9	1	8	40	1.5	800	Copper	Cross B	2	0.3	Superimposed	Concrete	Trefoil	613
10	1	6	20	3	800	Aluminum	Cross B	2	0.3	Adjacent	Concrete	Trefoil	417
11	1	6	40	1.5	800	Aluminum	Cross B	1.5	0.5	Adjacent	Buried	Flat	518
12	1	8	20	3	1600	Copper	Multi	2	0.3	Adjacent	Buried	Flat	372
13	1	6	40	3	1600	Copper	Cross B	1.5	0.3	Adjacent	Buried	Trefoil	574
14	1	8	20	3	800	Copper	Cross B	1.5	0.5	Superimposed	Buried	Flat	546
15	1	6	20	1.5	1600	Copper	Cross B	2	0.5	Adjacent	Concrete	Flat	1070
16	1	6	20	1.5	800	Copper	Multi	1.5	0.3	Superimposed	Concrete	Flat	415
17	2	8	20	1.5	1600	Aluminum	Cross B	1.5	0.3	Superimposed	Buried	Trefoil	795
18	2	8	20	1.5	800	Aluminum	Multi	2	0.5	Adjacent	Buried	Trefoil	382
19	2	8	40	3	800	Aluminum	Multi	1.5	0.3	Adjacent	Concrete	Flat	273
20	2	6	20	3	1600	Aluminum	Multi	1.5	0.5	Superimposed	Concrete	Trefoil	485
21	2	8	40	3	1600	Aluminum	Cross B	2	0.5	Superimposed	Concrete	Flat	540

(continued)

Table 1 (continued)

Run	Block	Duct type	Ambient temp	Thermal resistivity	Conduct gauge	Material conductor	Screen	Ground level	Duct distance	Location	Pipeline support	Disp	Ampacity
22	2	6	40	1.5	1600	Aluminum	Multi	2	0.3	Superimposed	Buried	Flat	367
23	2	8	40	1.5	1600	Copper	Multi	1.5	0.5	Adjacent	Concrete	Trefoil	607
24	2	6	40	3	800	Copper	Multi	2	0.5	Superimposed	Buried	Trefoil	251
25	2	8	40	1.5	800	Copper	Cross B	2	0.3	Superimposed	Concrete	Trefoil	613
26	2	6	20	3	800	Aluminum	Cross B	2	0.3	Adjacent	Concrete	Trefoil	417
27	2	6	40	1.5	800	Aluminum	Cross B	1.5	0.5	Adjacent	Buried	Flat	518
28	2	8	20	3	1600	Copper	Multi	2	0.3	Adjacent	Buried	Flat	372
29	2	6	40	3	1600	Copper	Cross B	1.5	0.3	Adjacent	Buried	Trefoil	574
30	2	8	20	3	800	Copper	Cross B	1.5	0.5	Superimposed	Buried	Flat	546
31	2	6	20	1.5	1600	Copper	Cross B	2	0.5	Adjacent	Concrete	Flat	1070
32	2	6	20	1.5	800	Copper	Multi	1.5	0.3	Superimposed	Concrete	Flat	415

Table 2 List of factors and their levels for the experiment

Factors	Labels	Low level	High level
Duct type	A	6	8
Ambient temp	B	20	40
Thermal resistivity	C	1.5	3
Conduct gauge	D	800	1600
Material conductor	E	Aluminum	Copper
Grounding	F	Cross B	Multi
Ground level	G	1.5	2
Duct distance	H	0.3	0.5
Location	I	Superimposed	Adjacent
Pipeline support	J	Buried	Concrete
Provision	K	Trefoil	Flat

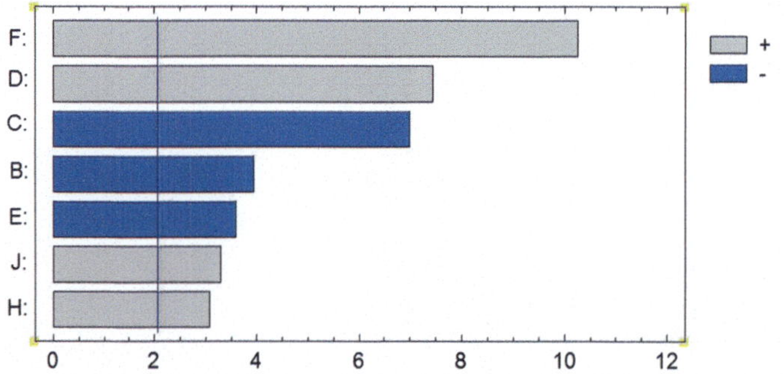

Fig. 10 Pareto plot of effects of the experiment

Table 3 Analysis of variance (ANOVA)

Source	*Sum of Squares*	*DF*	*Mean Square*	*F-Ratio*	*P-Value*
B: Ambient Temp	6.827E4	1	6.827E4	15.58	0.0006
C: Thermal Resistivity	2.142E5	1	2.142E5	48.89	0.0000
D: Conduct Gauge	2.433E5	1	2.433E5	55.53	0.0000
E: Material Conductor	5.628E4	1	5.628E4	12.85	0.0015
F: Grounding	4.613E5	1	4.613E5	105.30	0.0000
H: Duct Distance	4.104E4	1	4.104E4	9.37	0.0054
J: Pipeline Suport	4.728E4	1	4.728E4	10.79	0.0031
Total Error	1.051E5	24	4381.		
Total (corr.)	1.237E6	31			

R-square = 91.5 %
R-square (adjusted for d.f.) = 89.02 %
Estándar Error = 66.19
Mean absolute error = 44.11
Durbin-Watson statistic = 2.11 (P=0.5251)

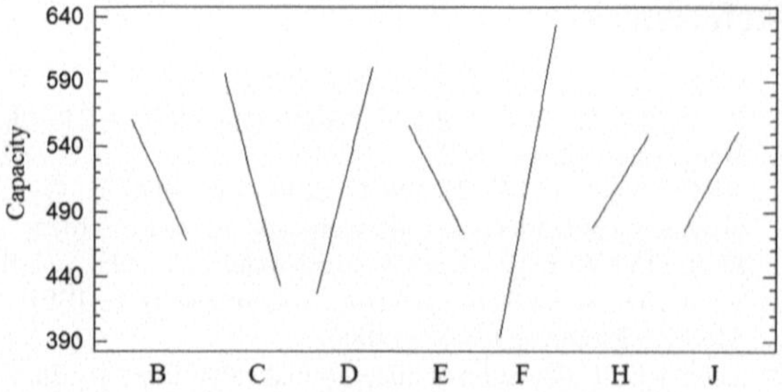

Fig. 11 Main effects plot for the experiment

Table 4 Optimal factor levels for AMPACITY

Factor	Low	High	Optimum
B	20	40	20
C	1.5	3	1.5
D	800	1600	1600
E	Aluminum	Copper	Aluminum
F	Cross B	Multi	Multi
H	0.3	0.5	0.5
J	Buried	Concrete	Concrete

is no indication of serial autocorrelation in the residuals with a significance level of 5%. On the other hand, the normality, independence and nonconstant variance assumption is fulfilled.

Table 4 show the factors levels combination, in which the conduction capacity, is maximum in the indicated region.

5 Conclusion

In this chapter, an experimental procedure to optimize the performance of capacity conduction has been proposed. A linear model was adjusted as a suitable approximation for the relationship between capacity conduction and the set of eleven independent variables. The R-Square indicates that the model adjusted explains 91.5% of the variability. The ANOVA shows that the selected lineal models adequately represented the data obtained, seven effects presented a P-value less than 0.05, indicating that are significantly different from zero with a confidence level of 95.0%. Grounding (F) has the strongest effect. The main effects plot for the experiment has been presented. Finally, the factors levels arrangement that optimize the performance of conduction capacity were determined.

References

1. Royal Society of Chemistry: IUPAC Compendium of Chemical Terminology. IUPAC, The Gold Book, Cambridge (1997)
2. ANCE: Nmx - J - 142, 2 - ANCE - 2011 - Conductores - Cables de energía con pantalla metálica, aislados en polietileno de cadena cruzada o a base de etileno - propileno para tensiones de 69 kV hasta 115 kV - Especificaciones y métodos de prueba, p. 2011. ANCE, México (2011)
3. León, F.D.: Major factors affecting cable ampacity. In: IEEE Power Engineering Society General Meeting, Montreal, Canada (2006)
4. León, F.D.: Calculation of underground cable ampacity. In: CYME International T&D, Quebec, Canada (2005)
5. Cruz-Salgado, J., Alonso-Romero, S., Zitzumbo-Guzmán, R., Domínguez-Domínguez, J.: Optimization of the Tensile and Flexural Strength of a Wood-PET Composite Optimización de la resistencia a la tensión y flexión de un compuesto de madera PET. Ingeniería, Investigación y Tecnología, pp. 105–112 (2015)
6. Jiju, A.: Design of Experiments for Engineers and Scientists, p. 73. Elsevier (2014)
7. Montgomery, D.C.: Design and Analysis of Experiments, 8th ed. Wiley (2012)

Analysis of Transport Logistics Operations at a Link in a Reverse Supply Chain that Values Used Cooking Oil

Benito Sánchez-Lara, Efraín Medina-Toribio, Reyna Gayosso-García, and Mayra Elizondo-Cortés

Abstract Used cooking oil is a so-called special handling waste whose improper handling causes severe environmental and social impacts. The Secretaría del Medio Ambiente of Mexico City reported that more than two million liters were collected during 2017. Used cooking oil requires differentiated treatment for final disposal or, better yet, to be valued through reverse supply chains. Its recovery implies its recycling and re-use. After a physicochemical process, it is incorporated as raw material for the production of biodiesel, soap, and animal food, among other products. In reverse chains, transport logistics operations are carried out necessary for their recovery. The empirical work presented was carried out in a company linking a used cooking oil supply chain. Elements of some logistics operations were analyzed by a company that values used cooking oil. The result of the analyses translates into the formulation of recommendations for the company.

Keywords Reverse supply chain · Used cooking oil · Logistics transport operations

---

B. Sánchez-Lara (✉) · E. Medina-Toribio · R. Gayosso-García · M. Elizondo-Cortés
Departamento de Ingeniería de Sistemas, Facultad de Ingeniería, Universidad Nacional Autónoma de México, Mexico City, Mexico
e-mail: blara@unam.mx

E. Medina-Toribio
e-mail: efrainmedina10@hotmail.com

R. Gayosso-García
e-mail: rgayossog@yahoo.com.mx

M. Elizondo-Cortés
e-mail: mayra.elizondo@comunidad.unam.mx

A. Ochoa-Zezzatti et al. (eds.), *Technological and Industrial Applications Associated With Industry 4.0*, Studies in Systems, Decision and Control 347,
https://doi.org/10.1007/978-3-030-68663-5_16

1 Introduction

Used cooking oil is an organic compound that is generated in room houses, industrial, commercial, and service establishments associated with cooking, frying, and food preparation activities.

NOM-161-SEMARNAT-2011 refers to used cooking oil as a special handling residue [1]. Special handling residues are a type of urban solid waste; they are produced by large generators, i.e., individuals or legal entities (enterprises) that generate ten or more tons per year. Besides, it is waste that does not meet the characteristics to be considered hazardous. The above standard sets out the criteria for classifying them and determining which ones should be subject to a management plan.

Used cooking oil is a polluting agent of urban waters, when draining causes affectations to the sewage system and makes wastewater treatment difficult; this increases the costs of maintenance and operation of the sewer system. When used cooking oil is spilled into soils and water bodies, it causes harmful effects due to its low biodegradability. Also, it alters the physicochemical and biological conditions of soil and water, which leads to loss of biodiversity.

Used cooking oil has been collected in Mexico City since 2010. Figure 1 shows the liters collected per year.

From Fig. 1: it is notorious that the collection of used cooking oil has grown exponentially; this could make think about the success of management plans; however, with no data about how much is generated, it is not possible to calculate what the magnitude of the achievement is. Once the used cooking oil is generated, it is necessary to reduce its impact, even considering that the highest volume generated is valorized at the end. By reference, the Secretaría del Medio Ambiente y Recursos Naturales (SEMARNAT) reported that in 2012 84 million tons of 14 different types of special handling residues were generated [2]. In this context, emerge the following

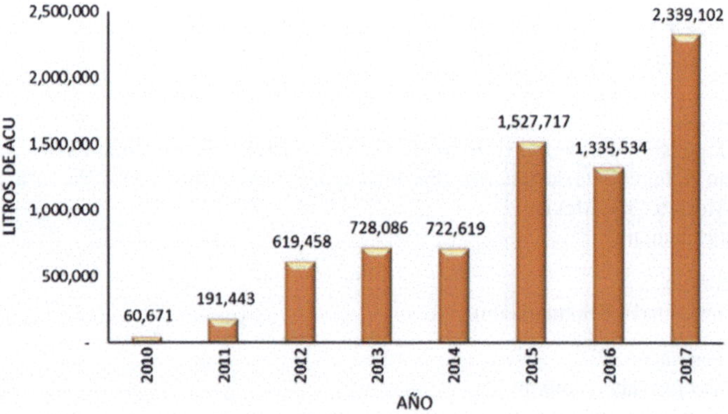

Fig. 1 Used cooking oil collected between 2010 and 2017 in Mexico City. *Source* SEDEMA (2017) Solid Waste Inventory, 2016

questions: Which enterprises valorizes used cooking oil? How do transport logistics operations carry out in the enterprises that valorizes used cooking oil? How is the collected used cooking oil transport? In which conditions is the collected used cooking oil transport? Which risk implies its transportation? This research addresses these questions.

There are many agents and activities involved in the operation of reverse supply chains; also, the relationships between these elements often generate some complexity. However, one way to gain an understanding regarding the handling of used cooking oil that is not finally available is the empirical work on reverse supply chains that valorizes it. This document presents an analysis of elements of some logistics transport operations; the study focuses on a used cooking oil company that valorize it. This company is one of the links of a reverse supply chain. The analysis touches on elements of three logistics transport operations: the definition and process of the appropriate mode of transport for the transport of used cooking oil, the use of vehicle capacity, and transport risk. The results of the analysis of the first two logistics operations allow an extensive risk analysis. The work is sectioned, as described below. The introduction set out the definitions of special handling residue and used cooking oil, provides collection data, and sets the objective of the work. The second section, of methodology, describes the conceptual and methodological framework for carrying out empirical work and presents the case study company.

In the third section, the analysis and their results are displayed. Finally, in the fourth section, some conclusions are offered.

2 Reverse Supply Chains and Logistics Transport Operations

Agents and processes for retrieving materials from a supply chain, when the product is defective or when it becomes waste, are links of a reverse supply chain. From a maintainable approach, reverse supply chains are elements of efficiency, drive economic value creation, and do not emphasize social and environmental impacts. The reverse supply chains are made up of processes aimed at reuse and final disposition [3]. It is determined that in addition to reuse, reverse supply chains include recycling processes [4]. For their part, authors in [5] note that the procedures involved in the reverse supply chains focus on recovering the surplus-value of the waste for a given market. That is, recovering the materials from the waste for which there is a market.

If a product is recycled, it is incorporated into a supply chain as raw material, thus extending the life cycle of the product and reducing the volume of waste, this constitutes a closed circuit [6].

Figure 2 illustrates a closed-loop supply chain that includes reverse supply chains and reverse logistics. For López-Parada in [6], some of the activities included in the

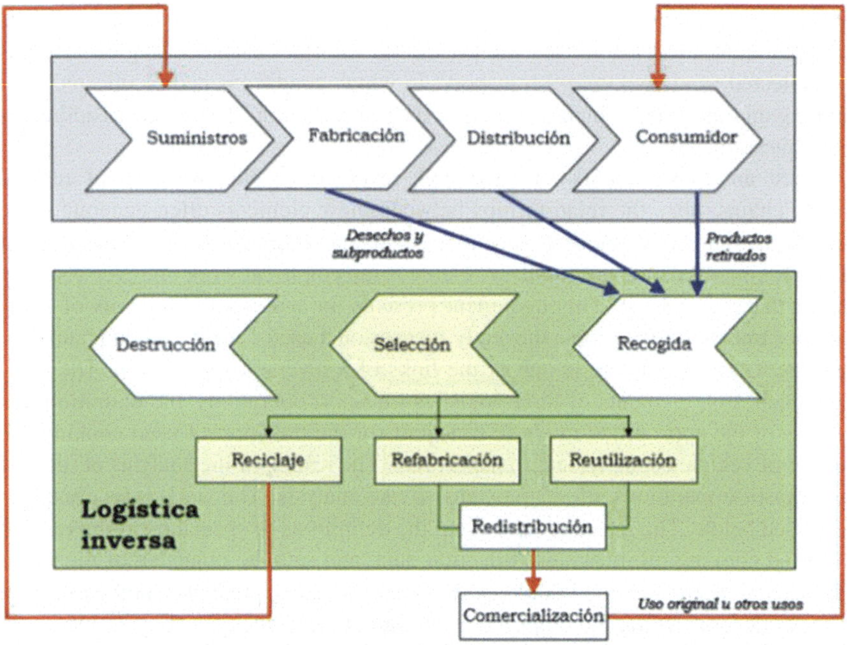

Fig. 2 The closed circuit of the supply chain. *Source* López-Parada (2017)

reverse supply chains are acquisition, transportation, inspection, handling, distribution, storage, and marketing. It is notorious that the operations indicated are the same as those carried out in direct supply chains, but are carried out on waste and not on products and performed by other economic operators, these agents are links of the reverse supply chain.

With a process approach, reverse supply chains involve a process for collecting waste and other to select between those to be finally available and those that can be reused, recycled and, or remanufactured. Also, through ad hoc marketing and distribution channels, the materials are incorporated into the same or another supply chain.

Reverse supply chains process the waste recovery; as mentioned, reverse supply chains are focused on assigning economic value to waste to be placed in a buying and selling market or a supply chain. The Norma Ambiental para el Distrito Federal (NADF-024-AMBT-2013) defines valorization as actions that allow recovering the remaining value or calorific value of the materials contained in the waste [7].

The urban solid waste daily generated in Mexico City is valued at approximately 38.73% [8]. The potential for valorization is an area of opportunity for new economic operators. In particular, the recovery of used cooking oil involves recycling and reuse; some of the products derived from such recovery are biodiesel, soap, animal feed, lubricants, paints, varnishes, and cosmetic products. A comparative analysis between countries on the valorization of used cooking oil is being prepared. Unfortunately,

Mexico is omitted, and although the reasons are unknown, it is likely because of the unavailability of the information [9].

Based on the preceding paragraphs, it is notorious that describing the different processes involved in reverse supply chains would require extensive study. For this reason, this empirical work is restricted to logistics, in particular logistics transport operations. Logistics is a discipline that studies how goods, people, or information use time and distance resources efficiently [10]. Logistics includes the different operations necessary to maintain a productive activity, including the associated with special handling residues to transport, so-called transport logistics operations. The transport logistics operations addressed in this document are the definition and operation of the mode of transport for used cooking oil transportation, the use of vehicle capacity and transport risk, are three out of eight transport logistics operations.

Generally speaking, the reverse supply chain link company was considered as a case study that provides data to the possible development of constructs and theories on contemporary phenomena, in this case, reverse supply chain management that includes waste recovery. The research is exploratory; that is, there is no rigorous and well-controlled experimental design.

The analysis was carried out on a company that obtains used cooking oil from its suppliers located in various states of the Mexican Republic. Then, they transport it to its plant in the metropolitan area of the Valley of Mexico. The characterization of the company is detailed in [11]. The study is focused on the link in the reverse supply chain. In the absence of some point-data from the company, reference data or estimates were considered.

3 Analysis of Transport Logistics Operations of Used Cooking Oil

3.1 The Mode of Transport Definition and Operation

For the movement of load, supplies, goods, or waste, there are two alternatives: multimodal and intermodal transport. The first involves the use of different modes of transportation. In contrast, the second consists of the use of a single cargo unit, especially containers that can be moved in various modes of transport. The selection depends on aspects such as:

- Harnessing transport capacity,
- Price-to-volume ratio (Economy of scale),
- Shipment size,
- Profile of the carrier or transportation enterprise, and
- Relationship between mode of transport and value of what is transported.

For example, high-value products require faster transport; low-value products resort to multimodal transport. The last case corresponds to the transport of used cooking oil.

The case study company carries out the transport of the used cooking oil that is collected with two types of vehicles. Considering that the used cooking oil is a special handling residue, three types of vehicles were analyzed for the transfer: tanker truck, pipe, and truck. Tanker trucks include a compartment equipped with break-waters that prevent the effects of sudden movements of the load and are equipped with suction and eviction mechanisms of the contents. Besides, they have equipment for the transfer of the load, the calibration of the tank, the automatic stop of the racking pump, among others. See Fig. 3 [12].

There are different axle configurations and lengths of tanker trucks. Table 1 shows some settings considering vehicle weight, wheelbase, and recommended load capacity [13].

On the other hand, the pipe is a variant of the tanker truck. In general, pipes are associated with the transport of water for irrigation, gasoline, kerosene, liquid chemicals, and others, its use is regulated in many industries. It has a cylindrical or ellipsoidal section, horizontal axis, with shells or bottoms domed at its ends, equipped with a valve, conductions, and loading and unloading devices.

Fig. 3 Tanker. *Source* Rigual [13]

Table 1 Code truck configurations

Vehicle weight (t)	Wheelbase (mm)	Recommended load capacity (t)
6	2500	Up to 2500
7.5	2800–3500	Up to 3000
12	3000–3800	Up to 6000
16	3600–4100	Up to 8000
18	3900–4200	Up to 10,000
26	3900–4500	Up to 13,000

Source Rigual [13]

Fig. 4 Trailer truck tautliner or curtain

Regarding the truck were considered two alternatives: tautliner trailer truck or curtain truck and open platform truck. The tautliner or curtain trailer is similar to a wagon, except that it facilitates access to the load with its sliding curtains. Access to charge is facilitated with the heavy-duty back door and curtains on both sides; with aluminum poles in the corners of the rear and rails on the ceiling are placed the laminated tarps. Figure 4 corresponds to a tautliner trailer truck.

The open platform truck can be used if the transported material is collected in individual containers with equal or different capacity; this is feasible for the used cooking oil. The platform can be fully or partially open. Perhaps it is the most usual truck, versatile but commonly used for transporting bulk materials and containers.

In the case of used cooking oil transport, multimodal transport and intermodal transport is considered feasible. For now, the case study company can maintain multimodal transport by considering the size of the shipment that depends on the offer and how it consolidates the load. Intermodal transport should be considered whether there was the possibility of implementing a cargo consolidation strategy, i.e., if the number of suppliers of used cooking oil and other materials that can be transported in the same containers was expanded [14].

3.2 Management of the Vehicle Capacity

This logistic transport operation refers to how the load is handled in modes of transport, i.e., in the vehicle. In general, you can select from two alternatives: an LTL (Less Than Truckload) scheme, that is, a consolidated load, and a full truckload (FTL) scheme, i.e., truck or full load.

The LTL scheme can be chosen when loads from different customers and with different volumes are combined into the same truck that goes to the same destination. This scheme is thought of for shipments that do not occupy a vehicle in its entirety. Partial load carriers operate by sharing the capacity of the trailer among multiple customers or suppliers. A typical partial-cargo carrier collects shipments from various customers and takes them to a central terminal. The load is segregated depending on the number of destinations. The charge can be loaded and unloaded before its final delivery according to the zoning that is done. Consolidated loads are called over-sized loads.

The FTL scheme refers to a complete load, i.e., full load is required for shipping to be economically viable. FTL implies that the cargo occupies a vehicle in full; either the volume of the charge takes up the available space, the weight is the maximum allowed, or because the load owner prefers to send the load even if the truck is not fully occupied.

The analysis to define the form that management should take advantage of the capacity of the vehicle was based on the characteristics of the used cooking oil. The features of the used cooking oil are then widely described.

As mentioned in the first section of this article, the used cooking oil is a residue that causes serious contamination problems. Spilling used cooking oil into the sink or toilet is dangerous; it is difficult to degrade. Because it is less dense than water oils, sunflower, soybeans, olive, and corn, cooking oil forms a film that makes it difficult to pass oxygen which puts at risk the living beings of the ecosystem where it is poured. Used cooking oil like other waste is generated in small quantities in homes, and it is usually poured into drains; this makes it very difficult to manage and control.

One liter of used cooking oil contaminates up to 40,000 L of water and has 5,000 times more pollutant load than wastewater. The cost of depuration of used cooking oil can be 700 times higher than sewage treatment. One liter of used cooking oil has the following average composition: 85% oil, 10% water with traces of oil and organic matter, 5% of sludge whose structure is 60% oil, 30% organic matter, and 10% water. This composition results in a relative density of 0.91 [12].

The case study company, in addition to used cooking oil, collects other fatty residues to complete the load that allows it to operate, including fats of animal and plant origin, and waste from different industries. In the end, it transports a mixture of fatty residues that may result in residue liquids special handling.

Despite collecting different fatty residues, in general, full-load trips are scheduled. The company gathers from its suppliers the used cooking oil, and the other fatty waste only does ensure the full load, so each trip requires that the charge occupies a vehicle in full. The management is of the FTL type, and the amount takes up the available space of the vehicle. For economic reasons, it is not considered an LTL scheme. The movement of the load through the network that connects the origins with the destinations is push-type, that is, it depends on the pressure of the preceding links: the suppliers of used cooking oil and fatty residues, who want to get rid of the waste and who partially finance the acquisition activities.

No need to say, the transport risk assumed with the movement of used cooking oil determines full load trips compared to trips to different suppliers and with different volumes. Transport analysis is the topic of the next section.

3.3 Transport Risk Analysis

Moving a load presents three types of risks: those that would cause the shipment to suffer delays in the path between the origin point and the destination, which would cause the load not to reach its destination, in general, factors that alter the origin-destination nodes, and, or route, and those associated with the transported material.

This article considered only the transport risk, i.e., the risk associated with the journeys between the supplier companies and the case study company. For reasons of confidentiality, company names were omitted, and only integrated data is given.

Risk is a measure of the likelihood of adverse or severe events. For the transport of hazardous materials, the risk is the probability of accident occurrence [15]. Accidents can occur in an area where the population, infrastructure, road network or environment is exposed to the consequences of these. The risks are different over a road network and are associated with so called risk factors, including congestion, weather conditions, natural disasters, driver behavior.

On risk management, three questions need to be asked: how to prevent or reduce the consequences if the risk becomes a reality, how to recover, and how individuals, groups, and communities should organize and inform themselves to reduce risk and improve confidence to deal with it [16]. Different individual and organizational factors, prioritized in different ways influence risk management. Individually, factors range from risk perception to political affiliation; also organizational aspects, ranging from environmental protection to technical and technological issues.

The transport risk analysis for the case study company was developed based on the method of risk and vulnerability analysis. However, because it did not have company data on some risk elements, it was not possible to carry out the valuation, the qualification, and acceptance stage. Nevertheless, it was possible to illustrate these stages.

The method of risk and vulnerability analysis involves thirteen steps [17]. The steps include the definition of the system, the scope of the review and its applicability; identifying vulnerabilities and threats; the formulation of scenarios and valuation, qualification, and acceptance of risks. Once the analysis is done, management involves, in different timeframes, driving risk scenarios to acceptable levels.

With the support of Google Earth©, taking as input data the origin-destination collection pairs of the company, six routes were spatially located, and the territorial scope of the collection network was identified. With data from the Instituto Nacional de Estadística y Geografía (INEGI), in particular, data on land traffic accidents in urban and suburban areas, and using QGIS 2.1826, the accident ranges of cargo transport vehicles per municipality were spatially located, only on Route 6 of collection [18]. On the other hand, with data from the geostatistical framework of municipalities of Mexico of Instituto Nacional de Estadística y Geografía [19] and using Autocad Civil 3D and Global Mapper, the number of accidents at sea level and road sections was associated, this is shown graphically.

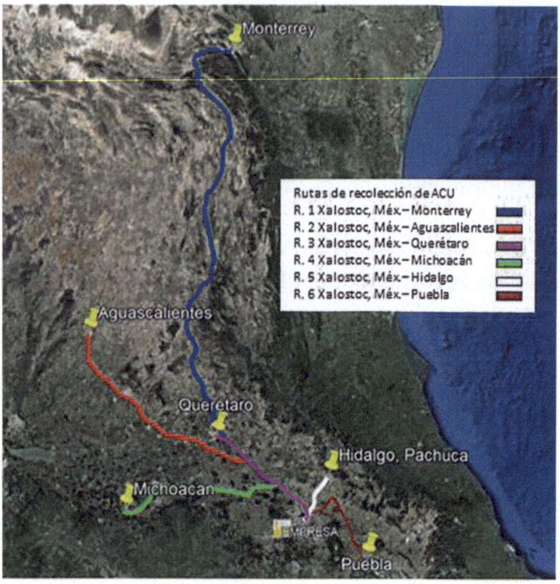

Map 1 Used cooking oil harvesting routes. *Source* Medina-Toribio [20]

Using as input the source pairs-collection destination, six routes were spatially located; this is shown in Map 1. The definition of routes considers, as the company's collection policy, transiting only on toll roads. Vehicle-occupied capacity was also taken into account.

Having considered the risk of transport, the elements on which the analysis was focused were: the load, i.e., the used cooking oil, the transport vehicle, the driver, and the tracks. These elements were called internal; factors outside the contextual conditions were called external factors, i.e., prevailing and, or emerging conditions, when transiting the different routes. Based on internal and external elements, threats, and vulnerabilities associated with transport risks were identified. Table 2 was used to associate quantifiable weaknesses, threats, and likely harms with transport risks and analysis elements.

Table 2 Framework for transport risk analysis

	Elements	Vulnerabilities	Threats	Measurable probable damage	Type of risk
Internal					Delayed on the journey
External					
Internal					Don't let the load reach its destination
External					
Internal					Associated with used cooking oil transported
External					

From the analysis framework, there are identified and described transport risks. In general, path risk is associated with internal factors. Vulnerabilities are associated with a lack of or inadequate vehicle maintenance: weight, volume and distribution load, lack of expertise, experience, and knowledge of the driver, as well as the status of the roads for transit. Threats result in accidents on the road or by inter-action with other vehicles, vehicle break downs, penalties for non-compliance with delivery times, loss of suppliers, and customers. If crime, social nonconformity, and climatic are considered as condition as external factors, the associated vulnerabilities are congestion, manifestations, and detentions, and the related threats are theft, extortion, and accidents.

On the other hand, the risk associated with transport considers as internal factors the used cooking oil and the volume that is transported, that is, the load. The vulnerability is given by density, weight, viscosity, and volume loaded, also by load distribution. Threats result in spills and their consequences. An external factor always present is the climatic condition that, for example, associated with inadequate packaging, causes difficulties in the subsequent treatment of used cooking oil and the consequent costs.

The scenario associated with the most identifiable transportation risk is a spill; however, there is no data on what the cost of remediation is. A benchmark was used to estimate the cost of environmental remediation for a used cooking oil spill: the 2010 British Petroleum Deepwater Horizon spill in the Gulf of Mexico. It resulted in an expense of $17.2 billion, expenditure per liter was estimated at $34; it must be noted that the damage to economic sectors such as fisheries and tourism was not quantified [21].

Finally, with the burden not reaching, the relevant elements are internal ones, and the vulnerabilities and threats are similar to the risk of delay. Still, these are maximized, i.e., they are the extreme cases; this remains true for external factors.

As mentioned in previous paragraphs, the extreme scenario is a spill of used cooking oil. The spill would make the three transport risks a reality and a myriad of consequences, from loss of load to failure of company reputation, this scenario is unacceptable and not tolerable. Other scenarios can be classified as acceptable, un-acceptable, tolerable, and non-tolerable. Scenario categories are reverted to the method of risk and vulnerability analysis.

As mentioned, due to the unavailability of accident data at the time of the case study company, it was not possible to develop in its entirety the method of risk and vulnerability analysis, remained outstanding: the risk profiles and the distribution pattern of these. Data on cargo vehicle accidents in Mexico were taken as a reference in 2018 to illustrate these steps. Table 3 shows the number of accidents that are disaggregated in fatal, non-fatal, and only caused by material damage. Based on this disaggregation, the three method scenarios of the method of risk and vulnerability analysis were defined: unacceptable, tolerable, and acceptable.

From the data in Table 3, the pattern of distribution of accident scenarios could be illustrated, that is, how accidents are distributed in scenarios. Table 4 contrasts accident percentage, grouped in scenarios, and the percentage that is considered as a reference for the assessment of such scenarios. It is identified that the acceptable

Table 3 Cargo vehicle accidents in Mexico in 2018

Type of accident	Number of accidents per cargo vehicle	Unacceptable scenario (fatal accidents)	Tolerable scenario (non-fatal accidents)	Acceptable Scenario (material Damage Only)
Collision with motor vehicle	19,410	141	1,261	18,008
Collision with pedestrian (hit)	366	60	306	0
Collision with animal	37	0	4	33
Fixed object collision	4,288	11	139	4,138
Rollover	802	32	143	627
Passenger fall	22	1	21	0
Exit the road	558	20	43	495
Fire	28	0	0	28
Collision with railway	66	2	10	54
Motorcycle collision	871	45	383	443
Collision with cyclist	203	19	113	71
Other	512	3	13	496
Total	**27,163**	**334**	**2,436**	**24,393**

Source INEGI [18]

Table 4 The pattern of distribution of accident scenarios

Scenarios	Number of accidents	Calculated distribution (%)	Reference distribution (%)
Acceptable	24,393	89.8	60 (minimum)
Tolerable	2,436	8.9	30 (maximum)
Unacceptable	334	1.2	10 (maximum)
Total	**27,163**	**100**	**100%**

accident scenario prevails, the tolerable and unacceptable scenarios are well below the reference data. That is, the highest number of accidents cause material damage. The scenarios are compatible to the method of risk and vulnerability analysis; however, it is possible to define more categories, it depends on the number of accident data, how they are typified and the assessment that organizations give to accidents. Reference values are also attached to the method of risk and vulnerability analysis.

To illustrate the accident ability on the routes that the used cooking oil collection follows, Map 2 shows the range of load vehicle accidents in the municipalities that includes Route 6. The municipalities with the most accidents are the origin and destination of the route (Fig. 5).

A better illustration of the risk of transport by route would involve having available data on topography, slopes, climatic conditions, conditions of the rolling layer of the pavement, unsafe areas, among others. However, from obtaining the contours and topographic profile over Route 6, Fig. 5 shows how the number of accidents is associated with height above sea level and road sections.

Figure 5 identifies that the number of accidents is not associated with higher-level road sections over the sea; it seems to be more associated with interaction with other

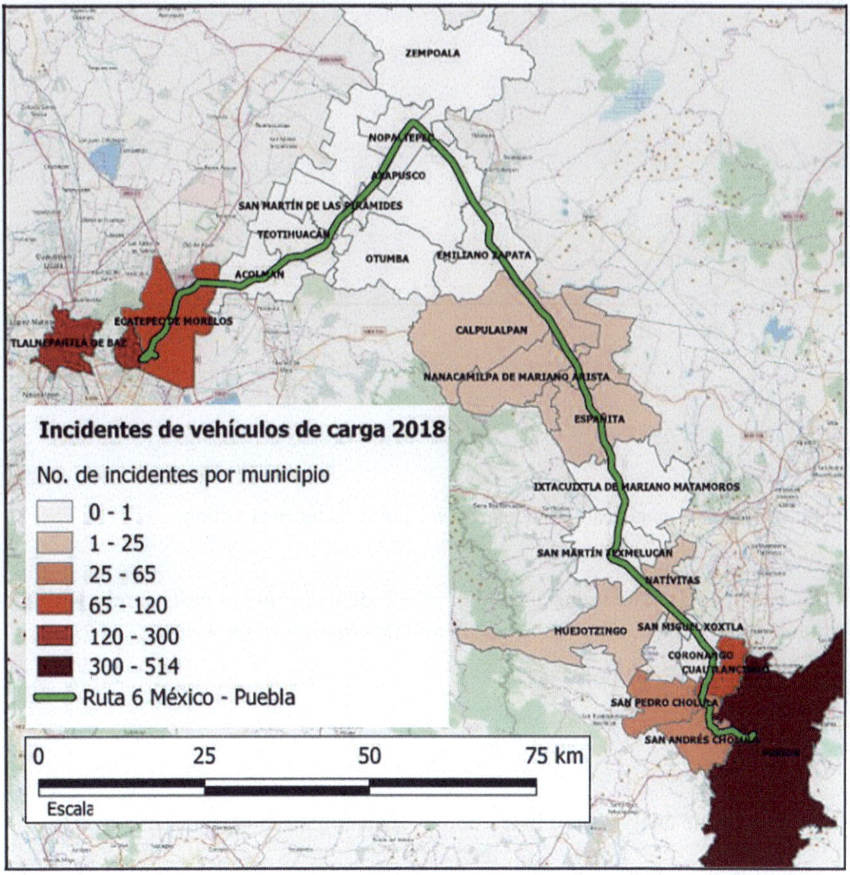

Map 2 Range of cargo vehicle accidents per municipality on Route 6 collection of used cooking oil

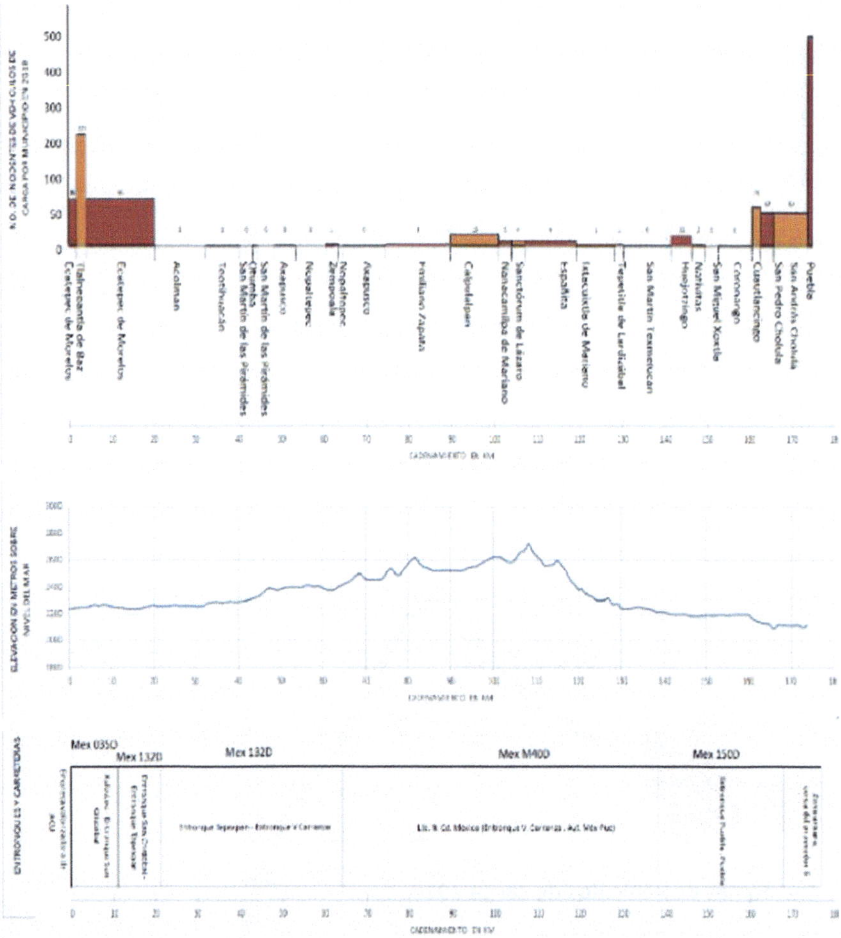

Fig. 5 Accidents by municipality and topographic profile of the road sections

modes of transport. The highest number of accidents occurs in parts surely with the greater vehicular flow, that is, the origin and destination of the route.

4 Results

The results of the analyses are listed below, focusing on those on transport risk.

1. Multimodal transport is suitable for the transport of used cooking oil given the way the load is consolidated: a supplier a trip. The valuator company makes a trip only if the supplier consolidates load, which implies storage costs for

the supplier. Intermodal transport is economically feasible if the number of suppliers of used cooking oil and other materials that can be transported in the same containers, for example, plastic or metal barrels, is expanded.

2. Programming full cargo trips are used to calculate the capacity of the vehicle, that is, an FTL scheme, the tanker is the convenient vehicle to mitigate the risk. The convenience of a LTL scheme involves the use of trailer tautliner or curtain with standard measuring containers that allow the handling and transfer of the vehicle–vehicle and vehicle-warehouse load. It also means a higher number of suppliers or more significant amounts of used cooking oil and other fatty residues collected. Under this scheme, the transport risk is significantly increased.
3. The risk of delay in the path is associated with the factors considered internal. They are relevant: the maintenance of the vehicle; the weight, volume, and distribution of the load; the expertise, experience, and knowledge of the driver; and the state of the tracks. The extreme scenario is that the charge does not reach its destination with the consequences that this implies.
4. The risk of the load not reaching its destination is also associated with internal factors. This risk is the one that gives rise to the extreme scenario of the delay in the path, i.e., that the load does not arrive.
5. As for the risk associated with transport, although internal factors remain relevant, the amount transported is most relevant, this is, the load itself. Load characteristics such as density, weight, viscosity, volume, and distribution take on importance. The extreme scenario is a spill of used cooking oil. The spill would bring the three transport risks to life and a myriad of consequences. A spill is an unacceptable and non-tolerable scenario.
6. From Table 3, under the information restrictions already mentioned, the acceptable scenario is where the higher number of accidents are located. As a percentage, this scenario is well above the referring data. As mentioned, the scenario categories of the method of risk and vulnerability analysis are resumed; however, the acceptability assessment can be changed concerning the preferences of each organization or company.
7. Two relevant conditions of the transport risk analysis are accidents or incidents that occur by vehicle-vehicle interaction and by the topographical conditions of the road that constitutes the route. Map 2 is identifiable that the spaces with the most accidents are the origin and destination of the path; that is, the places where the most vehicle-vehicle interaction is provided. From Fig. 3, it is identifiable that the accidents would most likely occur in areas of higher topographic profile above sea level.

The analyses were limited by the availability of data of the company case study. In particular, a review of transport risks and beyond transport, between a pair origin-destination, implies the availability of data associated with the recovery operations of used cooking oil, such as acquisition, classification, and selection. The required data to extend the scope to other logistics transport operations is topography, slopes, route length, safe parking spaces, emergency, and rapid response services, weather conditions, pavement conditions, unsafe areas, among others.

5 Conclusions

Logistics transport operations are elements of analysis of logistics that considers transport as an axis; its importance, however, remains in the behavior of the supply chain. The transport logistics operations has systemic behavior.

The analyses carried out are linked. That is why some decisions can be made considering the possible relationships between (1) the used cooking oil as a load, (2) the vehicle in which it is transported and how it takes advantage of its capacity, and (3) the risk that is assumed when carrying used cooking oil. Starting from a company condition, for example, the availability of vehicles limits how it takes advantage of its capacity, although this may not be true if the offer of used cooking oil is vast, and containers that optimize the space of the vehicle are used.

Because used cooking oil is a waste grouped in the so-called special handling residue, and its collection and transport are activities with inherent risks, transport risk analysis takes importance in reverse supply chains that valorize those residues. That situation also occurs with other types of waste, such as the hazardous ones and even with urban solids.

With the above, it is recommended that the company case study and other companies of the same sector or that transport material with tacit risks focus on risk responses. The mode of transport and how the capacity of vehicles are used are decisions linked to the risk assumed when transporting used cooking oil. Responses to each risk should be differentiated. The company already deals with the delay of its cargo; this is natural in the transport activity. It may have already had to deal with the fact that the shipment has not arrived, surely has solved it by modifying the transformation process, with its economic and operational consequences. But, the risk that results in a spill, although it is the risk with the least probability, must be a scenario for further study, a discharge for the company and society is unacceptable and intolerable. Knowing this to prevent it are the practical implications of this work.

In the previous context, other areas of work in logistics and supply chains come into play, including humanitarian logistics and chain fluidity studies.

References

1. Norma Oficial Mexicana, NOM-161-SEMARNAT-2011. Secretaria de Medio Ambiente y Recursos Naturales (SEMARNAT), Diario Oficial de la Fede-ración, 01 febrero de 2013
2. Secretaria de Medio Ambiente y Recursos Naturales (SEMARNAT): Diagnóstico básico para la gestión integral de los residuos. http://biblio-teca.semarnat.gob.mx/Documentos/Ciga/libros 2009/CD001408.pdf (2012)
3. Guide, V.D.R., Jayaraman, V., Linton, J.D.: Building contingency planning for closed-loop supply chains with product recovery. J. Operat. Manage. **21**(3), 259–279 (2003)
4. Kinobe, J.R., Gebresenbet, G., Vinnerås, B.: Reverse logistics related to waste management with emphasis on developing countries—a review paper. J. Environ. Sci. Eng. B formerly Part of J. Environ. Sci. Eng. **1**, 1104–1118 (2012)

5. Nakashima, K., Gupta, S.M.: Modeling supplier selection in reverse supply chains. In: Gupta, S.M. (ed.) Reverse Supply Chain: Issues and analysis, p. 401. Taylor & Francis Group, New York (2013)
6. López-Parada, J.: Tesis. Incorporación de logística inversa en la ca-dena de suministro y su influencia en la estructura organizativa de las empresas Universidad de Barcelona. Barcelona, España. http://www.tdx.cat/bitstream/handle/10803/1493/03.JLP_3de10.pdf;sequence=4 (2010)
7. Norma Ambiental NADF-024-AMBT-2015, Secretaría del Medio Am-biente, Gaceta Oficial de la Ciudad de México, 08 de julio de 2015
8. Secretaría de Medio Ambiente: Inventario de residuos sólidos, Ciu-dad de México, 2016 (2017)
9. Ribau-Texeira, M., Nogueira, R., Nunes, L.M.: Quantitative assessment of the valorisation of used cooking oil in 23 countries. Waste Manage. **78**, 611–620 (2018)
10. Robusté, A.: Logística del Transporte, Universitat Politécnica de Catalunya. Iniciativa Digital Politécnica (2005)
11. Reyes-Pardo, S.A., Medina-Toribio, E., Sánchez-Lara, B.: Carac-terización de cadenas de suministro inversas. El caso de una cadena para el manejo del aceite comestible usado. Memorias del Congreso Internacional de Logística y Cadena de Suministro (CiLOG2018) Cd. Mx., Octubre. ISBN 978-607-96403-1-6 (2018)
12. González-Canal, I., González-Ubierna, J.: Aceites usados de co-cina. Problemática ambiental, incidencia en redes de saneamiento y coste del trata-miento de depuradoras. Consorcio de Aguas de Bilbao Bizkaia. http://residusrecur-sos.cat/uploads/activitats/docs/20170427092548.pdf (2015)
13. Rigual: Ficha técnica de cisternas de aceites usados. https://www.ri-gual.es/producto/maquinaria-energia/cisternas-de-reparto/7 (2020)
14. Reyes-Pardo, S.A.: Aprovechamiento de la capacidad del transporte en cadenas de suministro inversas: el caso del aceite comestible usado, Avances de Tesis de Maestría, Posgrado en Ingeniería, UNAM (2020)
15. Garbolino, E., Tkiouat, M., Yankevich, N., Lachtar, D.: Transport of dangerous goods. In: Garbolino, E., Tkiouat, M., Yankevich, N., Lachtar, D. (eds.) Springer Netherlands, Dordrecht. http://doi.org/10.1007/978-94-007-2684-0 (2012)
16. Greenberg, M.R.: Explaining Risk Analysis. Routledge, London. https://doi-org.pbidi.unam.mx:2443/10.4324/9781315647579 (2017)
17. Bedoya-Arias, D., Villalba-Salazar, N.: Gerencia de riesgos en una empresa de transporte de carga masiva. Revista Universidad EAFIT **40**, 18–27 (2004)
18. INEGI: Estadísticas de accidentes de tránsito terrestre en zonas ur-banas y suburbanas. https://www.inegi.org.mx/programas/accidentes/de-fault.html#Datos_abiertos (2018)
19. INEGI: Marco geoestadístico de municipios de México. https://www.inegi.org.mx/temas/mg/ (2019)
20. Medina-Toribio, E.: Operaciones logísticas de transporte de una ca-dena de suministro inversa de un residuo de manejo especial: el caso del aceite co-mestible usado, Tesis de Maestría, Posgrado en Ingeniería, UNAM (2019)
21. El Economista: ¿Cuánto cuesta el derrame de BP a la naturaleza mexicana? https://www.eleconomista.com.mx/internacionales/Cuanto-cuesta-el-derrame-de-BP-a-la-naturaleza-mexicana-20170424-0102.html (24 de abril de 2017) (2017)

The Transformation of Supply Chains in the Circular Economy from International Experiences to the Mexican Cases

Mariana Hernández-González, Benito Sánchez-Lara, Mayra Elizondo-Cortés, and Luisa Fernanda Diego-Villegas

Abstract In recent years, with the intention of reducing the demand for virgin raw materials and the generation of waste, it has been promoting the transition from a linear production model which is constituted by extraction, production and waste operations, to a circular one in which the resources are maintained as much as possible in the production system. This transition has been called Circular Economy (CE), is involving important and diverse challenges for different economic agents, in particular, supply chains have had to rethink and redesign their processes and practices. The present work, based on a systematic review of the specialized literature, compiles a set of experiences in different countries regarding the transition to a circular model, in particular collects practices that are carried out in different links of the supply chains. The analysis of the circular practices was carried out taking as criteria its level of circularity and compliance with the objectives of the CE. The findings from the analysis of circular practices are challenges, circumstances and the scope of the transition to circular models, opportunities and challenges are also identified in the supply chains. Associated with the analysis, based on data from the National Statistical Directory of Economic Units, regions with potential were identified for the shaping of circular supply chains. The potential is valued by the number and industrial vocation of the economic units in the different sectors.

Keywords Circular economy · Circularity · Circular supply chains · Circular practices

---

M. Hernández-González · B. Sánchez-Lara (✉) · M. Elizondo-Cortés · L. F. Diego-Villegas
National Autonomous University of Mexico, Mexico City, Mexico
e-mail: blara@unam.mx

M. Hernández-González
e-mail: marianahzgz@comunidad.unam.mx

M. Elizondo-Cortés
e-mail: mayra.elizondo@comunidad.unam.mx

L. F. Diego-Villegas
e-mail: luudvillegas@comunidad.unam.mx

A. Ochoa-Zezzatti et al. (eds.), *Technological and Industrial Applications Associated With Industry 4.0*, Studies in Systems, Decision and Control 347,
https://doi.org/10.1007/978-3-030-68663-5_17

1 Introduction

Circular economy (CE) is an emerging topic which has attracted the attention of researchers, governments, communities and private initiative, because it is presented as an alternative to reduce the environmental impact of productive activity through a change in the mode of operation of a company, in which objectives such as productivity, efficiency and corporate responsibility are preserved [1]. CE has gained followers and support among different knowledge fields, being an alternative model to change production and consumption, from the current mode of operation [2].

Because a completely accepted definition of CE is not available, this document uses one based on reported works in [3–6], and therefore, CE will be understood as a restorative economic R[1] system, untying global economic development of finite resources, with the objective of achieving sustainable development operating in micro, meso and macro analysis levels. Where micro level corresponds to products, enterprises and consumers, meso level to industrial parks and other groups of organizations and macro level to a city, region or country [6].

The R frameworks application on the different levels, requires compromise and environmental awareness from the actors involved, being the government, private initiative and society the most remarkable ones. In addition, particularly in micro and meso levels, organizational and financial boundaries must be overcome [7]. In [8] it is indicated, that introduction to circular model into the company requires monitoring transition progress from linear economy to circular economy, which requires to identify effects generated by the product during its whole lifecycle, as well as all the other activities implemented by the companies. Thereby, it would be achieved qualitative aspects to be part of the methods and tools used to determine the business general success under circularity.

Although it is an innovative and increasingly studied topic, it still ambiguous how the organizations have to move to a new circular economic system. In some cases, clean production, efficiency increasing and costs reduction, is what have forced companies to look for CE practices [9].

However, it has not to be loss of sight that moving away from linear paradigms challenge ingrained patterns that can be an important obstacle to good practices application, this is the importance of having studies that transcend from theoretical plane to the practice.

Literature about application and effective CE operation can be divided on studies which deal regional level implementation on a specific geographic place, and in those that apply it on a sector, product or material [10]. In other words, practical cases are dialed through industries and geographies, attached by a material likely to form material supply loops into the company or in a supply chain (SC). To [11], the successful application of CE requires rethink the production processes, distribution, sales, consumption and waste, to create cooperation systems among the CE cases which are currently operating.

[1]Good practices of sustainability: refuse, rethink, reduce, reuse, repair, refurbish, remanufacture, repurpose, recycle and recover.

As regards the sectoral study, a commonly explored line of research, it is to prioritize the application of CE to sectors whose waste is a social and environmental problem due to its abundant generation, while in the case of waste whose recovery is feasible. In this work we chose to follow this trend and select four types of materials that make up urban solid waste (USW), whose recovery processes are already in operation such as glass, plastic, metals, paper and cardboard.

This work is in addition to these efforts and pursues two main objectives, on the one hand, to analyze circular practices that allow to identify the scope of transition to CE and on the other hand, to identify potential regions to shape circular SC, based on quantity and kind of company's activity on each entity. This approach intends to explore the circumstances in which circular practices are presented and which challenges are revealed from those experiences. For its part, the identification of regions is made as a starting point to note regions which have minimum condition to contain companies on different links of the SC.

In the following section are mentioned some considerations reported on literature about actions, on international level, focused on applying CE and the topics about SC applications, after that, on the third section, the methodological, the steps that allowed to deal the objectives of the study are presented. On a fourth section are presented the results of the literature review and the identification of companies, to finish with a conclusion section.

2 Circular Economy and Supply Chains

SC are key to CE application because they involve going beyond the limits of the company to obtain from suppliers, eco-friendly materials; and from customers, their cooperation to post-consume products valorization [7]. Nowadays, CE has tried to integrate itself into the SC on different geographic areas, for example, the European Union, the United States, Japan and China, through the application of numerous initiatives to high technology sectors [12].

While each SC has its own particularities in terms of scale and structure, it is possible to make a generalization of them, under the CE, defining the materials flow and its passage through the different links that form it. In such a way, it can be observed five general links of the SC: i. material sourcing, ii. Both parts and final product manufacturing iii. Distribution and sales, iv. consumption and use, and finally v. service suppliers such as transport, repair, recycling and all related to product handling, once it has been discarded by de consumer [10, 13].

The initiatives aimed to introduce CE into SC has been made in different analysis level, from the R frameworks in some of its links with the purpose of shape loops of material reintegration, to formation of nets in which the SC share specific objectives. The most widespread formation are the eco- industrial parks, ecological or environmental SC, and close cycle SC [7]. In all of them it is intended to close resource flows (energy, raw materials, water) that allow increasing its utilization, while reducing waste and environmental impacts. These objectives are reinforced in [14], where the authors point that in addition, it is required a systemic change, together with non-technological changes and in the diverse actors, because an increasing innovation by itself will not achieve a decoupling between economic growth and environmental impacts, and only with development and the adoption of more radical and systemic ecological innovations, it would be possible to achieve the sustainable transformations and in long term.

On the other hand, it has to be mentioned that certain circular practices of product recovery such as reuse, remanufacture and recycling, have become more common. Its application can allow mitigation of negative environmental impacts, increasing competitively, reducing production costs along the entire lifecycle, comply with the current regulations and promote sustainable development [15]. These strategies involve the existence of a close loop system for which is necessary enterprises to compromise themselves to think about the product lifecycle from a holistic approach.

Figure 1 shows the expected structure for a closed SC and which are the R-framework practices that are usually applied at the different links, highlighting that practices such as Rethink and Reduce are performed when supplying, designing and manufacturing the product, because these practices seek to modify and reduce the materials used in the manufacture of the product, in addition to thinking about the product lifecycle. On the other hand, practices such as Recycle and Recover are carried out at the end of life, when waste recovery services are used once discarded by the consumer.

The group that concentrates more practices while applying itself in more links of the SC, is where Re-use, Repair, Refurbish, Remanufacture and Repurpose are used, involving links ranging from manufacturing to disposal of waste. This is because these Rs seek to take advantage of products already manufactured for general loops that allow to prolong their service life or that of the parts that make up them.

Fig. 1 9R framework practices inscribed in a closed-cycle supply chain

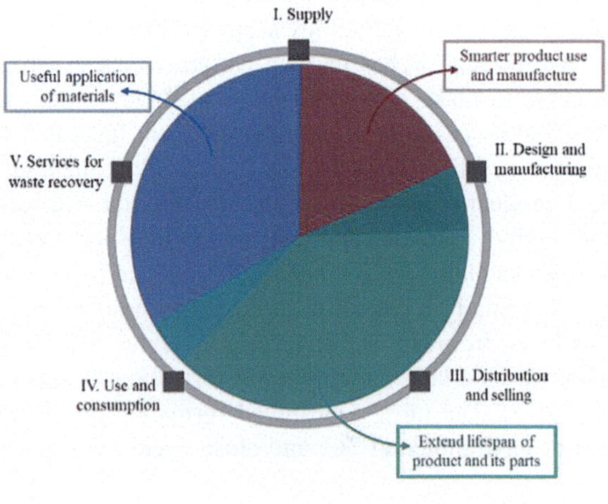

It should be mentioned that this does not mean that a link such as manufacturing cannot perform practices such as Recycle, there are cases where different Rs are implanted in the same organization. The goal of this figure is to show how all Rs can be applied on the same SC.

The implementation of this initiatives has highlighted different challenges to SC redesign, the authors of [13] group them into seven categories: finance, market, product, legislation, chain management, technology and consumer behavior.

Stand out among them, challenges related to the product and chain management. On this matter it is mentioned that we should be careful when making incremental improvements in productive processes since, by making them more efficient, major production could be encouraged, reducing environmental benefits.

Concerning to chain management, it has to be considered that SC redesign hardly relapse in only one company, because this has limitations on its vertical integration and production control, in view of that, it is clear the need of having information about the involved actors on the different moments of productive process, as well as the kind of materials used and generated on those processes. As part of this investigation, a first approach to information available on Directorio Estadístico Nacional de Unidades Económicas (DENUE) was made, obtaining as a result the number of economic units (EU) related with a determined material and because of its vocation and geographic location could shape a SC.

3 Methodology

This work has two main objectives; the first one is to analyze circular practices that allow identifying the scope of transition to CE, as well as opportunities and challenges to SC. This objective is dealed through literature review, following the five steps methodology purposed by [16], which process is described below:

3.1 Literature Review

1. *Question formulation*: As already mentioned, this review focuses on circular practices, seeking to know which scope has its application had on the transition process to CE? in addition to identifying which opportunities and challenges do this practices involve to shaping SC enrolled into CE?
2. *Data harvesting*: The research was made using the Biblioteca Digital de la Universidad Nacional Autónoma de México,[2] which shelter more tan 200 data bases with about 22,000 magazines, among which are found: *Journal of Cleaner*

[2]https://bidi.unam.mx/.

Production,[3] *Resources, Conservation & Re-cycling,*[4] *Journal of Industrial Ecology,*[5] *Waste Management*[6] y *Sustainability,*[7] which concentrate almost 60% of the overall publications about CE [17]. The search terms were *"circular economy" AND "practice case"* and "economía circular" AND "caso", using three filters: complete text, English or Spanish language, in the period between 2010 and 2020.

3. *Selection and evaluation of studies*: Since most practical cases only show partial application of what has been previously defined as CE, it was considered to include as circular practice those which met three criteria: to show an initiative executed in an organization with the purpose of applying the CE, to meet with some of the three circularity principles, and to fulfill some of the CE purposes. The research yielded a total to 72 registers, from which 45 were extensively reviewed since the reading of the summary denoted relationship with the topic of interest. From this review 15 articles were discarded since they were not practical applications, finally the analysis was carried out on the remaining 30 articles.
4. *Analysis and interpretation*: A comparative analysis was chosen, which consist of three elements: comparators and variables, observations and data [18]. Were defined as comparators and variables:

   4.1. *Circularity level.* Established by priority order of R framework strategies (refuse, rethink, reduce, reuse, repair, refurbish, remanufacture, repurpose, recycle and recover) purposed by [19] in which are defined the three circularity levels described below in Fig. 2.
   4.2. *Stakeholder.* Referring to the agents involved in the practice, three main agents are distinguished in the CE: companies, society and government. These agents can be involved in the practice individually or in collaboration, what is sought is to distinguish whether their participation exists.
   4.3. *Circularity principles.* In [4] Ellen MacArthur Foundation points three circularity principles:
      a. *Preservation principle.* It refers to the search for efficient use of natural and renewable resource flows.
      b. *Optimization principle.* The greater rotation of the products and its components is sought, achieving a greater use cycle and therefore a longer lifecycle.
      c. *Efficiency principle.* It involves reducing externalities of natural resource utilization processes and seeking synergies between the different agents.

[3]https://www.journals.elsevier.com/journal-of-cleaner-production.

[4]https://www.journals.elsevier.com/resources-conservation-and-recycling.

[5]https://onlinelibrary.wiley.com/journal/15309290.

[6]https://www.journals.elsevier.com/waste-management.

[7]https://www.mdpi.com/journal/sustainability.

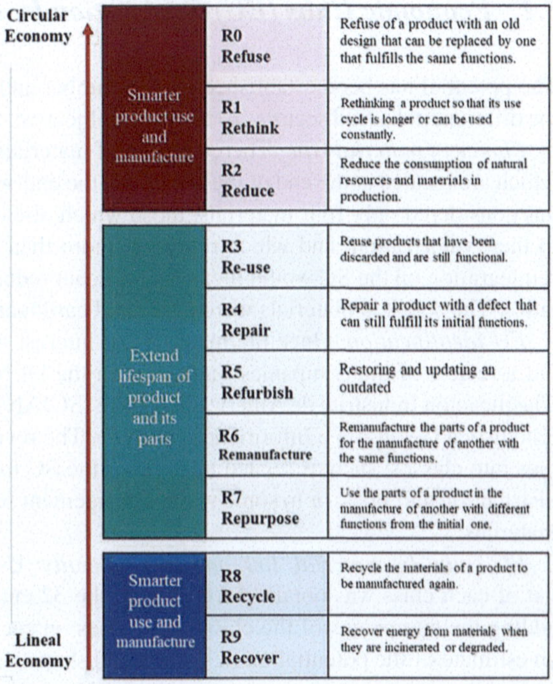

Fig. 2 Circularity levels

4.4. *Purpose of the practice.* As these are circular practices, it seeks to fulfill the CE purposes, which according to [6] are those shown in Fig. 3:

To complete the comparative analysis, the 30 chosen documents were considered as observations and as data the subcategory in each case, in such a way that, for each document were identified the correspondent subcategories, allowing its comparison.

5. *Report and use of results:* Are shown in the results section.

As mentioned at the beginning of this section, this work has two objectives, being the second one the identification of potential regions to shape circular SC.

Fig. 3 Circular economy purposes

3.2 *Economic Units (EU) Exploration by Kind of Material*

The potential has been established by the number and industrial vocation of EU on the different industrial sectors. To fulfill this objective, the following steps were done:

Choice of material type: There are a lot of materials subject to valorization, all of which are located at the end of the direct SC, the end-waste. However, this document has considered only four materials, those which already have companies dedicated to their valorization, and which represent more than 75% of USW [20], and their reintegration on the SC would mean a significant reduction of pollution problems in cities. The selected materials were paper and cardboard, plastics, metals and glass.

EU Identification: Once the materials of interest were identified, it was carried out a review of the companies registered on the DENUE based on the Sistema de Clasificación Industrial de América del Norte (SCIAN) used by Instituto Nacional de Estadística Geografía e Informática (INEGI). The review allowed grouping companies into classes, such as related to some of the SC links: sourcing, manufacturing, distribution and sales, or to some waste management service; for each of the selected materials.

EU count by material, link and federal entity: Using DENUE records the EU list of each class was obtained for each of the 32 entities of the country. Thus, by adding the companies of the classified classes on each link for different materials, an estimate of the potential unities that could shape a circular SC can be obtained.

Finally, are shown those entities where EU can be found for all links of the SC and for the different materials under study. This establishes a starting point for the identification of areas with high potential to form circular SC, based on the proximity between the companies of the sector, which provides a similar context under which it operates and, being entities with the highest percentage of EU, also allows for the diversification of allies.

4 Analysis of Descriptive Statistics

In the following part, is presented the analysis of descriptive statistics, prepared in accordance with the two stated objectives: the literature review of CE practices, and the exploration related to companies in Mexico.

Distribution of the CE practice sample. As already mentioned, the literature review was carried out on a 30 article sample, Figs. 4 and 5 set out how the practices recorded in these articles are distributed both regionally and by industry. While it is recognized that is not a representative sample due to the bias that the process of collecting documents may entail, this distribution does follow situations already expressed on literature as determinants to circular practices application. With regard to geographical distribution, stand out the 47% of this practices located on the European continent.

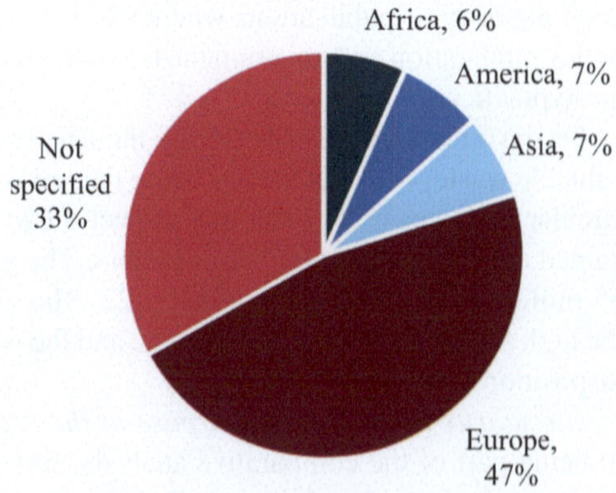

Fig. 4 Percentage of circular practices by geographical region

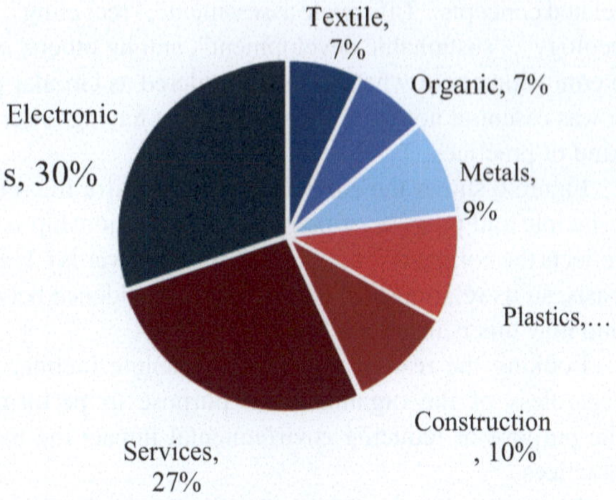

Fig. 5 Percentage of circular practices by industry

This can largely respond to the stimulus of a highly regulated and explicit CE-oriented, besides to the commitment acquired by the European Union to the amendment to its waste regulation, the Waste Framework Directive, which incorporated objectives related to: USW recycling, reuse of the same, selective collection of organic waste, modification of regulations throughout the region on producer responsibility and the setting of targets to reduce waste generation by 2030. Associated whit this, the commitment was enhanced by initiatives such as "An European strategy for plastic in a circular economy", which seeks to improve profitability and quality of plastic recycling, reduce plastic waste and boost investment and innovation to find circular solutions [21].

In this respect, stands out the low circular practices percentage founded in the Asian continent (6.7%). However, a possible explanation is the filter generated by the search terms and the selected period, in the time between 2010 and 2020, thus, while it is known that China has formally adopted the CE model since 2002, in the large majority of publications where Chinese case is reported, they focus on public policy publication and government-led industrial reconversion and not, on micro and meso practices [7].

On the other hand, distribution by industry or economic sector, addresses the main valuable waste, with electronics being the industry that groups the largest number of circular practices (30%). On this respect it can be mentioned that this industry has gained relevance for a number of factors: The growth of e-waste generation, almost 45 million tonnes per year globally [22], the great diversity of materials involved, the high added value in such products, and the pollutants that derived from their final disposition.

Circularity principle and purpose of the practice: These categories, in addition to being part of the comparative analysis, served as a filter for the selection of the articles to be considered; its importance is that in both recognize the circular nature of the practices. While CE is a relatively recent concept, it has been nurtured by other related concepts: "Life cycle assessment", "recycling", "cradle to cradle", "industrial ecology", "sustainable development", among others; and this similarity may lead to a confusion about what can be considered as circular practice [10]. For this reason, it was essential not only to define CE but having criteria to distinguish it from other kind of practices.

Figure 6 shows the comparative analysis for the relationship between circularity principle and purpose of practice. This relationship is interesting since the purpose reflects the company´s intention to transit towards CE and the principles of CE are its basis, so its relationship indicates the coincidence between the company's intention and how this is inscribed in the CE.

Looking the results, there is an absolute inclination to optimization principle, regardless of the organization´s purpose to perform the practice, besides, it is the purpose of reducing environmental impact the most frequent motivator of the practices.

At the same time, stands out that 50% of the cases mentioned the search of reducing environmental impact, applying practices that led to a greater product lifecycle.

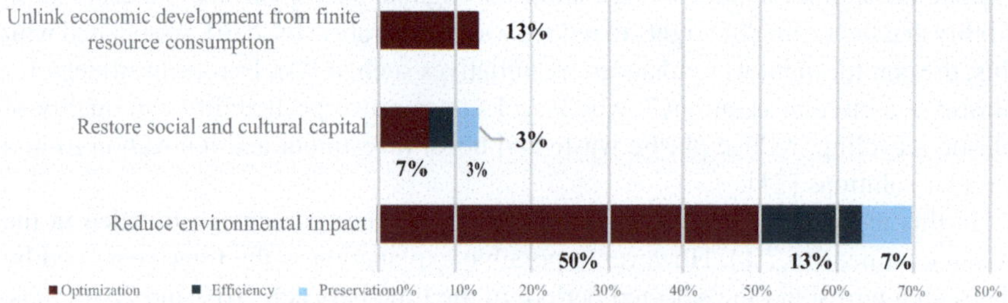

Fig. 6 Percentage of circular practices by it purpose and the present circularity principle

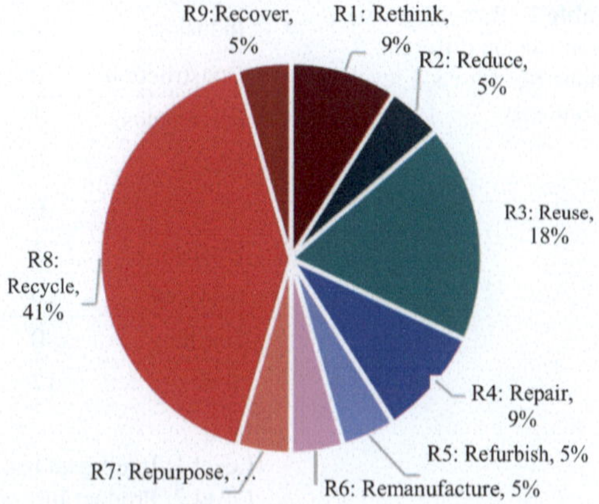

Fig. 7 Distribution of cases with the principle of optimization according to the applied R

This result is further explored in Fig. 7, where the reported cases consistent with the optimization principle and their location in the R frame are observed. Looking Fig. 7, it can also be said that recycling is the most frequent action, thus causing that, together the 5% represented by the action of recovery, as a result 46% of cases with low circularity level, where the practice aims to take advantage of some of the material that made up a particular product. This value contrasts with 14% of practices located at the highest level of circularity, where it seeks to rethink a product so that its cycle of use is longer or reduce the consumption of natural resources. In this sense, it should be thought about whether to comply with an optimizer to transit to CE [23].

On the other hand, it is interesting to mention that practices corresponding to 14% of higher circularity, were developed in micro and small companies, showing that the belief that circular practices require high financing, is not a rule [24].

Circularity level: The scale purposed by [19] based on an extended R framework, which, groups into three broad categories was used to place the company at a higher or lower circularity level. In this regard, it has to be mentioned that the authors include on their proposal the evaluation of type of transition: focused on socio-institutional change, focused on technological innovation, or focused on socio-institutional change but with a decisive institutional component. In this way, the authors seek to integrate contextual elements into the evaluation as consumer participation and local legislation are taken into consideration. This responds to the level of circularity being related with different variables and not all of them depend on the actions of the company itself.

Table 1 shows the percentage of practices in each level of circularity and in each industry. Fort the level: "Smart product use and manufacturing", the percentages are low except for the service sector, this responds to the most commonly used models at this level are those where product ownership is shared or where second hand markets are generated [25], so strategies focused on customer service are required or high duration products and such strategies cannot be applied to industries such as construction or those related to organic products.

Table 1 Percentage of practices on different industries for each level of circularity

Industry	Level 1 (%)	Level 2 (%)	Level 3 (%)
Construction	0	7	3
Electronics	3	13	13
Metal	0	7	3
Organic	0	3	3
Plastic	3	0	7
Services	10	7	10
Textile	0	3	3
Total	17	40	43

Note
Level 1: Intelligent use and manufacture of products
Level 2: Product life extension
Level 3: Useful application of materials

It is important to remember that, besides the sector influence, there are other factors that affect the choice of the practice, inscribed in an R frame-work such as technology and customer consumption habits, however, these factors are not part of the study [26].

Stakeholder: The environmental and economic problems that generate the need to propose a new economic system, such as CE, are of such magnitude that require the collaboration of agents that, when working in a coordinated way, generate and develop feasible proposals to be implemented. With this in mind, one of the categories of this comparative analysis refers to stakeholders, who are explicitly mentioned as essential participants in circular practices. Table 2 shows the percentage of practices involving consumer, government, other economic units, or instead, those

Table 2 Percentage of the practices by stakeholders involved depending on the purpose of the practice

Purposes of the practice/ stakeholders	Company (%)	Company-consumer (%)	Company-government (%)	Company-company (%)
Purpose A	7	0	0	7
Purpose B	47	10	3	10
Purpose C	0	3	0	0
Purpose D	0	7	3	3
Total	53	20	7	20

Note
Purpose A: Unlink economic development from finite resource consumption
Purpose B: Reduce environmental impact
Purpose C: Achieve sustainable development
Purpose D: Social and natural capital restoration

which worked independently. These percentages are associated to the purpose of the practice.

Of these results, 53% of the practices carried out without the collaboration of another agent, in this regard it is worth mentioning that, within the organization, the commitment, technical knowledge and understanding of circularity, by senior management, has shown an important effect on the adoption of CE [27]. Taking into account the previously mentioned, the government, consumer or other business participation can be considered as non-crucial.

Apart from the already mentioned agents, the authors of [24] expose the importance of having multidisciplinary knowledge to support the development of effective innovations in the practice, that if not found into the organization can be obtained by linking with other agencies such as research and development centers.

Figure 8 shows the cases found where the company collaborated with another link of the CS for the realization of circular practice, in addition to showing the level of

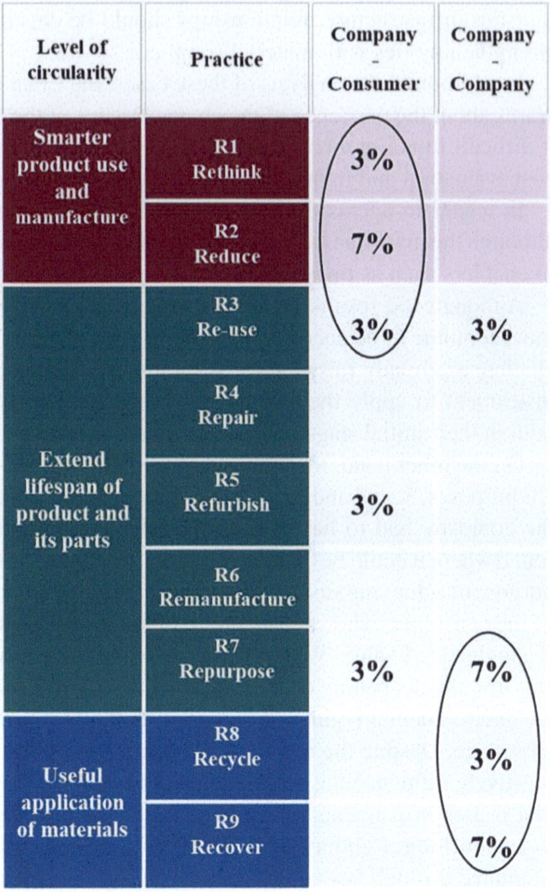

Fig. 8 R framework practices applied in collaboration with another SC agent

circularity regarding the practice of the R framework that was carried out. It notes that 40% of the practices included the consumer or another company, with the same number of cases in both relationships.

On the other hand, it is observed that for Company–Consumer relations, practices tend to have a higher level of circularity, this may seem confusing if it is recalled that the actions of greater circularity occur in the design phase of the product, however these initiatives refer to situations where the company modified its business model, in collaboration with its client, to provide the service that was usually provided with a product. This kind of practices are growing in different industries but it should be emphasized that they are only possible when we have the acceptance of the customer, situation that is not always possible, especially if they are environments where the ownership of the product is very ingrained to consumption habits.

With regard to collaboration between enterprises, the opposite is the case, practices are concentrated in low levels of circularity, in a number of cases where a provider of transport, waste management and treatment services is used.

While such relationships are of great importance for the formation of a reverse logistics infrastructure, relationships should be developed on both sides of the SC, strengthening ties with material suppliers.

In addition to the analysis of these cases, the Ellen MacArthur Foundation in [3], warns about the existence of the characteristics of the agents involved that can make it difficult to adopt circular practices, including: misaligned incentives, short-term cost evaluation and internal company habits.

In regard to agents outside the company, the mentioned characteristics prevail, although the main ones are: the lack of infrastructure for reverse logistics, the absence of enablers such as financing, legislation and consumer habits.

Although the results show low collaboration between companies, this linkage must continue to be encouraged, among other reasons for its potential to meet some challenges already raised such as the financial aspect [13], considering that initial investment to apply the CE practices may represent a challenge to organizations, both in their initial stage and in their implementation.

On the other hand, the analysis of the practices showed that in order to fulfill the CE purposes, social and natural restoration, and to achieve sustainable development, the company had to have the participation of another agent, since no cases were found where it could be done in isolation. This fact was observed particularly, with the purpose of achieving sustainable development. The practice required the participation of the consumer, who was already familiar with the sustainability concept and gave it a high added value. With regard to government participation, it is observed that it was low and deepening on the particular case, it is known that it focused on verifying the environmental regulation accomplishment, since this was seen as a competitive advantage. Despite the low participation, the idea that government intervention can positively influence the induction of circular practices is widespread, to the extent that it uses instruments such as contracts and environmental standards in addition to establishing a climate conductive to cooperation [28]. Specifically, CE can be promoted through laws, policies and strict governance [29].

5 Challenges and Opportunities

In summary, the main challenges presented in implementing circular practices were:

- Systemic change. Achieving systemic change that outweighs incremental improvements, which, while innovative, have not allowed economic growth to be decouple from environmental wear.
- Eco-design: Product design has shown to be a stage with high potential to encourage the application of circular practices, but at the same time the few changes that this has had, it has become in a difficult obstacle to overcome is stages such as remanufacturing and repair.
- Vertical integration: To get a redesigned and oriented to circularity SC, it takes more than one company that integrates it, to initiate the change, due to limitations because of its vertical integration and limited control to its own production.
- Shared information systems. It has been given few cases where development of systems has allowed the dissemination of information on required materials, production processes and waste generated among the members of the chain.
- Processes for waste management. In particular, for the management of electronic waste, it is necessary to improve the procedures and conditions under which it is recycled, due to the emission of pollutants released in its final disposal.
- Studies by industry. It is necessary to find the right type of practices and strategies for each industry, considering that there are strategies that cannot be applied in all industries, such as the lengthening of the lifecycle in organic products.
- Incentives. Some of the difficulties in this area are the short-term cost assessments, which do not prioritize those solutions with benefits in the long term, in addition to the internal habits of each organization too ingrained in traditional production practices.
- Infrastructure for reverse logistics. In several regions it is still an underdeveloped link, with few options for waste management, transport and treatment, as well as limited information flows.
- Community and government. The absence or little participation of other actors such as consumers whose participation, interest and knowledge, around CE and sustainable development, is indispensable to carry out some practices of the highest level of circularity. While lack of funding or inconsistent legislation by governments makes the transition more difficult.

On the other hand, important opportunities could also be identified with potential to bring significant changes in traditional productive systems and orient them towards circularity; among the main ones is:

- Regional and industrial diversity. There are initiatives around the world advancing into CE application and cases in different industries show the wide range of possibilities for valorization.
- Frequent practices. Some of the practices under the R framework are increasing their use, such as reuse, recycling and repurpose, opening the way to the rest of the practices.

- Companies of various sizes. While there are cases where a big company generates a circularity dynamic in the SC to which it belongs, an important percentage of the cases with a high level of circularity is developed in small companies, giving evidence that high amounts of investment are not indispensable and, on the other hand, the reduced and flexible structure of these organizations is an advantage.
- Service-oriented models. The popularity and diversification of this type of models, where the customer does not maintain ownership of the product and only makes use of the service that it offers, allows to take advantage of the recent and growing, behavior change of the consumer, opening up the possibility of applying these models in other sectors not yet explored.
- Experience and commitment. In much of the cases reported here, the knowledge and commitment on the part of senior management had a significant impact on the application of circular practices, so the willingness shown to make modifications to the company, already represents by itself a challenge overcome to achieve success.
- Scientific and technological research. Beyond the activities that can be carried out by the business sector, there is a growing scientific and technological research that supports the development of functional innovations that can be leveraged by the industry to the extent that the relevant links are established.

6 Companies by Material, Link and Entity

This section shows the results of the EU analysis, whose activity relates to one of the four selected materials, paper and cardboard, plastic, metal and glass. Table 3 makes a classification of companies and its class codes. In this regard, it is worth clarifying that this codes corresponds to SCIAN, which stablishes a category for each company according to its economic activity, so that this system has 20 sectors which are divided into subsectors, branches, sub-branches and finish with 1084 classes on the most specific level. This groups of classes is where are selected those, which by

Table 3 Total count of class and eu codes considered by link and material

Link material	Paper and cardboard		Plastic		Metal		Glass	
	CC	EU	CC	EU	CC	EU	CC	EU
Supply	8	44	2	38	7	157	1	23
Manufacturing	18	7,661	17	37,109	33	98,922	9	3,091
Distribution and selling	10	809,703	6	31,037	4	22,359	4	38,859
Waste recovery	7	2,399	8	5,601	5	11,814	7	714
Total	43	819,907	33	73,785	49	133,252	21	42,687

Note CC Class code
EU Economic unit

their activity, correspond to each material and in turn to a SC link, remembering that generally 4 links are being considerate: materials supply, manufacturing, distribution and sales, service providers, which include transport, repair, recycle and everything related to waste management.

Among the values in Table 3 stands out the low number of companies on the supply link, in contrast with units dedicated to distribution and sale. There are several explanations for this behavior, which address market characteristics, size of the company or profitability of each activity; however, this huge disparity can be seen as an opportunity to encourage companies' creation, which, with waste management or post-consume products, are added to material supply, taking advantage of the learning and developments in logistics field already implemented by distribution companies.

The total values are broken down in Table 4, where federative entities with highest number of EU are presented, classified in some of the 4 mentioned links, for each material. Due to the extent of this count, it was decided to include only the 5 entities with the highest number of companies, so that, each box in the table includes a

Table 4 Federative entities with the highest eu concentration by link and material

Link/Material	Paper and cardboard		Plastic		Metal		Glass	
Supply	25%	Michoacán	13%	CDMX	15%	Sonora	22%	Sonora
	18%	Chihuhua	13%	N. L	8%	Zacatecas	22%	Veracruz
	11%	Oaxaca	11%	Gto	8%	CDMX	13%	N. L
	11%	Puebla	11%	Jalisco	8%	Jalisco	9%	Coahuila
	9%	Jalisco	11%	Edo. Méx	8%	Edo. Méx	9%	S. L. P
Manufacturing	8%	Edo. Méx	10%	Edo. Méx	11%	Edo. Méx	15%	Edo. Méx
	8%	CDMX	10%	Gto	7%	Jalisco	13%	CDMX
	7%	Jalisco	9%	Michoacán	7%	Michoacán	13%	Jalisco
	7%	Veracruz	9%	Jalisco	7%	Veracruz	9%	N. L
	6%	Puebla	7%	Veracruz	7%	Puebla	7%	Puebla
Distribution and selling	16%	Edo. Méx	14%	Edo. Méx	11%	Guerrero	16%	Edo. Méx
	8%	CDMX	9%	CDMX	11%	CDMX	9%	CDMX
	7%	Puebla	8%	Jalisco	9%	Jalisco	9%	Jalisco
	6%	Veracruz	6%	Puebla	8%	Edo. Méx	6%	Puebla
	6%	Jalisco	5%	Michoacán	6%	N. L	6%	Gto
Waste recovery	22%	Edo. Méx	20%	Edo. Méx	19%	Edo. Méx	11%	Edo. Méx
	17%	CDMX	9%	CDMX	8%	CDMX	8%	Jalisco
	7%	Jalisco	9%	Jalisco	7%	Jalisco	7%	Puebla
	6%	Puebla	9%	Gto	6%	N. L	7%	CDMX
	6%	N. L	7%	Puebla	5%	Puebla	6%	B.C

descending list of 5 entities, including the percentage of companies that each entity owns, relative to the totals in the table above.

It is observed the constant presence, in almost each box, of entities such as Estado de Mexico, Mexico City and Jalisco. This result is expected due to certain characteristics of these entities such as their population density and urbanization degree, which makes them attractive points to companies' installation. On the other hand, in very specific boxes, entities that reinforce its vocation are observed, such as Sonora and Zacatecas, distinguished by their mining orientation.

In order to detect entities that could allow SC formation, Fig. 9 shows the maps where it can be observed the entities of Table 4, showing those that have companies in all links.

The case of glass stands out because no entity was found with companies in all links, with the supply link missing for this material, in entities such as Mexico City, Estado de Mexico, Jalisco and Puebla that do house companies in the rest of the links. This may be an indication of the complications of establishing SC, with respect to glass within a single entity. How-ever, it has to be mentioned that only the companies with the most representative class codes are being represented, in the case of glass only one class is available (CC: 212324, silica mining); In addition, only entities with the highest EU percentage were considered. Having a more detailed review, it is Estado de Mexico that includes all the links because it has a companies' registration in the supply link located in this place.

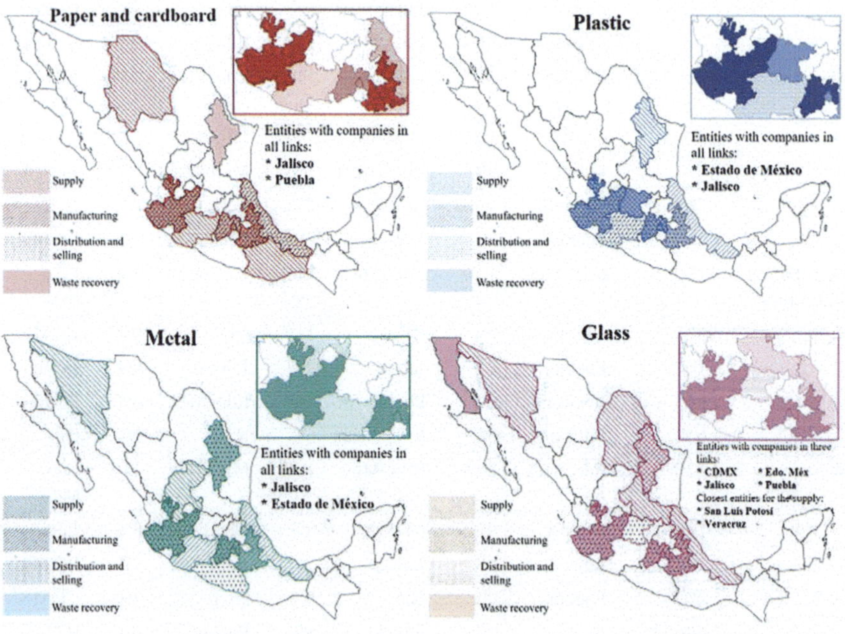

Fig. 9 Entities with highest number of EU in each link and entities with EU in all links

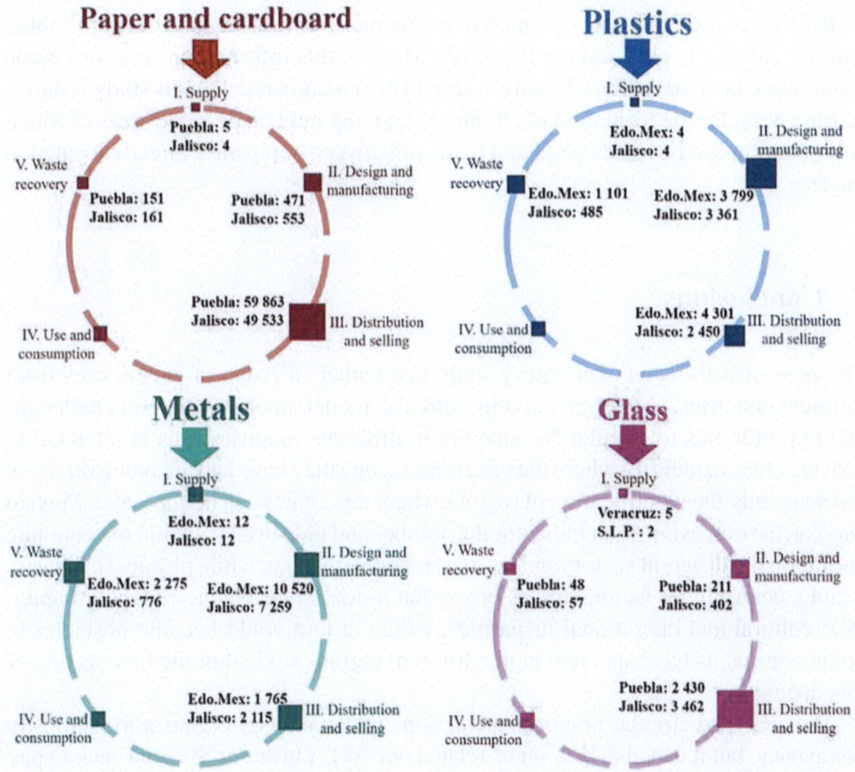

Fig. 10 Number of companies registered in entities with the greatest potential, for each SC link of each material

On the other hand, Jalisco is the entity where companies related to all materials and links are located.

Figure 10 shows the companies registered with activities related to each link of the SC for the different materials, this count includes only the federal entities where there is activity in all links, except the case of the glass for which not a single entity that included all the links was found, being necessary to address the nearest entity.

This figure highlights two general aspects, on the one hand, for most materials there are companies within the same entity in which a circular SC could be formed, being the link of the supply of materials the one with the smallest number of companies. This is beneficial considering geographical proximity not only implies a reduction in transport costs, but also allows collaboration under the same contextual conditions, facilitating understanding and information flows.

On the other hand, it is noted that, for most materials, there is a link with a high number of companies, with manufacturing and distribution being the most robust. This situation can be considered an advantage insofar as it is the links with the knowledge and experience necessary to create synergies in the rest of the chain.

It also stands out the concentration of companies from various sectors in just three entities, Jalisco, Puebla and the Estado de Mexico, this information serves to guide future work in these regions. Finally, it should be remembered that this study is only a starting point for the formation of circular SC, giving light on possible areas in which collaboration can be re-searched and fostered between companies already located in the area.

7 Conclusions

The present work is an exploratory study that gathers a group of experiences from different countries, about transition to a circular model, in or-der to point challenges and opportunities to circular SC shaping in different countries, this is intended to provide clues to identify where the efforts and scope they have had are being directed. Subsequently the identification of regions where this chains can be formed in Mexico was carried out, using as an indicator the number and industrial vocation of economic entities in the different sectors and entities, recognizing that, while physical closeness is not a determining factor, studies prove that it does contribute to reducing regulatory, cultural and operational disparities, which in turn could become obstacles to collaboration, as has happened in the different regions studied in the first section of this document.

The analyzed circular practices are actions undertaken by organizations (mostly companies, but it can also be a set of related network, cluster or SC) that seek to pay the CE, making modifications to its operation, thus representing an explicit effort to move towards circularity. It is recognized that these are much focused actions, which importance does not relapse in the impact that they, on their own, can have in the construction of a circular economic system, but in the effect that they can generate at scale when more organizations join to this task.

As it is an exploratory study, it has limitations; regarding to literature review, stands out the number of considered articles, the result of very specific search terms that leave out other works with a related topic, however, the obtained results show a coincidence with what is reported in literature addressing theoretical and practical CE questions, thus pointing that quantity is not a factor on invalidating the results.

Regarding to EU identification, it was chosen only those companies classified under determined class codes that were directly related to transformation and handling of the selected materials, recognizing that a SC has greater diversity of participants with activity in different sectors. On the other hand, the disaggregation, shown in the present document, is done only at state level, since it is a starting point to identification of areas with a specific industrial vocation, evidenced by the proliferation of companies, considering that, the higher the number of companies, the greater the options available to create connections.

A focused analysis on the identified entities, which maps the registered companies, will provide more information about the physical proximity between them. It is also recommended that future work integrate another factors that contribute to circular

SC formation and explore cultural, legislative and technological conditions in each region.

The consumer role should not be overlooked, although, this work does not address it, its low participation of circular practices implementation, is a reflection of the lack of study and development on this area. In general, it is recognized the existence of cultural obstacles, particularly related to participation on second hand markets or for the acquisition of remanufactured products, the lack of knowledge about the product content and its manufacturing processes and the preference for product ownership, are some of the challenges to be overcome. The impact of this challenges is no less, considering that a low involved consumer with this issues, contributes to uncertainty about quantity, quality and period of waste or return of products or materials likely to be reintegrated into the chain, either by being re-manufactured or recycled. Challenging and reconstructing these patterns of thought, is a shared responsibility among all the stakeholders.

The shape of structures such as supply chains, stablished in a climate of trust and cooperation between the different actors is a slope that would potentially facilitate the exchange of technical knowledge, more effective communication mechanisms and more integrated processes that can contribute mitigating certain challenges of circular economy. The analysis and outreach exercises are in addition to the efforts to achieve the integration of networks that have more reaching, compared to that of companies work-ing individually and thus, to stablish models to be followed by other indus-tries by demonstrating the value of commitment among all actors.

References

1. Aranda-Usón, A., Portillo-Tarragona, P., Scarpellini, S., Llena-Macarulla, F.: The progressive adoption of a circular economy by businesses for cleaner production: an approach from a regional study in Spain. J. Clean. Prod. **247**. Recuperado de (2020). https://doi.org/10.1016/j.jclepro.2019.119648
2. Millar, N., McLaughlin, E., Börger, T.: The circular economy: swings and round-abouts? Ecol. Econ. **158**, 11–19 (2019)
3. Ellen MacArthur Foundation: Towards the Circular Economy, Vol. 2: Opportunities for the Consumer Goods Sector. Recuperado de (2013). https://www.ellenmacarthurfounda-tion.org/publications/towards-the-circular-economy-vol-2-opportunities-for-the-consumer-goods-sector
4. Ellen MacArthur Foundation: Hacia una economía circular: motivos económicos para una transición acelerada. Recuperado de (2015). https://www.ellenmacarthurfoundation.org/as-sets/downloads/publications/Executivesummary_SP.pdf
5. Lett, L.A.: Las amenazas globales, el reciclaje de residuos y el concepto de economía circular. Revista Argentina De Microbiología **46**(1), 1–2 (2014)
6. Kirchherr, J., Reike, D., Hekkert, M.: Conceptualizing the circular economy: an analysis of 114 definitions. Resour. Conserv. Recycl. **127**, 221–232 (2017)
7. Masi, D., Day, S., Godsell, J.: Supply chain configurations in the circular economy: a systematic literature review. Sustainability **9**(9), 1602 (2017)

8. Janik, A.: The analysis of existing indicators, methods and tools for measuring circularity in the context of their future development. In: International Multidisciplinary Scientific GeoConference: SGEM: Surveying Geology & Mining Ecology Management, vol. 17, pp. 43–50 (2017)
9. Groslambert, S.: Circular economy of buildings: Leroy Merlin case study (2019)
10. Kalmykova, Y., Sadagopan, M., Rosado, L.: Circular economy–from review of theories and practices to development of implementation tools. Resour. Conserv. Recycl. **135**, 190–201 (2018)
11. Gardetti, M.A.: Introduction and the concept of circular economy. Circular Economy in Textiles and Apparel, pp. 1–11. Elsevier (2018)
12. De Angelis, R.: Business Models in the Circular Economy: Concepts, Examples and Theory. Springer (2018)
13. Bressanelli, G., Perona, M., Saccani, N.: Challenges in supply chain redesign for the circular economy: a literature review and a multiple case study. Int. J. Prod. Res. **57**(23), 7395–7422 (2019)
14. Beltramello, A., Haie-Fayle, L., Pilat, D.: Why New Business Models Matter for Green Growth, No. 2013/1. OECD Publishing, Paris, France (2013)
15. Badurdeen, F., Aydin, R., Brown, A.: A multiple lifecycle-based approach to sustainable product configuration design. J. Clean. Prod. **200**, 756–769 (2018)
16. Randolph, J.: A guide to writing the dissertation literature review. Pract. Assess. Res. Eval. **14**(1), 13 (2009)
17. Abadia, L.G., Galvão, G.D.A., Carvalho, M.M.D.: Economia circular: um estudo bibliométrico. ENEGEP 2016-Contribuições da Engenharia de Produção para Melhores Prá-ticas de Gestão e Modernização do Brasil (2016)
18. Lor, P.J.: International and Comparative Librarianship: Concepts and Methods for Global Studies. Walter de Gruyter GmbH & Co KG (2019)
19. Potting, J., Hekkert, M.P., Worrell, E., Hanemaaijer, A.: Circular economy: measuring innovation in the product chain (No. 2544). PBL Publishers (2017)
20. Semarnat: Informe de la Situación del Medio Ambiente en México, sección Residuos. Recuperado de (2016). https://apps1.semarnat.gob.mx:445/dgeia/informe15/tema/pdf/Cap7_Residuos.pdf
21. COM: Comisión de Medio Ambiente, Salud Pública y Seguridad Alimentaria. "Una estrategia europea para el plástico en una economía circular". Parlamento europeo. Recupe-rado de (2018). https://eur-lex.europa.eu/legal-content/ES/TXT/?uri=CELEX:52018DC0028
22. Chaudhary, K., Vrat, P.: Circular economy model of gold recovery from cell phones using system dynamics approach: a case study of India. Environ. Dev. Sustain. **22**(1), 173–200 (2020)
23. Howard, M., Hopkinson, P., Miemczyk, J.: The regenerative supply chain: a frame-work for developing circular economy indicators. Int. J. Prod. Res. **57**(23), 7300–7318 (2019)
24. Bek, D., Lim, M.: The circular economy: a key approach for addressing strategic business challenges in supply chains. Soc. Bus. **8**(1), 95–102 (2018)
25. Van Loon, P., Delagarde, C., Van Wassenhove, L.N.: The role of second-hand markets in circular business: a simple model for leasing versus selling consumer products. Int. J. Prod. Res. **56**(1–2), 960–973 (2018)
26. Lieder, M., Asif, F.M., Rashid, A., Mihelič, A., Kotnik, S.: A conjoint analysis of circular economy value propositions for consumers: using "washing machines in Stock-holm" as a case study. J. Clean. Prod. **172**, 264–273 (2018)
27. Sehnem, S., Jabbour, C.J.C., Pereira, S.C.F., de Sousa Jabbour, A.B.L.: Improving sustainable supply chains performance through operational excellence: circular economy approach. Resour. Conserv. Recycl. **149**, 236–248 (2019)
28. de Oliveira, M.C.C., Machado, M.C., Jabbour, C.J.C., de Sousa Jabbour, A.B.L.: Paving the way for the circular economy and more sustainable supply chains. Manag. Environ. Qual. Int. J. (2019)
29. Govindan, K., Hasanagic, M.: A systematic review on drivers, barriers, and practices towards circular economy: a supply chain perspective. Int. J. Prod. Res. **56**(1–2), 278–311 (2018)

Nanostores' Density and Geographical Location: An Empirical Study Under Urban Logistics Approach

Raul Soto-Peredo, Benito Sánchez-Lara, and Mariana Gómez-Eguiluz

Abstract In urban areas, nanostores are perhaps the main marketing channel for Fast-Moving Consumer Goods. However, nanostores are not usually considered in the design of distribution strategies, so relevant aspects that generate different impacts on their environment are neglected. This empirical study considers that the goods distribution to nanostores in the urban context are the cause of externalities and that therefore it is necessary to analyze the elements that affect this operation to facilitate the design of new distribution strategies under an urban logistics approach with the aim of seeking a balance between: habitability, mobility, sustainability, and resilience, while still fulfilling the purposes of the distribution logistics operation. This study relies on geographic information systems in order to look for similarities or dissimilarities based on the spatial distribution of some relevant elements, such as the number and geographic location of nanostores, the road infrastructure on which supply is supported, the location of a certain type of customers (millennials), as well as the relationship with other services located in proximity. As a result of the empirical study, despite not achieving conclusive results, the evidence will make it possible to establish new hypotheses for future research with better and more robust support.

Keywords Urban logistic · Nanostores · Spatial distribution · Externalities · Last mile

R. Soto-Peredo · B. Sánchez-Lara (✉) · M. Gómez-Eguiluz
National Autonomous University of Mexico, Mexico City, Mexico
e-mail: blara@unam.mx

R. Soto-Peredo
e-mail: raul.soto@uacm.edu.mx

M. Gómez-Eguiluz
e-mail: marianage.28@comunidad.unam.mx

A. Ochoa-Zezzatti et al. (eds.), *Technological and Industrial Applications Associated With Industry 4.0*, Studies in Systems, Decision and Control 347,
https://doi.org/10.1007/978-3-030-68663-5_18

1 Introduction

The traditional retail in Mexico has different names: neighborhood stores, proximity stores, small independent retailer, among others. Blanco and Fransoo [5] began to called them as nanostores.

Despite to many common identifiable characteristics in nanostores [29], their wide product assortment difficult characterize them. However, most of these economic units primarily sells Fast-Moving Consumer Goods (FMCG). A usual nanostore classification distinguishes among urban and rural stores. But there is no taxonomy or categorization related to their customer's characteristics and/or the nanostores' context [14].

Nanostores, from the economic point of view, fulfill two identifiable functions: (1) as source of employment (formal, informal and as self-employment) in micro businesses structures [18, 21] and (2) as the major distribution channel for FMCG. In Mexico nanostores represents about 48–70% of FMCG sales levels [17], in 2013 staff in this sector represent 28% of total national employed personnel. From the social point of view, supports the social texture established from the relationship between nanostores and its customers [13]. Also, from this relation, some strengths arise from the FMCG distribution that make nanostores attractive to customers, these are [5, 13, 14]:

- Product presentation versus price. Nanostores often make sales of individual products or in "small" presentations, for their clients this means requiring less cash for the purchase. It seems that they are price competitive, however, in net terms the products can become more expensive.
- Shopping experience. Usually, nanostores' owners know their clients personally and provide them with informal loans. Additionally, nanostores are neighborhood or community meeting points.
- Proximity. Nanostores save their clients times, both to get the products and to get to the store. Nanostores serve small groups of neighbors.

On the other hand, nanostores as brick and mortar stores will endure as part of physical location for the FMCG sales [10]. Not just because of the market share or the purchasing power of their customers, but for how nanostores fit the needs of customers, providing "personal touch" (achieving emotional proximity) and allowing physical proximity to where customers live or work [7, 11, 14].

According to Aithal et al. [1], nanostores' location decision is based on chance, availability, and opportunistic behavior. This approach is opposite of that used by large organized retailers because location decision results from considering elements such as freight generation, commodity flow, vehicle loading, vehicle design, cost, traffic flow, land use, building and site design, and pollution level [3]. This nanostore location decision leads to serving pedestrians and residents of the neighborhood [8]. Therefore, can be inferred that pedestrians have any other interest in neighborhood but housing. Consequently, the services located in the neighborhood, related to land use, supply customers to the nanostores.

Into a FMCG point of view, according to Murphy and Enis [26], FMCG look for a market saturation and, consequent, into a distribution strategy focused in moving high volume of products and prioritizing cost minimization. This approach excludes nanostores from all the logistics decision-making process because the relationships between distributors and retailers is given by contracts and high volume of exchanged products [30]. Also because those distributors are responsible for selecting the attributes of the vehicle to be used based on: type and quantity of products transported, the total number of trips, driving restrictions and other operating rules [4]. This is the last mile distribution approach.

Another urban distribution approach is based into a customer stereotype. The Nielsen Company [32] recognize, into a worldwide scope, a customer who is looking for convenience, has a busy lifestyle, is permanently online and moving. This customer is looking for more available time or to make better use of it into an environment characterized by: fast urbanization, smaller households, crowded public transport, gender roles evolution, generational needs and technology embracement. In order to fit into this type of customer, aligned to their mindset [32], the distribution looks for improve the level of service required, for example, through delivery time, time windows or online orders. As consequence, this shape some operational characteristics as fleet size, vehicle type and capacity, and other operating rules in order to fulfill this kind of customer and their needs.

The millennial population is part of this customer stereotype. They are recognized as the newest workforce, it is estimated that by 2025 the millennial population represent 75% of the workforce [16]. Also, according to Kelly et al. [25], the millennials prefer avoid mass channels distribution and are more careful about what to buy and where to buy it. In 2018, 33% of the U.S. millennial population chose traditional channels [20].

Nevertheless, nanostores' customer in Mexico does not fulfil that stereotype although neither the nanostores' consumer profile is well-defined but show other relevant features as [15]:

- does frequent and low-volume purchases,
- looks for small presentations related to a lower spend amount of money,
- prefers nanostores close to home with product assortment,
- embraces cash purchases, and
- usually makes to the nanostores on foot and does not have an own car.

Despite not observing a strong relationship between millennial population and nanostores' clients, the millennial population is of interest because they are an identifiable group in statistics population and because they share some characteristics with the nanostores' customer, such as saving time in the purchasing process, achieved from nanostores' proximity.

Finally, some externalities from nanostores' operation under urban logistics approach are [29]:

- traffic congestion, related to nanostores' procurement activities and clients purchasing;

- reduced mobility, due to transport infrastructure capacity, transport goods vehicles and loading and unloading goods;
- risk from handling goods in public spaces and public space appropriation;
- pollution, associated to distribution activities; and
- wealth, nanostores are job generators and constitute the major group of companies in Mexico as micro and small companies.

Thus, nanostores can be studied from different approaches and taking into account different elements of analysis. Is notorious their economic and social relevance. Their organizational structure, the way they are managed, how they compete with each other, how they survive as individual entities and how they relate to their context as groups, are work lines to explore and explain.

However, the approach used for this nanostores study observe balance between four Urban Logistics elements [17, 31]: mobility, to improve freight distribution without interruptions; sustainability, to reduce negative externalities under an economic, social and environmental approach; habitability, to improve freight distribution safety, reliability and protection, as well as handling products and waste, and to improve air quality, to develop safer communities and healthier lifestyles; and resilience, to stimulate response capacity to disruptions, such as disasters or supplies shortages.

This article shows the results of an empirical study on nanostores as the major distribution channel for FMCG. The study focusses on a random group of Basic Geostatistical Areas (BGA) from Mexico's Valley Metropolitan Area (MVMA). The study describes the nanostores by their number and geographic location as a relationship resulting from the transport infrastructure, the millennial population density and proximity to other services, such as education, health, sports or recreational facilities, and cultural facilities. In addition to the findings, the work offers inferences and future work lines.

The article is structured in four sections. In the introductory section, the background of the Mexican nanostores is given, emphasizing their economic and social importance. The second section describes the methodological framework for the empirical study and the tools used for data representation. From a Geographical Information Systems (GIS) perspective, this is a study with elements of spatial distribution and association. The third section shows BGA maps that allow contrasting by number, location, and spatial distribution of nanostores. The fourth section focuses on the results and their discussion, the findings and work lines. Finally, a set of conclusions is offered.

2 Empirical Study Design

This work is an investigation based on empirical evidence, acquiring knowledge through recording direct and indirect experience. This work is part of a body of research findings about the interaction between nanostores and the urban context and how this interaction affects urban logistics.

Table 1 BGA set studied

Municipality or county	Identifier number (ID)
Chiconcuac de Juárez	1503000010044
Ecatepec de Morelos	150330001014A
Benito Juárez	0901400010187
Benito Juárez	0901400010350

The study examines as cases a set of BGAs located in the MVMA. An BGA [23] is a geographical extension with three defining attributes: it is a territory perfectly recognizable, because it is delimited by identifiable and enduring topographic features; generally, it is homogeneous in the geographical, economic and social dimensions and its extension is such that it can be journey over by a single person. The BGA can be more or less urbanized. The most urbanized can vary in size between 20 and 80 blocks, the least urbanized can contain one or more districts with less than 100,000 inhabitants, depending in both cases on the housing density. The BGA were considered comparable zones in base of two statistical attributes: they are primary sampling units and organized information units. The collected data were gathered from the BGAs set presented in Table 1.

The considered BGAs have features that could distinguish them but for the study's purposes were used to improve contrast. A not completely homogeneous set was sought trying to find similarities or dissimilarities in the most notable way possible.

Figure 1 shows the studied BGAs. The BGA in Chiconcuac de Juárez municipality, in the State of Mexico, is an urban zona with proximity to rural areas and has a low population density. This is where the town hall is located. The BGA in Ecatepec de Morelos, is another municipality in the State of Mexico were the town hall and multiple services are located, its main land use is services. The BGA in Benito Juárez county, in Mexico City, is also considered as urban.

The restriction criteria for the BGAs selection were that the selected BGA were classified as urban, have registered 2,500 inhabitants or more or were located the town hall. The reason for use urban BGAs is because most freight distribution has an urban scope.

This study has elements of spatial distribution and spatial association. The spatial distribution [9] considers that a set of the same type entities has a concentrated or distributed space behavior, their pattern could be random, dispersed or regular way. The spatial distribution in statistical terms represents frequency, class intervals or density of an entity in the geographical space.

The spatial association looks for similarities based on the spatial distribution [9]. The association between two or more variables can be verified using the Exploratory Spatial Data Analysis (ESDA). ESDA allows identification atypical locations or outliers, discover patterns and suggest different spatial regimes and other forms of spatial instability [27].

The used spatial association in this study considered the number and geographical location of nanostores related to the transport infrastructure, the millennial population

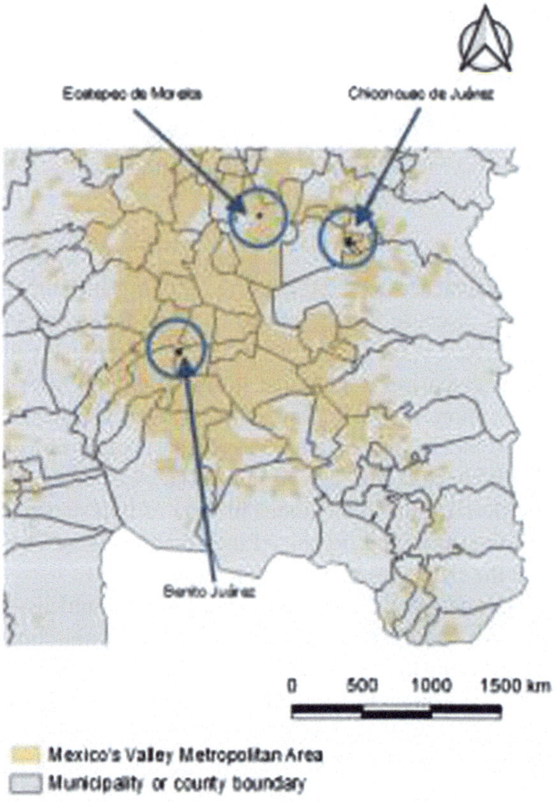

Fig. 1 BGA's geographic location

group, and other services location. The analysis was performed with QGIS software version 3.10.

The references about the millennial population group recognize different age ranges. But it was considered that millennial group is made up of people who were born between 1980 and 2000. The reason on focus on millennial group is because this group is assumed as the possible major customer of FMCG and, at the same time, the nanostores as the major distribution channel for FMCG [5, 32, 33]. Therefor a relationship is assumed between nanostores and millennial population. In this study the National Institute of Statistic and Geography 2010 National Population and Housing Census [22] was the primary information source.

The census data is disaggregated and georeferenced in BGA tier. The population data are stratified by age and gender. In particular, since the millennial population were between 10 and 30 years old at the moment of the census, the population strata considered were: 12–14, 15–17 and 18–24 years, also recognized as the population born between 1986 and 1998.

Transport infrastructure is a second variable for the study. It is relevant considering that freight transportation in the urban context coexists naturally with other transport systems and modes. Furthermore, the nanostores operations as sourcing, warehousing, and freight distribution are also factors that arises congestion and at the same time are affected by it.

The 2019 National Geostatistical Framework [24] was the source information for the transport infrastructure analysis. This framework includes georeferenced data, about state, municipal and BGA limits, in addition to roads in the national territory, insular territory, block polygons, road axes, green areas, rivers, railways, town hall, among other data.

The roads are classified into arterial, and collectors and local roads. Arterial roads are intended to improve continuous traffic flow or get traffic control by traffic lights. Collectors and local roads allow access into neighborhoods, also, these roads enable the flow of non-continuous vehicular traffic and parking is not restricted.

The third variable are other services located in the studied BGAs. The services considered in the study were those georeferenced available in the 2019 National Geostatistical Framework [24], such as medical assistance centers, educational facilities, sports or recreational facilities, cultural facilities, and temples.

At last, nanostores' information was gathered from the 2019 Economic Units National Statistic Directory (EUNSD) [21]. The data includes economic activity classification, name, location and labor size among other data from active economic units in Mexico. In EUNSD, nanostores are classified as 461110, known as "retail in grocery and miscellaneous stores". For the study, registered economic units in July 2010 were considered to provide consistency with the millennial population.

This study recognized that the information used is from different years, the population and nanostores information dates from 2010, meanwhile infrastructure and other services information dates from 2019. For the analysis, it was assumed that the BGAs, roads, and other services location information keep unchanged (constant) compared to 2010.

3 Results Analysis

The study has two types of findings: one corresponds to the exercise of association of variables using Geographic Information System (number of nanostores, type of road in the BGA and the millennial population group) and the other that correspond to the comparative analysis of urban BGAs.

In relation to the findings and association analysis, Table 2 summarizes the data and some statistics on the variables considered.

The data reveal the collectors and local roads as the prevailing road type where nanostores are located in every studied BGA. In other side, despite Ecatepec de Morelos register the highest nanostores number, at the same time, has the lowest millennial population ratio per nanostore. The opposite is present in the 0901400010187 BGA (Benito Juárez), has the lowest nanostores number and the highest millennial population ratio per nanostore.

The Chiconcuac de Juárez's area influences a different behavior to the rest of the BGAs. While a greater number of nanostores entails a smaller number of blocks

Table 2 BGAs' collected data and primary statistics

BGA name	Chiconcuac de Juárez	Ecatepec de Morelos	Benito Juárez	Benito Juárez
BGA ID	1503000010044	150330001014A	0901400010187	0901400010350
BGA area (km^2)	1.37	0.32	0.23	0.37
Blocks (number)	53	26	26	27
Millennial population (inhabitants)	1,144	369	830	389
(%)	25	18	15	12
Dominant class interval of millennial population	0–16	9–18	0–31	0–11
Blocks in dominant class interval of millennial population (%)	33	39	76	48
Road type where nanostores are located	Collectors and local roads	Collectors and local roads	Collectors and local roads	Collectors and local roads
Nanostores (number)	13	18	8	14
Nanostores density (number/km^2)	9	56	35	38
Blocks per nanostore	4.1	1.4	3.3	1.9
Nanostore per capita (population over 18 years old)	5 (per 100 inhabitants)	12 (per 100 inhabitants)	2 (per 100 inhabitants)	5 (per 100 inhabitants)
Average distance between nanostores (m)	194	58	126	76
Nanostore average coverage radius (m)	183	74	102	64
Millennial population ratio per nanostore	88	21	104	28

served, as well as a smaller average distance between nanostores and a smaller coverage area, Chiconcuac de Juárez has the opposite behavior.

Figures 2, 3, and 4 show the information about the transport infrastructure and the nanostores' geographical location. Millennial population per block is shown in Figs. 5, 6, and 7.

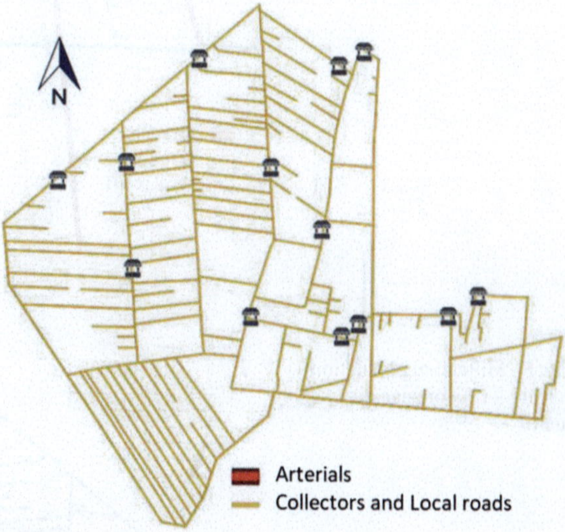

Fig. 2 Nanostores location—Chiconcuac de Juárez

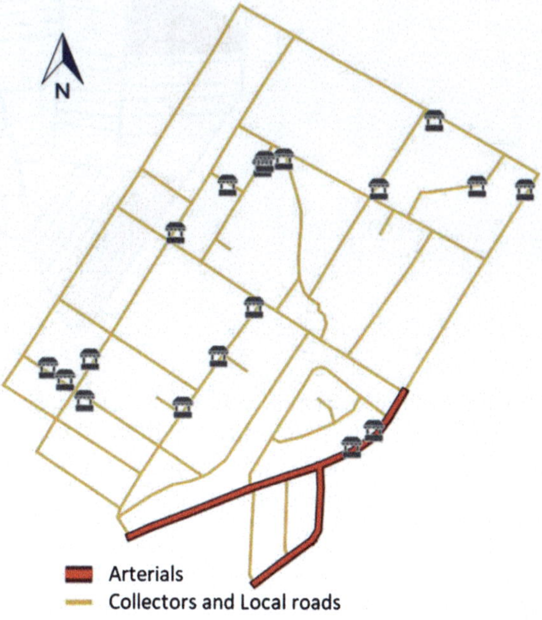

Fig. 3 Nanostores location—Ecatepec de Morelos

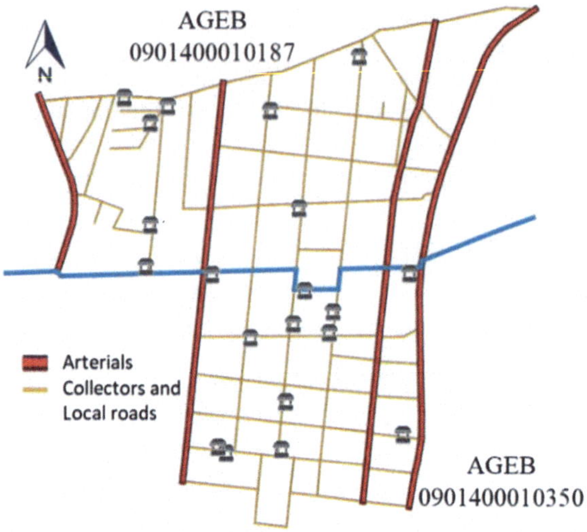

Fig. 4 Nanostores location—Benito Juárez

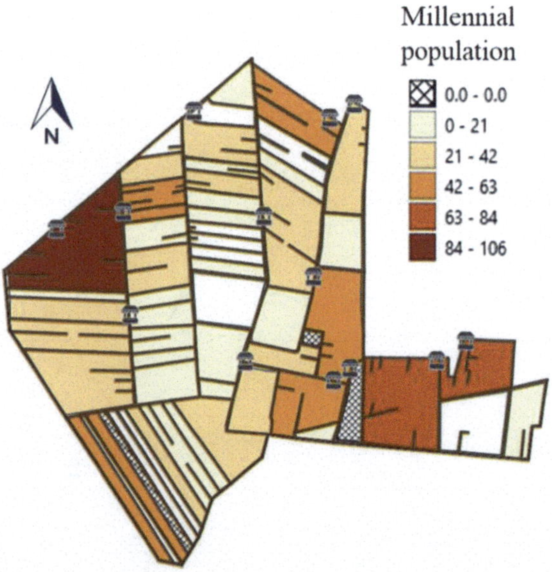

Fig. 5 Millennial population density—Chiconcuac de Juárez

In terms of transportation infrastructure, Chiconcuac de Juárez stands out for not having arterial roads. Benito Juárez's BGAs show the most extended arterial roads but no nanostore is located in those roads. It is only in Ecatepec de Morelos where two nanostores are located on arterial roads, these represent only 11% of nanostores in the BGA.

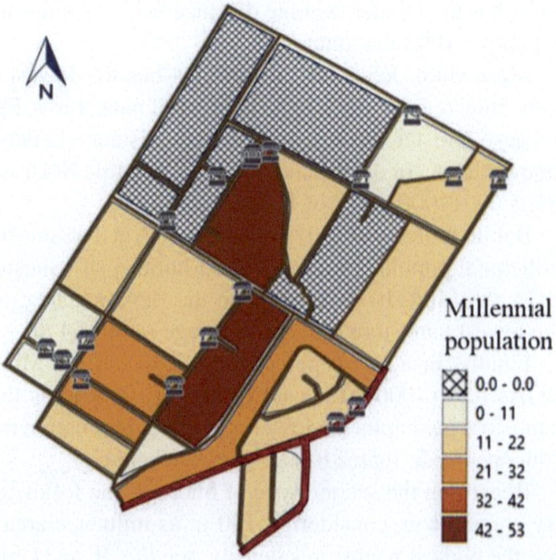

Fig. 6 Millennial population density—Ecatepec de Morelos

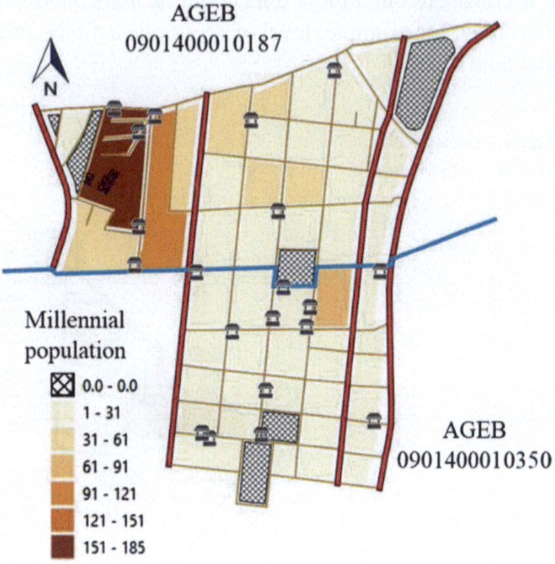

Fig. 7 Millennial population density—Benito Juárez

The Chiconcuac de Juárez's BGA stands out, by far, for register the largest millennial population (1,144 inhabitants) and area (1.37 km^2), Fig. 5. Using Voronoi polygons and estimating as average coverage radius, the covering radius by each nanostore is estimated in 183 m. This data is notorious even though this BGA has the largest number of blocks, that is, the BGA has the largest area. It is also remarkable that the

BGA has the greater average distance between nanostores but only 5 nanostores per capita per 100 inhabitants.

Meanwhile, Ecatepec de Morelos has the lowest millennial population, at the same time register the largest number of nanostores, Fig. 6. This BGA is exceptional because register the smallest average distance between nanostores than any other studied BGA. In other hand, the blocks in this BGA is similar to those accounted in BGA 0901400010187 of Benito Juárez.

Benito Juárez's BGA 0901400010187, at the same time, register the second largest millennial population and the least number of nanostores, Fig. 7. In other side, the BGA 0901400010350, also from the Benito Juárez, is distinguished by registering the lowest nanostore average coverage radius 64 m.

Finally, in a broad perspective, Ecatepec de Morelos and both Benito Juárez BGAs (0901400010187 and 0901400010350) has the same behavior. The higher nanostore per capita the lower millennial population ratio per nanostore. Meanwhile Chiconcuac de Juárez has an opposite behavior.

Regarding the second type of findings, the following paragraphs are formulated. By comparison, considering 100 m as influence area, Ecatepec de Morelos' BGA has the highest nanostores density, see Fig. 9, and Chiconcuac de Juárez's BGA the lowest, see Fig. 8. It is not possible to propose some type of conclusive association, it is considered that these data could be associated with variables not considered in this study, for example, level of economic activity, population socio-economic level and land use.

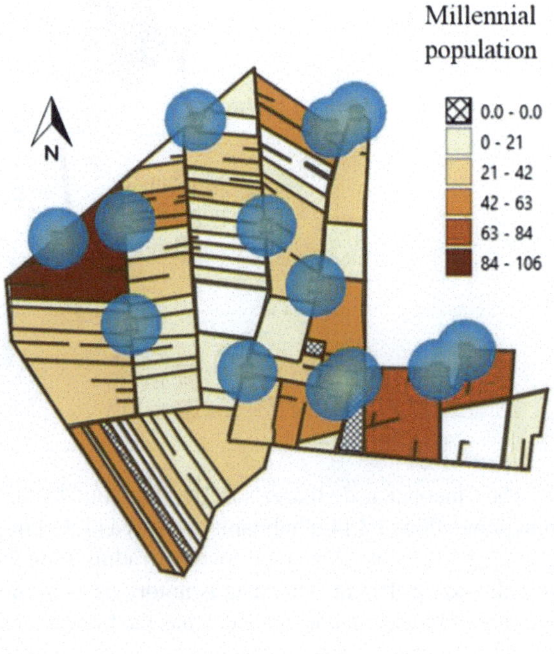

Fig. 8 Nanostores density—Chiconcuac de Juárez

In general, Figs. 8, 9, and 10, reveal a random behavior between nanostores' location and millennial population density. But confirm the nanostores absence on arterial roads.

Although Ecatepec de Morelos and both Benito Juarez's BGA has arterial roads, this condition is not observed as a decisive factor for nanostores densification.

At last, the services located in the studied BGAs are shown in Figs. 11, 12, and 13.

Fig. 9 Nanostores density—Ecatepec de Morelos

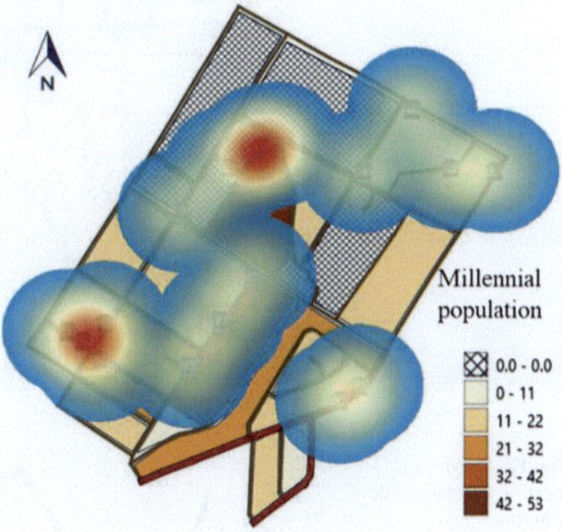

Fig. 10 Nanostores density—Benito Juárez

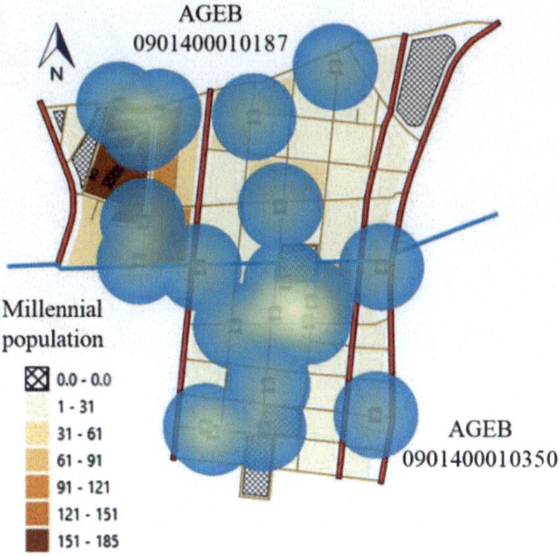

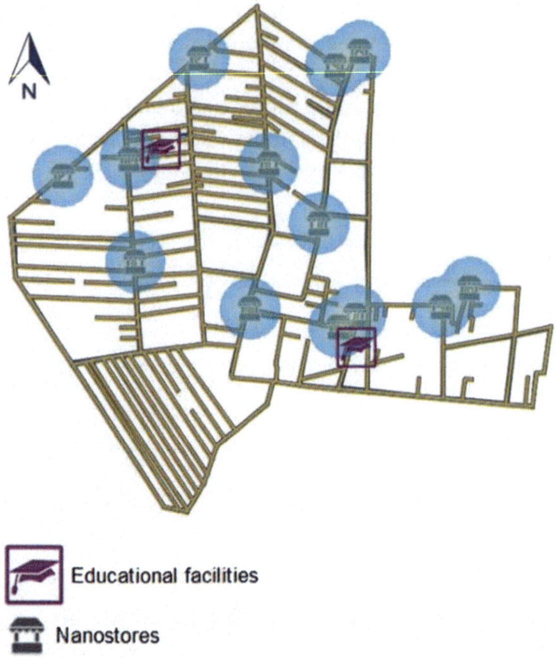

Fig. 11 Services location—Chiconcuac de Juárez

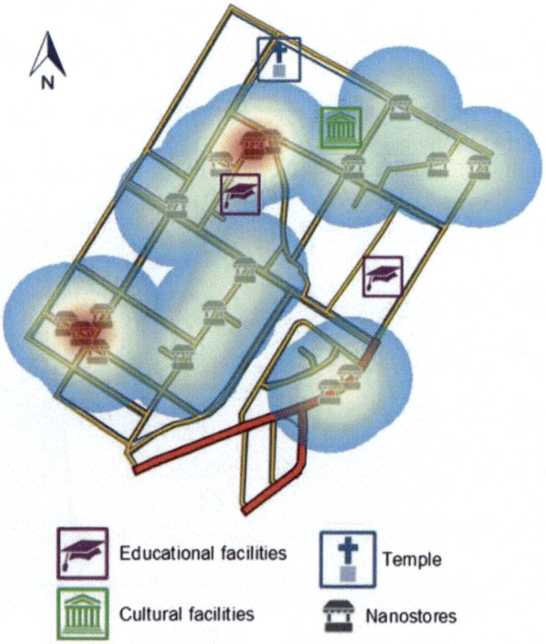

Fig. 12 Services location—Ecatepec de Morelos

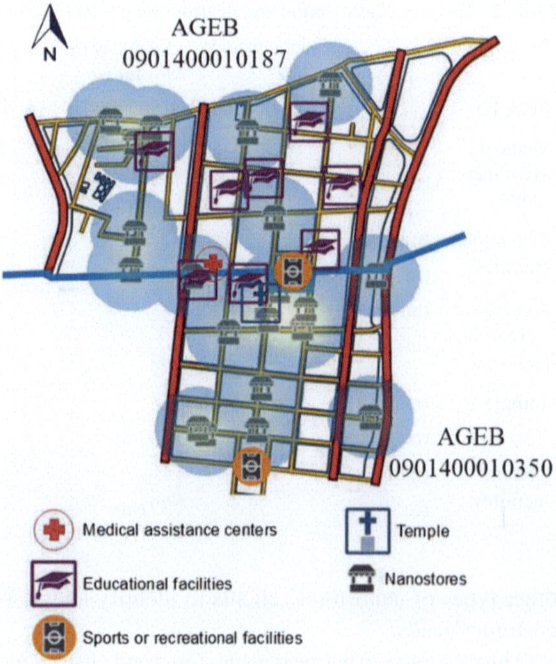

Fig. 13 Services location—Benito Juárez

The services types and number are different on each studied BGA. Based on this information and assuming a pedestrian walking at 5 km/h speed, moving according to the available roads along the urban layout, the average distances to get from a given service to any nanostore is the indicated in Table 3.

In global terms, a travel of 447 m (5.4 min) is needed from any service to reach a nanostore. But Chiconcuac de Juárez stands out doubling that travel distance.

Educational facilities are found in all BGAs, sports or recreational facilities as well as temples are present on 50% of BGAs and medical assistance centers or cultural facilities are registered on 25% of BGAs.

Despite the reduced services facilities existences, educational facilities have the shortest travel distance in Ecatepec de Morelos, meanwhile sports or recreational facilities have the closest travel in Benito Juárez's BGA 0901400010187. As to Benito Juárez's BGA 0901400010350, the temple and educational facilities have the closest proximity to a nanostore.

4 Findings, Limitations and Future Work

Based on data, millennial population is a specific market segment for FMCG, but an association cannot be fully accepted. Millennials does not explain by itself nanostores number or density in the studied BGAs. It is necessary to expand the study of

Table 3 Services classification and distribution per BGA, and proximity to nanostores

BGA name		Chiconcuac de Juárez	Ecatepec de Morelos	Benito Juárez	Benito Juárez
BGA ID		1503000010044	150330001014A	0901400010187	0901400010350
Medical assistance centers	(number)	–	–	1	–
	(m)	–	–	397	–
Educational facilities	(number)	2	2	6	2
	(m)	939 m	231 m	535 m	394 m
Sports or recreational facilities	(number)	–	–	1	1
	(m)	–	–	318 m	639 m
Temple	(number)	–	1	–	1
	(m)	–	503 m	–	361 m
Cultural facilities	(number)	–	1	–	–
	(m)	–	771 m	–	–

other types of nanostores' clients to identify and associate consumption habits and customer needs.

The relationship between nanostores and customers require a deep insight, considering a broader socio-economic strata analysis, as well as income, age and consumption habits. But the importance that the millennial population will assume cannot be underestimated, since it will be the most relevant workforce in the near future.

It is important to note that excluding Chiconcuac de Juárez, the number of nanostores in BGAs is inversely proportional to average distance between nanostores. That different behavior in Chiconcuac de Juárez is reinforced by the service and nanostore density. This lead to consider the existence of different BGAs which only common characteristics is to be located in the urban context.

Hence, it is important to emphasize that the studied BGAs were classified as part of the urban context but is not conclusive that this only characteristic is enough to make comparisons from any random BGA in the urban context.

Another kind of association is between nanostores location and transport infrastructure. From this association were expected a higher nanostore densification nearby a higher road hierarchy. However, this does not take place in the studied BGAs, it is more evident that nanostores are located in lower tier road hierarchy. This leads to believe that nanostores respond to neighbors' needs, in other words, the nanostores' target customer group are those residents in proximity. Being more precise, the target customers are located on the near blocks where nanostores are located.

Therefore, the nanostores location is not a function of roads hierarchy and neither nanostores density a function of millennial concentration. This allows to infer the following idea, the FMCG distribution strategy to nanostore is strongly related to wholesalers and distributors strategy rather than own nanostores' strategies. This idea agrees with Boulaksil and Belkora [8], according to the authors, the nanostores are highly fragmented and unorganized, provides little or no information to support decisions systems, and assume a passive roll into the placing order system. Under those characteristics can be recognize at least four distribution channels [5], all of them depends on manufacturer, wholesalers, or distributors but none on nanostores.

The observed BGAs are a very limited sample, the same as the variables used to study some type of association. Future work must extend the number of BGA, consider different socio-economic tiers as different nanostores' customers profiles, land use and a large transportation infrastructure perspective, as loading/unloading bays proposed by Cedillo-Campos [11].

Therefore, in addition to expanding the sampling BGAs, it is relevant for further investigation knowing or verifying aspects such as:

- Nanostore data and location, in order to identify the distribution and spatial concentration of the nanostores and to validate the available information. Because the information used about nanostores comes from 2014 Economic Census [21], significant changes associated with the permanence of these businesses are to be expected, considering that in Mexico 75% of the new companies close their operations within two years [2], so changes are to be expected between each economic census.
- Context and proximity. Validate information available is relevant to have an updated and reliable context. Besides, assuming the environment as a factor that drives the consumption of a certain set of goods and, consequently, defines the profile of a customer segment of the nanostores, can be used to characterize BGAs and/or the nanostores. It is expected that nanostores close to recreational areas sell a greater number of beverages (bottled water, soft drinks, juices, flavored drinks, etc.); those that are located near school zones sell mainly sweets, snacks and chips; and in those located in residential areas, the products consumed are more diverse.
- Competition. The nanostores contend between themselves. But wholesaler have had a growing interest to adopt smaller formats (similar to nanostores) to offer greater proximity to customers which has driven a greater competition, together with the proliferation of convenience stores.

Finally, other variables that have not been considered in this study but that are undoubtedly relevant because they are directly associated with nanostores and their operation are:

- Labor and physical characteristics of the nanostores. With regard to the labor characteristics, it is expected that the economic units of interest will be micro-enterprises, so there should be no more than five workers in each one of them. On the other hand, with regard to its physical characteristics, it pursue to characterize the accessibility that customer has with respect to the products, as one variable to qualify the shopping experience; at the same time, the sales floor and storage as indicators of the economic unit's capacity to carry out its operations, both sales and supply; finally, equipment as an indicator of the level of financial leverage and the capacity of the business unit to offer and diversify its services.

- Assortment and product presentation. Due to nanostores sell products in presentations (packages) of small sizes in order to adapt to the needs of their customers (adjustment dimension) [28], it is necessary to verify if nanostores choose to sell the fractionated products, in smaller portions than the original packaging of the product [7].

5 Conclusions

According to the data collected in this study, a direct relationship between the presence of high tier hierarchical road and the nanostores' location and distribution is not observed. Despite the fact that higher tier road promotes transit and flow of potential customers, the nanostores' location do not suggest to respond to this condition. The presence of primary road does not seem to be decisive criteria in choosing the site for a nanostore. One possible explanation is the parking restrictions on primary roads, this implies difficulty in accessing potential customers and merchandise providers.

The absence of nanostores on or nearby the primary roads could be more a reflection of the nanostores customers' own characteristics. Because in general, the typical nanostores' customers do not have their own car and arrive at the nanostore on foot. So, it is necessary to inquire more about the nanostores' behavior and their own operations, in particular supply operations and related actors, to get greater knowledge about how nanostores achieve satisfy their customer needs.

Into the behavior between millennial population density versus nanostores density, there is no clear correlation between these variables. For the Ecatepec de Morelos BGA, it is observed that the block with the highest millennials density match the area with the highest nanostores concentration, but it is important to note that this same block is surrounded by a cultural facilities, a cathedral, parks, government facilities and other businesses, so it is to be expected an atypical behavior. Even looking for the same pattern in other blocks it is observed that is not repeated in the area. On the contrary, in the same BGA, other areas with high nanostores density do not match with high millennial density. Therefore, it is not possible to affirm existence of a correlation or a random behavior between millennial population and nanostores density.

In the other side, the average distance from any service to reach a nanostore allows to infer a possible correlation between the service type and nanostores density, this is a sort of nanostore specialization. This idea can lead to a definition of a taxonomy of nanostores in Mexican context, which, based on their operating characteristics, improve the design of distribution strategies that continue to fulfill the purposes of this logistics activity but, at the same time, improve the fulfilling of some purposes of the urban logistics.

The literature indicates that in the distribution process the nanostores are not considered [4, 12, 19, 30], either as links or as participants in the supply chain, but as destination. So, it is estimated that the elements that make them different are left aside because environmental and social aspects are not considered. Therefore, it is necessary to identify the attributes of the nanostores that make the distribution strategies sustainable and integrate the very nature of the nanostores.

Finally, taking into account gathered data, the studied BGAs are not different nor contrasting by themselves. Hence, for subsequent empirical studies, this is considered to support a random BGA selection. In other words, the urban BGA category can be considered as the appropriate level of analysis for empirical studies. But, a more detailed characterization must be made, considering both BGAs and nanostores, if the study purpose is to optimize the use of limited financial resources or improve performance rates. As Ge et al. [19] indirectly do it to select the channel distribution to nanostores as market size function.

As the propose of this article is to aim into the nanostores relationship with customer profile, vial infrastructure, and proximity to other services, another relevant relationship have been sidelined. Therefore, description and decisions about distribution channel are explained by Blanco and Garza [6], Garza Ramirez [18] and Ge et al. [19]. Meanwhile into the stakeholder participation Cedillo-Campos [11, 12] has a broader explanation. And into nanostore relevance, characteristics and social functions ideas can be confronted in D'Andrea et al. [14], Booz Allen Hamilton [7], and Boulaksil and Belkora [8].

References

1. Aithal, S., Aithal, R.K., Maurya, H.: Exploring channel conflict in an emerging economy: the small retailer's perspective. Int. J. Retail Distribut. Manage. **45**(10), 1061–1078 (2017). https://doi.org/10.1108/IJRDM-03-2016-0037
2. Almada, A., Eaton, A., Flores, B., Mondragón, C., Ortiz, C., Jacob, E., de Anda, R.: El libro del fracaso. The Failure Institute, México (2014)
3. Anand, N., Van Duin, R., Quak, H., Tavasszy, L.: Relevance of city logistics modelling efforts: a review. Transport Rev. **35**(6), 701–719 (2015)
4. Antún, J.: Distribución Urbana de Mercancías: Estrategias con Centros Logísticos, vols. IDB-TN-167. Banco Interamericano de Desarrollo (2013)
5. Blanco, E.E., Fransoo, J.C.: Reaching 50 million nanostores: retail distribution in emerging megacities. Working paper, BETA Research School, Eindhoven University of Technology (2013)
6. Blanco, E.E., Garza, J.: Multichannel distribution strategies in Latin America. In: 4th World Conference on Production & Operations Management (P&OM), (July) (2012)
7. Booz Allen Hamilton: Creando valor en canales minoristas para consumidores emergentes. Coca-Cola Retailing Research Council—América Latina (2003)
8. Boulaksil, Y., Belkora, M.J.: Distribution strategies toward nanostores in emerging markets: the Valencia case. Interfaces **47**(6), 505–517 (2017). https://doi.org/10.1287/inte.2017.0914
9. Buzai, G.: Capítulo 7. Análisis Espacial con Sistemas de Información Geográfica. Sus cinco conceptos fundamentales. En G. Buzai, Geografía y Sistemas de Información Geográfica. Aspectos conceptuales y aplicaciones. GESIG (2009)
10. Calvillo, E.: Deloitte. Recuperado el 07 de Marzo de 2018, de https://www2.de-loitte.com/mx/es/pages/dnoticias/articles/futuro-tiendas-fisicas-mexico.html (2018)
11. Cedillo-Campos, M.: LogistiX-Lab: Primer Laboratorio de Logística Urbana en Tiempo Real de América Latina. Presentación disponible en línea: https://www.youtube.com/watch?v=O0w9k88kdCg. Premio InterTraffic Latinoamerica, Ciudad de México, México (2019)

12. Cedillo-Campos, M., Fransoo, J.: Distribución urbana de mercancías aplicando la inteligencia colectiva en logística. Revista Ingeniería Civil, Colegio de Ingenieros Civiles de México, No. 599, Julio, pp. 8–12 (2019)
13. Coronado, S.: Proximidad y precio, factores que atraen clientes: Euromonitor. El Financiero (4 de Abril de 2018)
14. D'Andrea, G., Lopez-Aleman, B., Stengel, A.: Why small retailers endure in Latin America. Int. J. Retail Distribut. Manage. **34**(9), 661–673 (2006). https://doi.org/10.1108/09590550610683184
15. Escamilla, R., Freire, R., Mejía, C.: Nanostores: as bajo la manga en estrategias de suministro exitosas. Énfasis Logística, 36–37 (2019)
16. European Field Marketing Partners: How Millennials are influencing FMCG consumer trends. Recuperado el 01 de Marzo de 2020, de https://www.efmp.com/articles/how-millennials-are-influencing-fmcg-consumer-trends/ (2019)
17. Fahimnia, B., Bell, M., Hensher, D., Sarkis, J.: Green Logistics and Transportation. A Sustainable Supply Chain Perspective, 1 edn. Springer International Publishing. http://link.springer.com/10.1007/978-3-319-17181-4 (2015)
18. Garza Ramirez, J.: Distribution Strategies in Emerging Markets: Case Studies in Latin America Master of Science in Engineering and Management Massachusetts Institute of Technology By. Massachusetts Institute of Technology (2011)
19. Ge, J., Honhon, D., Fransoo, J.C., Zhao, L.: Supplying to Mom and Pop: traditional retail channel selection in megacities. Manu. Service Operat. Manage. 23–31 (February) (2020). https://doi.org/10.1287/msom.2019.0806
20. Helms, G., Kaufman, N., Mehta, N., Taylor, N., Scaglione, C., Wirth, J.: Millennials on Millennials. The Nielsen Company (2018)
21. Instituto Nacional de Estadística y Geografía: Directorio Estadístico Nacional de Unidades Económicas. Obtenido de Inicio/Servicios/Directorio Estadístico Nacional de Unidades Económicas: http://www.beta.inegi.org.mx/app/mapa/denue/ (noviembre de 2019)
22. Instituto Nacional de Estadística y Geografía: Inicio/Programas/Censo de Población y Vivienda 2010. Retrieved from Censo de Población y Vivienda 2010: https://www.inegi.org.mx/programas/ccpv/2010/default.html#Datos_abiertos (2020)
23. Instituto Nacional de Estadística y Geografía: Inicio/Programas/Micronegocios 2008/Glosario. Retrieved from Glosario: https://www.inegi.org.mx/app/glosario/de-fault.html?p=ENAMIN 2008 (2020)
24. Instituto Nacional de Estadística y Geografía: Inicio/Servicios/Mapas. Retrieved from Marco Geoestadístico, septiembre 2019: https://www.inegi.org.mx/app/biblioteca/fi-cha.html?upc=889463776079 (2020)
25. Kelly, G., Kopka, U., Küpper, J., Moulton, J.: The New Model for Consumer Goods. McKinsey & Company (2018)
26. Murphy, P., Enis, B.: Classifying products strategically. J. Market. **50**, 24–42 (Julio de 1986)
27. Oxoli, D., Prestifilippo, G., Zurbarán, M.: Enabling spatial autocorrelation mapping in QGIS: The Hotspot Analysis Plugin. Geoingegneria Ambientale e Mineraria **LIV**(2), 45–50 (2017)
28. Sapien Nitro: The bottom line on experience: measuring return in the age of storytelling. Boston, Massachusetts, Estados Unidos (2016)
29. Soto-Peredo, R., Sánchez-Lara, B., Gayosso, R.I.: Distribución detallista de mercancías de consumo de rápido movimiento: caracterización de una unidad de análisis. Ci-LOG 2019 (2019)
30. Taniguchi, E., Thompson, R.G.: City Logistics. Mapping the Future. CRC Press, Boca Raton, Florida (2015). https://doi.org/10.1007/978-1-4614-3628-7
31. Taniguchi, E., Thompson, R., Yamada, T., van Duin, R.: City Logistics. Network Modelling and Intelligent Transport Systems. Pergamon. Emerald Group Publishing Limited (2001). https://doi.org/10.1007/s13398-014-0173-7.2
32. The Nielsen Company: The millennial mindset. Unlocking a generation of potential (2016)
33. Nielsen Company: La búsqueda de conveniencia. México, México (2018)

Blockchain Model Implementation to Select the Best Bid in an Industrial Supply Chain

María Inés Borunda-Aguilar, Iván Juan Carlos Pérez-Olguín, Alberto Ochoa-Zezzatti, Erwin Adan Martinez-Gomez, and Gilberto Ochoa

Abstract This paper presents an intelligent model which uses Artificial Intelligence within the Industry 4.0. It is of utmost importance to be able to choose the best supplier in order to identify a competitiveness model associated with the reduction of direct and indirect costs. The present investigation describes a novel *Blockchain* that uses big data for adequately determining critical factors of the purchasing department and then using a hashing model verifying and validating with real data from a manufacturing company as a components supplier within the automotive industry. This research focuses mainly on specifying the mathematical modelling involved in Blockchain and its implementation using innovative factors from the financial area.

Keywords Blockchain · Supply chain · Optimal bid using artificial intelligence

M. I. Borunda-Aguilar (✉) · I. J. C. Pérez-Olguín · A. Ochoa-Zezzatti · E. A. Martinez-Gomez
Department of Industrial Engineering and Manufacturing, Autonomous University of Ciudad Juarez, Ciudad Juarez, Mexico
e-mail: maria.borunda@uacj.mx

I. J. C. Pérez-Olguín
e-mail: ivan.perez@uacj.mx

A. Ochoa-Zezzatti
e-mail: alberto.ochoa@uacj.mx

E. A. Martinez-Gomez
e-mail: emartine@uacj.mx

G. Ochoa
ITESM, Campus Guadalajara, Jalisco, México
e-mail: gilberto.ochoa@tec.mx

A. Ochoa-Zezzatti et al. (eds.), *Technological and Industrial Applications Associated With Industry 4.0*, Studies in Systems, Decision and Control 347,
https://doi.org/10.1007/978-3-030-68663-5_19

1 Introduction

As part of a methodology to facilitate the decision-making process, a tool which makes use of Fuzzy Multiple Criteria Decision Making (FMCDM), is presented to help suppliers in the manufacturing industry respond to quotation requests, belonging these suppliers to a globalized supply chain. The proposed system offers criteria to fulfill requests for quotations and shows the best supplier alternatives when making comparisons regarding quality, cost, and payment terms. This allows the decision maker to consider all the elements and thereby select the best alternative so that both organization and customer obtain the best possible value (cost reduction, competitive price, among others). Guaranteeing the reliability of the choice through *Blockchain* and by an encryption or hash encryption algorithm (SHA256).

It is common practice, within globalized supply chains, to consider variation in sales prices, depending on manufacturing and logistics costs, required volume, or specific packaging requirements. Resulting in unique conditions for each customer negotiation, whether is a one-time purchase or long-term contract. This negotiation process is called Request for Quote (RFQ), whose information is fed by buyers, several engineering departments, as well as suppliers using buyers as an interface.

In addition to the sales prices, within the RFQ process, sales and payment conditions, delivery dates, product specifications and features are negotiated. While considering business objectives, such as production efficiency, profitability, and customer satisfaction.

Finally, for the reliability of the entered values into the system database that contains the estimates fed by the RFQ, it is proposed the use of *Blockchain,* which obtains the information entered by each department involved, as it is; raw materials, supplies, machinery and equipment; thus, building new integration methods based on the information consolidation in a decentralized database. In addition, for the multi-criteria decision-making process used, evaluations are considered for qualitative attributes by weighting them through a Likert scale, providing quantitative attributes for strategic reasoning techniques, suggesting alternatives that maximizes the plaintiff company's benefits.

2 Related Works

RFQ system is used to match the excess inventory of multiple suppliers with the expected customer's demand, the resulting matches are presented to the customers, when the product's cost is below the current market prices this system uses new integration methods with web services to obtain information on supply and demand [1].

From this integration models, [2] describes a holistic literature review of recent work on block supply chains, including both the financial and business perspective. In addition to discussing the barriers and challenges of the adoption of *Blockchain* in the supply chain, through a qualitative approach. However, it does not present an adaptation model for the implementation of *Blockchain.*

For [3] the value of these technologies (*Blockchain*) for supply chain management lies in four areas: greater visibility and traceability, digitization, and disintermediation of the supply chain, as well as improvement in security of data and smart contracts. Presenting two works in January and May 2019, in [4] industrial applications in different areas are analyzed, [5] adopts *Blockchain* in the supply chain, [6] discusses the value of *Blockchain* application in smart contracts, [7] proposed *Blockchain* applications for influencing existing laws and regulations, and [8] built a shared safety net system for data management.

Overall, the *Blockchain* technology enforcement in supply chains offers numerous opportunities for researchers to address innate issues of the technology (such as scalability) and of practitioners scientific community in general, to bring the technology in our everyday lives and closer to the understanding and acceptance of the general public [1]. The authors only realized an extensive study of the existing literature.

Nowadays the literature focuses on the benefits and usefulness of the implementation of *Blockchain*, especially in the supply chain. This is due to the movement sensitivity in some products which guarantees that they are transported with the appropriate measures and specifications, based on its needs, to ensure product quality and veracity.

The *Blockchain*'s traceability property is one of the features that attracted the attention of the business world [9]. Organizations are often part of large and complex networks in which transactions of products and services are made, from production and creation to consumer and utilization. The supply chain is made up of the network of organizations, such as companies, transport, retailers, wholesalers, suppliers, consumers and the activities and transactions between them.

Figure 1 shows a vision of the supply chain considering the product cycle, from the raw material to the customer. Basically, a supply chain seeks to satisfy consumer needs for a better-quality product or service [10] with the smallest amount of inventory for the producer or retailer.

However, it is still considered more beneficial to use *Blockchain* in business and finance, basically where there is economic flow. Leaving a wide research and implementation opportunity in industrial areas, as corporate clusters.

Therefore, after current job reviews available on the network, no consolidation was found in the Request for Quote processes with *Blockchain* used to feed the estimating systems within the industry. Hence the innovation of this document, where Blockchain presents a new growing and demanding market niche. Taking into account the use of security tools based on cryptography, through the hash functions 256, guaranteeing the reliability and veracity of the information during the Request for Quote process, by bearing out that each department information does not suffers any modification or alteration.

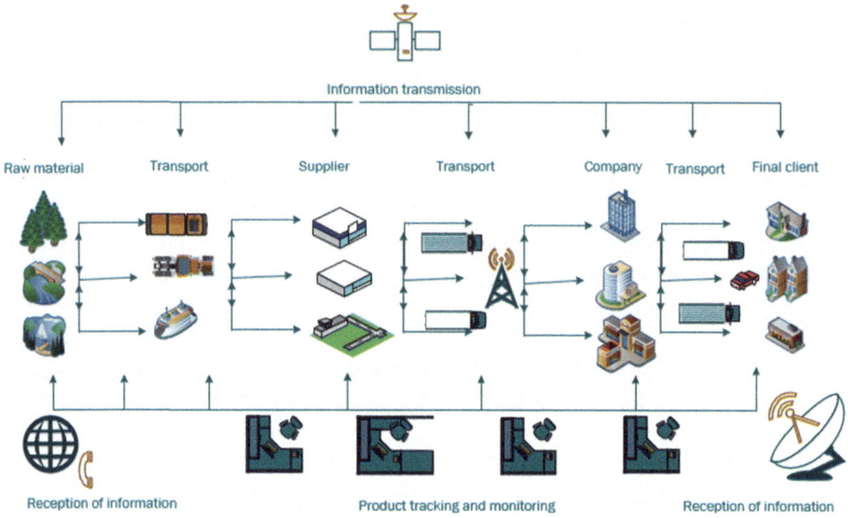

Fig. 1 Supply chain using *Blockchain* tracking monitoring

3 Proposed Model

The computer security model based on Blockchain, requires a long process of adaptation, modification, and awareness of the different departments involved and, therefore, of the personnel in charge, requiring a collaboration culture. Although, changing the organizational culture based on previously established processes creates a huge challenge in the industry, especially if some processes have been faulty over the years. Therefore, it is essential to count with the participation and openness of both management levels and key personnel from different departments. This project presents an opportunity for the industry's improvement, ensuring reliable information for assertive decision-making based on the company objective's achievement. Figure 2 shows a Blockchain model within a company, incorporating the previously security model in its design. As far as users are concerned, this is where the suppliers and products' selection begin with the MCDM model. The miners are replaced by a distributed network of virtual machines which are interconnected certifying each transaction reliability. At this point, the Hash 256 model is established.

The Blockchain provides an infrastructure where transactions take place in a transparent, secure and reliable manner, without the need for central organization. The blocks are validated by a virtual network, being all the strings identical, and stored in each decentralized computer sending them to the database. The distributed system allows data synchronization through the computer virtual network through a consensus mechanism. Validating the information contained in the previous block of the chain through a SHA 256 hash. Therefore, once the transaction is registered within the *Blockchain*, it cannot be manipulated, since a malicious act would manifest itself as an inconsistency between the *Blockchain* maintained individually, sending an alert

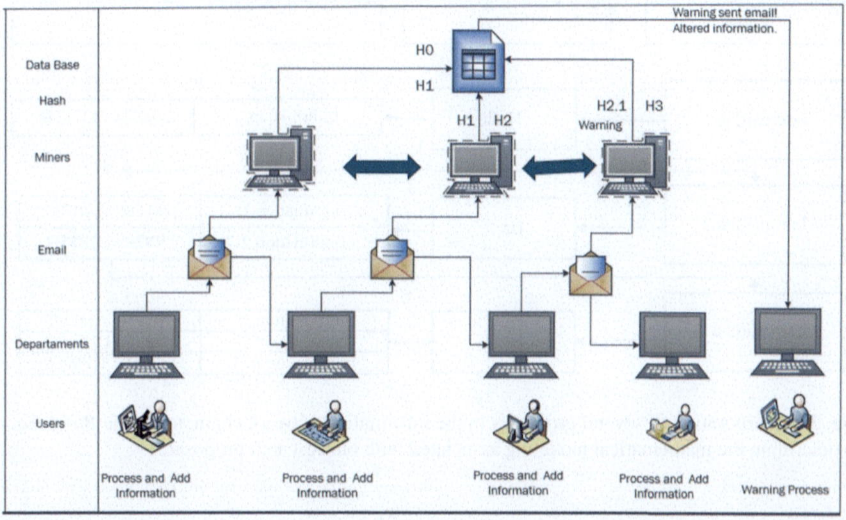

Fig. 2 Model for the estimation process based on *Blockchain* considering the critical financial factors

message to the central network of inconsistency in the information and reporting the machine that presents it, invalidating the process.

The fundamental feature of the model proposed consists in being a distributed system, making the *Blockchain* attractive for the supply chain implementation, since it implies a series of benefits [11], as is the management of transactions and information storage in different nodes, which should compare the immutability of information to ensure the reliability of this through a local consensus. There is no need for negotiation or reconciliation among the various participants in the supply chain. Fraudulent and adverse attempts are completely ruled out.

It is important to mention that for the Hash 256 choice, the creation of the nested Hash model and the *Blockchain* model, previous articles were created for their evaluation, and this document presents the incorporation of these in an alternate system, which incorporates from the selection of the product and suppliers to the supply chain in an information system for decision making.

The next step is to ensure that the selected suppliers are the ideal manufacturer for the company to incorporate in its manufacturing process. For this to be reached, the information must be reliable through the Blockchain system, where it will go through each department involved, validating the information with the SHA256 hash. At this point, the model's improvement and proposal is to have the hash nested as explained in the following paragraphs, because currently the use of the hash only encrypts one information at a time; e.g. each department information is entered and encrypted separately.

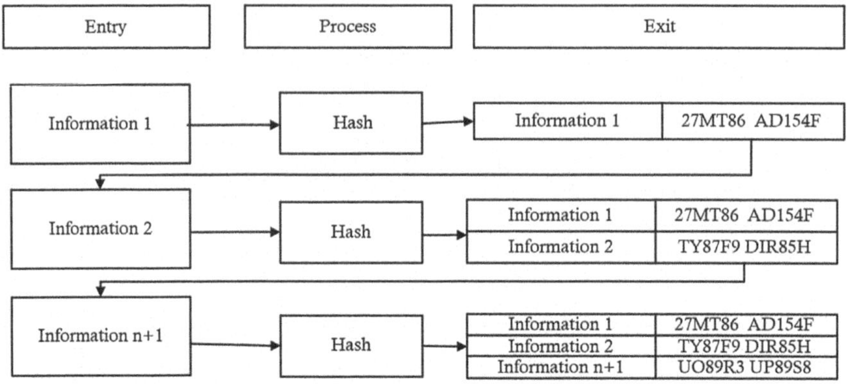

Fig. 3 Incorporation of several processes in the information network chain, using the *Blockchain* to determine the mathematical modeling associated with our research proposal

In this part of the model, the improvement is that currently the use of the hash only encrypts one information at a time, that is, the information of a certain department is entered and encrypted separately, however, we present the possibility of having the hash nested as explained below.

Figure 3 shows the interaction of departments A, B and C, where A sends information to B, B cannot modify it, but can add new information and transfer it to C, where C in turn can add information, but not modify the former, as it is protected by Hash cryptology. Creating a reliable ecosystem between suppliers and their customers [12]. This is achieved through a policy that focuses on chain transparency to ensure product traceability, where accurate data collection and secure data storage is required. Suppliers and the company must be synchronized through the use of the system, which will be used to transfer information to the different departments involved.

3.1 Content Modeling

Code proposal for Hash 256:

```
function out = sha256( msg )
% Initial Hash Values(8 constant 32-bit words).(§5.3.3)
default_hash = [
'6a09e667';
'bb67ae85';
'3c6ef372';
'a54ff53a';
'510e527f';
```

```
'9b05688c';
'1f83d9ab';
'5be0cd19'
];
% Constant value array (64 constant 32-bit words) to be used for the iteration t of the hash
computation.(§4.2.2)
K = [
'428a2f98'; '71374491'; 'b5c0fbcf'; 'e9b5dba5';
'3956c25b'; '59f111f1'; '923f82a4'; 'ab1c5ed5';
'd807aa98'; '12835b01'; '243185be'; '550c7dc3';
'72be5d74'; '80deb1fe'; '9bdc06a7'; 'c19bf174';
'e49b69c1'; 'efbe4786'; '0fc19dc6'; '240ca1cc';
'2de92c6f'; '4a7484aa'; '5cb0a9dc'; '76f988da';
'983e5152'; 'a831c66d'; 'b00327c8'; 'bf597fc7';
'c6e00bf3'; 'd5a79147'; '06ca6351'; '14292967';
'27b70a85'; '2e1b2138'; '4d2c6dfc'; '53380d13';
'650a7354'; '766a0abb'; '81c2c92e'; '92722c85';
'a2bfe8a1'; 'a81a664b'; 'c24b8b70'; 'c76c51a3';
'd192e819'; 'd6990624'; 'f40e3585'; '106aa070';
'19a4c116'; '1e376c08'; '2748774c'; '34b0bcb5';
'391c0cb3'; '4ed8aa4a'; '5b9cca4f'; '682e6ff3';
'748f82ee'; '78a5636f'; '84c87814'; '8cc70208';
'90befffa'; 'a4506ceb'; 'bef9a3f7'; 'c67178f2'
];
% First padd the input message to be a multiple of 512(bits).(§5)
[padded_msg,padded_len] = padder( msg );
% Split padded message to N (512-bit) blocks.(§6)
[M,total_blocks] = split2block( padded_msg,padded_len );
W = zeros( 64, 32 );
H = zeros( 8, 32 );
% Main SHA-256 compuation process.(§6.2.2)
for j = 1:8 % Load initial hash values at first iteration.
H(j,:) = hexToBinaryVector( default_hash( j , : ) , 32 );
end
for i = 1:total_blocks % For every block M(i).
% Step 1 - Prepare the message schedule.
for j = 1:64
if j >= 1 && j <= 16
W( j, 1:32 ) = M( i, 32*(j-1)+1:j*32 );
else
W( j, 1:32 ) = mod32add( sigma1( W( j-2, : ) ), W( j-7, : ) , sigma0( W( j-15, : ) ), W( j-16, : ) );
      end
    end
```

User creation code:

```
def Save():
  if (USERV.get() !="") & (PASSV.get() != "") & (Area.get() != "DEPARTAMENT"):
    crsr.execute("SELECT * FROM USER WHERE USUARIOS= '%s'" % USERV.get())
    ans= crsr.fetchall()
    valueUS=""
    for i in ans:
     valueUS= str(i[1])
    if (USERV.get()!= valueUS):
```

```
CountDB= open("Service/Firmas/Datos/CountDB.txt", "r")
CDB= int(CountDB.read())
CountDB.close()
cote=datetime.datetime.now().strftime("%m/%d/%Y")
User= USERV.get()
Password= PASSV.get()
str1= User+ Area.get()
HASH= hashlib.sha256(str1.encode())
HASH= HASH.hexdigest()
sql_command = """INSERT INTO USER VALUES (?, ?, ?, ?, ?, ?);"""
crsr.execute(sql_command, (CDB,User,Area.get(),Password,HASH,cote))
connection.commit()
NewNumDB=str(1+CDB)
CountDB= open("Service/Firmas/Datos/CountDB.txt", "w")
CountDB.write(NewNumDB)
CountDB.close()
messagebox.showinfo("Admin", "Usuario Agregado %s" % USERV.get()+"\n"+ HASH)
USERV.set("")
PASSV.set("")
Area.set("DEPARTAMENT")
else:
messagebox.showerror("Usuarios", "Usuario repetido")
else:
messagebox.showerror("Campos vacios", "Llene todos los campos")
```

Process, at this point the information is validated by each department and consolidated in an excel book:

```
NValue= "APPROVED"
DateAppr= datetime.datetime.now().strftime("%d/%m/%Y")
crsr.execute("UPDATE HASH SET DateFin = '"+DateAppr+"' where
SERDOC='"+SerDocCla+"'")
crsr.execute("UPDATE HASH SET ESTATUS = '"+NValue+"' where
SERDOC='"+SerDocCla+"'")
connection.commit() ACtualiza la fecha y el estatus en la tabla
messagebox.showinfo("Finish Request for quote", "Approved" )
elif (i[2]=="Div. Industrial Engineer"):
SerDoc=window.directory
SerDoc= SerDoc.split("/")
SerDocCla= SerDoc[-1]
DepUser= i[2]
UserFir= i[4]
crsr.execute("SELECT * FROM HASH WHERE SERDOC= '%s'" % SerDocCla)
ans= crsr.fetchall()
for i in ans:
if (i[1]==SerDocCla)&(i[8]=="")&(i[9]=="")&(i[34]=="")&(i[36]=="")&(i[35]=="IN
PROCESS"):
crsr.execute("UPDATE HASH SET USER2 = '"+USERV.get()+"' where
SERDOC='"+SerDocCla+"'")
```

```
            crsr.execute("UPDATE HASH SET F2 = '"+UserFir+"' where SERDOC='"+SerDocCla+"'")
            connection.commit()
            messagebox.showinfo("Approved", "Signed by %s" % USERV.get()+ "\n"+DepUser)
          else:
            messagebox.showerror("User", "No request signed or document not found")
          window.directory=""
          if (i[5]!="")&(i[7]!="")&(i[34]=="")&(i[35]=="IN PROCESS")&(i[36]=="")&(i[9]!="")&(i[11]!="")&(i[13]!="")&(i[15]!="")&(i[17]!="")&(i[19]!="")&(i[21]!="")&(i[23]!="")&(i[25]!="")&(i[27]!="")&(i[29]!="")&(i[31]!="")&(i[33]!=""):
            NValue= "APPROVED"
            DateAppr= datetime.datetime.now().strftime("%d/%m/%Y")
            crsr.execute("UPDATE HASH SET DateFin = '"+DateAppr+"' where SERDOC='"+SerDocCla+"'")
            crsr.execute("UPDATE HASH SET ESTATUS = '"+NValue+"' where SERDOC='"+SerDocCla+"'")

            connection.commit()
            messagebox.showinfo("Finish Request for quote", "Approved" )
      elif (i[2]=="Metal Tooling Engineer"):
        SerDoc=window.directory
        SerDoc= SerDoc.split("/")
        SerDocCla= SerDoc[-1]
        DepUser= i[2]
        UserFir= i[4]
        crsr.execute("SELECT * FROM HASH WHERE SERDOC= '%s'" % SerDocCla)
        ans= crsr.fetchall()
        for i in ans:
          if (i[1]==SerDocCla)&(i[10]=="")&(i[11]=="")&(i[34]=="")&(i[36]=="")&(i[35]=="IN PROCESS"):
            crsr.execute("UPDATE HASH SET USER3 = '"+USERV.get()+"' where SERDOC='"+SerDocCla+"'")
            crsr.execute("UPDATE HASH SET F3 = '"+UserFir+"' where SERDOC='"+SerDocCla+"'")
            connection.commit()
            messagebox.showinfo("Approved", "Signed by %s" % USERV.get()+ "\n"+DepUser)
          else:
            messagebox.showerror("User", "No request signed or document not found")
          window.directory=""
```

The results of the previous codes are the following figures. Figure 4 shows how when each user is registered in the system, a hash is immediately assigned, which in the union of his name + surname + department = hashSHA256, as a security prototype, (only for visualization reasons, it can be seen,) is stored in the distributed database to validate the following processes.

Figure 5 shows a different system where the user acquires the Excel document (where the information is stored for each department involved in the corresponding area) and adds the content that it requires, if it considers that everything is fine, it validates with the hash, transferring the document to the next department.

As can be seen, in Fig. 6, a tracking system is presented to define each process and its proper documentation, waiting on the validation of the final department for joint approval. At this point, the information from each department is considered as reliable, however in the following and last step the information is confirmed as valid. Using the multicriteria selection process, it can be determined which products is the most beneficial to be manufactured.

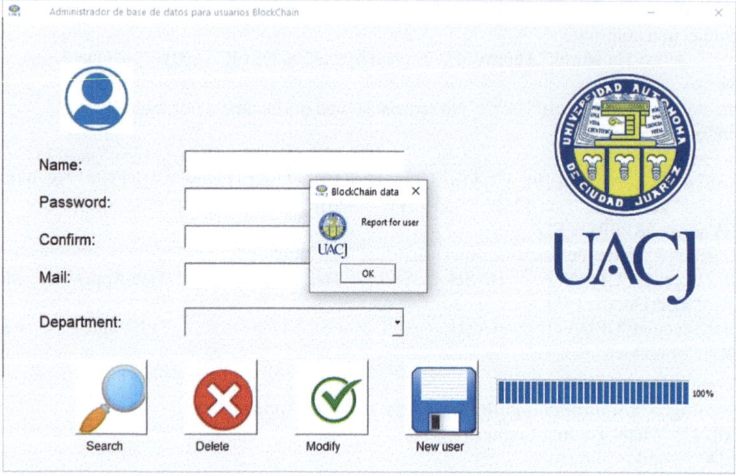

Fig. 4 SHA256 Hash mapping

Fig. 5 Validation by department

The model proposes the use of FMCDM, to select the best supplier and product. On the last step, the formulas, and the development of Manoj Mathew [13] is applied.

On the following section of this document, the development of the analytical tool would be described step by step, in order to show the results obtained.

A general multi-criteria decision problem with m alternatives A_i $(i = 1\ldots, m)$ and n criteria C_j $(j = 1, \ldots, n)$ can be expressed as follows:

Fig. 6 Tracking and validation system by department

$$D = [x_{ij}] \text{with} i = 1, ..., m \text{and} W = (w_j) \text{with} j = 1, ..., n. \quad (1)$$

D refers to the decision matrix, x_{ij} represents the input value for the alternative A_i, which are related the criterion C_j and W refers to the vector of weights, represented by w_j the weighting value of the criterion C_j. The classification of the criteria is:

Profit Criteria: where the best value for decision making is the highest value of x_{ij}.
Cost Criteria: where the best value for decision making is the lowest value of x_{ij}.

In the case of FMCDM the x_{ij} values are obtained from a combined decision matrix, which is developed with criteria of different scales, $x_{ij} = \tilde{x}_{ij}$.

$$\tilde{x}_{ij} = (a_{ij}, b_{ij}, c_{ij}) \quad (2)$$

where

$$a_{ij} = \min_k \{a_{ij}^k\}, b_{ij} = \frac{i}{k} \sum_{k=1}^{K} b_{ij}^k, c_{ij} = \max_k \{c_{ij}^k\} \quad (3)$$

The value of k represents the number of matrices, i represents the number of columns and j the number of criteria.

Similar procedure is applied for weighting values, $w_{ij} = \tilde{w}_{ij}$.

$$\tilde{w}_{ij} = \left(w_{j1}, w_{j2}, w_{j3}\right) \tag{4}$$

where

$$w_{j1} = \min_k\left\{w_{j1}^k\right\}, w_{j2} = \frac{1}{K}\sum\nolimits_{k=1}^{K} w_{j2}^k, w_{j3} = \max_k\left\{w_{j3}^k\right\} \tag{5}$$

In order to normalized fuzzy decision matrix is required to obtain the $\tilde{r}_{ij}$ values for each combination of alternative and criteria. The fuzzy decision matrix is normalized, by adding the benefit and the cost criterion, on each cell. The value of c_{ij}^* is defined by the maximum value of each cell. The cost is represented by the smallest value in each cell. The result is divided by each of the cell values. Keeping in mind that cost is the lowest, if it goes up, then values *abc* are changed to *cba.*

$$\tilde{r}_{ij} = \left(\frac{a_{ij}}{c_j^*}, \frac{b_{ij}}{c_j^*}, \frac{c_{ij}}{c_j^*}\right) withc_j^* = \max_i\left\{c_{ij}\right\} forprofitcriteria \tag{6}$$

$$\tilde{r}_{ij} = \left(\frac{a_j^-}{c_{ij}}, \frac{a_j^-}{b_{ij}}, \frac{a_j^-}{a_{ij}}\right) witha_j^- = \min_i\left\{a_{ij}\right\} forcostcriteria \tag{7}$$

Once the fuzzy decision matrix is normalized, the next step is obtaining the weighted normalized decision matrix ($\tilde{v}_{ij}$) using the fuzzy sets operators ($\tilde{A}_1 \otimes \tilde{A}_2$) to perform the operations.

$$\tilde{v}_{ij} = \tilde{r}_{ij} x \tilde{w}_{ij} \tag{8}$$

$$\tilde{A}_1 \otimes \tilde{A}_2 = (a_1, b_1, c_1) \otimes (a_2, b_2, c_2) = (a_1 * a_1, b_1 * b_2, c_1 * c_2) \tag{9}$$

Calculate the fuzzy positive ideal solution (FPIS) and the fuzzy negative ideal solution (FNIS).

$$A^* = \left(\tilde{v}_1^*, \tilde{v}_2^*, \ldots, \tilde{v}_n^*\right) with\tilde{v}_j^* = \max_i\left\{v_{ij3}\right\} \tag{10}$$

$$A^- = \left(\tilde{v}_1^-, \tilde{v}_2^-, \ldots, \tilde{v}_n^-\right) where\tilde{v}_j^- = \min_i\left\{v_{ij1}\right\} \tag{11}$$

Subsequently, calculate the difference between each alternative and fuzzy positive ideal solution and the difference between each alternative and the fuzzy negative ideal solution.

$$d\left(\tilde{x}, \tilde{y}\right) = \sqrt{\frac{1}{3}[(a_1 - a_2)^2 + (b_1 - b_2)^2 + (a_1 - c_2)^2]} \tag{12}$$

To calculate the difference between the values of A^* and the values of A^- and calls them $d_i^* y d_i^-$ calculate the distance between each alternative of FPIS and FNIS. Then, determine the difference of the values. For example, for d_i^* the values of each row for each candidate are added. And by the vertex method, the distance between each alternative and the positive ideal solution is calculated, and the negative ideal solution is calculated as:

$$d_i^* = \sum_{j=1}^{n} d(\tilde{v}_{ij}, \tilde{v}_{ij}^*) \tag{13}$$

$$d_i^- = \sum_{j=1}^{n} d(\tilde{v}_{ij}, \tilde{v}_{ij}^-) \tag{14}$$

For all criteria, each alternative distance from the best alternative (d_i^*) and the worst alternative (d_i^-). Subsequently, the proximity coefficient is calculated CC_i for each alternative. And it is determined how close each coefficient is CC_i for each alternative. Having the results obtain the best alternative through the following formula.

$$CC_i = \frac{d_i^-}{d_i^- + d_i^*} \tag{15}$$

4 Results

Pythagorean Fuzzy Sets, the model presents decision-making through multiple criteria, such as the selection of suppliers, which belong to 42 countries in Europe and Asia, considering cost and distance as selection factors (considering the industry is in the American continent). These factors are weighted with the [14] Likert Scale (also called summary evaluation method) for which the scale from one to seven was considered, with seven being the most important. In addition to the level of raw material capacity weighted quantitatively and having as a restriction from zero to twelve.

Among the alternatives, the manufacture of seven products is also considered, we must define which product should be made and which countries are going to be the raw material suppliers. The seven industrial washers must cover the twenty-seven factors that define their manufacture, such as: Load capacity, Energy consumption, Water consumption, Wash cycle, Detergent, detergent viscosity, Electric shock, Motor, Rotor, Fuzzy Logic, Design, Cost, Warranty, Repair cost, Lifetime, Ecosystem Plugin, Drying time, Use of collision detection system, Complexity of use of technology, Rinsing cycle, Financing system, Months without interest, Symbolic capital of the product, Ease of spare parts, Adaptability to Additive Manufacturing, Treatment of complex fabrics: Terlenka. It should be mentioned that among the restrictions there are factors to be maximized such as load capacity and to minimize energy consumption, depending on each criterion specific feature.

In Table 1 the criteria are described in the upper part and on the left the seven production alternatives. Subsequently, the resulting tables are presented, following the steps of the previously presented model (Table 2).

All criteria to maximize—is a condition of the TOPSIS method. Therefore, the minimization criteria had to be converted (Tables 3, 4, 5 and 6).

The fuzzy Pythagorean data do not require standardization, so only the weights are multiplied with each criteria value (Table 4).

Finally, Table 7 shows the best alternative which according to Fuzzy Multiple Criteria Decision-Making is option A.

5 Conclusions

Because of the synergistic and immutable characteristics of the technology, the adoption of *Blockchain* in global supply chains is one of the most promising recent applications. So, it has quickly positioned itself in the industrial, financial, and technological worlds. Basically, in any entity that wants to exchange assets. The *Blockchain* provides the infrastructure for these activities to be carried out in a transparent, secure, and reliable way.

The distributed system shields the information once registered, so it provides the basis for tracking the product at any point in the supply chain giving veracity, speed, and performance.

So, the research provides further dissemination of *Blockchain*'s technology within the supply chain for decision making in an RFQ system. Through the incorporation of models and tools the user can obtain the best option of suppliers and product, with the safety and security provided through the MCDM model, storing the information in a transparent manner guaranteeing the reliability in the information through the Blockchain model. Thus, consolidating the supply chain and providing timely and accurate information and location of the product or ra material.

Finally, it should be noted that *Blockchain*'s technology provides the opportunity for traceability and origin of inputs or products, fighting fraud, increasing trust management based on reliability, transparency and integration devices to incorporate in global supply chains; as well as, ensuring the use of vehicles with the require specifications for product transfer, such as drugs or food.

Table 1 Decision matrix table

	Capacity of load	Energy consumed	Water consume	Detergent	Washing cycle	Drying cycle	Electrical downloading	Engine	Rotary	Fuzzy Logic
A	22	57	18.7	0.875	62	57	22.7	54.6	27.4	6.4
B	21	58.4	16.7	0.5965	60	48	24.3	59.2	38.4	6.7
C	19	59.2	14.8	0.6441	54	47	24.8	57.4	36.2	6.3
D	17	64.3	22.1	0.7097	59	57	26.9	59.4	32.8	6.5
E	20	55	23.4	0.7966	58	54	23.4	57.4	36.4	6.8
F	24	58.4	26.2	0.8657	44	57	24.6	59.2	50.8	6.4
G	22	55	27.8	0.9219	61	63	27.9	56.4	37.9	6.2
Criteria Type	Direct	Indirect	Indirect	Indirect	Indirect	Indirect	Direct	Direct	Direct	Direct
Weights	2	2	1.5	1.6	1.2	1.8	2.2	2.4	1	2
Ideal	24	55	14.8	0.5965	44	47	27.9	59.4	50.8	6.8
Worst	17	64.3	27.8	0.9219	62	63	22.7	54.6	27.4	6.2
	Design	Cost	Warranty	Cost of repair	Lifetime	Ecosystem PlugIn	Viscosity	Use of collision detection system	Rinse cycle	Financing system
A	6.6	5500	22	22.7	10	6.2	34	7	28.2	0.2941
B	6.1	6200	17	24.7	7	6.3	36	6.7	29.3	0.5714
C	6.4	6350	24	23.8	9	6.7	37	6.4	30.7	0.4231
D	6.8	5792	23	24.7	10.2	6.4	38	6.3	31.2	0.6591
E	6.7	5874	24	27.2	10.7	6.3	44	6.4	30.5	0.5957
F	6.2	5962	20	23.8	10.4	6.8	52	6.2	29.8	0.6667

(continued)

Table 1 (continued)

	Capacity of load	Energy consumed	Water consume	Detergent	Washing cycle	Drying cycle	Electrical downloading	Engine	Rotary	Fuzzy Logic
G	6.9	5470	17	21.9	10.1	6.2	57	6.8	27.9	0.4583
Criteria Type	Direct	Indirect	Direct	Indirect	Direct	Direct	Direct	Direct	Indirect	Indirect
Weights	3	2	1	2	2	3	2	1	2	2
Ideal	6.9	5470	24	21.9	10.7	6.8	57	7	27.9	0.2941
Worst	6.1	6350	17	27.2	7	6.2	34	6.2	31.2	0.6667

Table 2 Pythagorean fuzzy sets

	Capacity of load	Energy consumed	Water consume	Detergent	Washing cycle	Drying cycle	Electrical downloading	Engine	Rotary	Fuzzy Logic
A	−5	7.3	9.1	0.04688	0	6	0	0	0	−0.2
B	−4	5.9	11.1	0.32538	2	15	−1.6	−4.6	−11	−0.5
C	−2	5.1	13	0.27781	8	16	−2.1	−2.8	−8.8	−0.1
D	0	0	5.7	0.2122	3	6	−4.2	−4.8	−5.4	−0.3
E	−3	9.3	4.4	0.12526	4	9	−0.7	−2.8	−9	−0.6
F	−7	5.9	1.6	0.0562	18	6	−1.9	−4.6	−23.4	−0.2
G	−5	9.3	0	0	1	0	−5.2	−1.8	−10.5	0
Normalization	11.3137	17.9416	20.7227	0.4991	20.445	25.8844	7.4666	9.1804	31.0871	0.8888
	Design	Cost	Warranty	Cost of repair	Lifetime	Ecosystem PlugIn	Viscosity	Use of collision detection system	Rinse cycle	Financing system
A	−0.5	850	−5	4.5	−3	0	0	−0.8	3	0.3725
B	0	150	0	2.5	0	−0.1	−2	−0.5	1.9	0.0952
C	−0.3	0	−7	3.4	−2	−0.5	−3	−0.2	0.5	0.2436
D	−0.7	558	−6	2.5	−3.2	−0.2	−4	−0.1	0	0.0076
E	−0.6	476	−7	0	−3.7	−0.1	−10	−0.2	0.7	0.071
F	−0.1	388	−3	3.4	−3.4	−0.6	−18	0	1.4	0
G	−0.8	880	0	5.3	−3.1	0	−23	−0.6	3.3	0.2083
Normalization	1.3565	1485.895	12.9615	9.163	7.6223	0.8185	31.3369	1.1576	5.1186	0.5057

Table 3 Normalized decision matrix

	Capacity of load	Energy consumed	Water consume	Detergent	Washing cycle	Drying cycle	Electrical downloading	Engine	Rotary	Fuzzy Logic
A	−0.4419	0.4069	0.4391	0.0939	0	0.2318	0	0	0	−0.225
B	−0.3536	0.3288	0.5356	0.6519	0.0978	0.5795	−0.2143	−0.5011	−0.3538	−0.5625
C	−0.1768	0.2843	0.6273	0.5566	0.3913	0.6181	−0.2813	−0.305	−0.2831	−0.1125
D	0	0	0.2751	0.4251	0.1467	0.2318	−0.5625	−0.5229	−0.1737	−0.3375
E	−0.2652	0.5183	0.2123	0.251	0.1956	0.3477	−0.0938	−0.305	−0.2895	−0.6751
F	−0.6187	0.3288	0.0772	0.1126	0.8804	0.2318	−0.2545	−0.5011	−0.7527	−0.225
G	−0.4419	0.5183	0	0	0.0489	0	−0.6964	−0.1961	−0.3378	0
	Design	Cost	Warranty	Cost of repair	Lifetime	Ecosystem PlugIn	Viscosity	Use of collision detection system	Rinse cycle	Financing system
A	−0.3686	0.572	−0.3858	0.4911	−0.3936	0	0	−0.6911	0.5861	0.7367
B	0	0.1009	0	0.2728	0	−0.1222	−0.0638	−0.4319	0.3712	0.1883
C	−0.2212	0	−0.5401	0.3711	−0.2624	−0.6108	−0.0957	−0.1728	0.0977	0.4817
D	−0.516	0.3755	−0.4629	0.2728	−0.4198	−0.2443	−0.1276	−0.0864	0	0.015
E	−0.4423	0.3203	−0.5401	0	−0.4854	−0.1222	−0.3191	−0.1728	0.1368	0.1404
F	−0.0737	0.2611	−0.2315	0.3711	−0.4461	−0.733	−0.5744	0	0.2735	0
G	−0.5898	0.5922	0	0.5784	−0.4067	0	−0.734	−0.5183	0.6447	0.412

Table 4 Weighted normalized decision matrix

	Capacity of load	Energy consumed	Water consume	Detergent	Washing cycle	Drying cycle	Electrical downloading	Engine	Rotary
A	−0.8839	0.8138	0.6587	0.1503	0	0.4172	0	0	0
B	−0.7071	0.6577	0.8035	1.0431	0.1174	1.0431	−0.4714	−1.2026	−0.3538
C	−0.3536	0.5685	0.941	0.8905	0.4696	1.1126	−0.6188	−0.732	−0.2831
D	0	0	0.4126	0.6802	0.1761	0.4172	−1.2375	−1.2548	−0.1737
E	−0.5303	1.0367	0.3185	0.4015	0.2348	0.6259	−0.2063	−0.732	−0.2895
F	−1.2374	0.6577	0.1158	0.1802	1.0565	0.4172	−0.5598	−1.2026	−0.7527
G	−0.8839	1.0367	0	0	0.0587	0	−1.5322	−0.4706	−0.3378
Ideal	0	1.0367	0.941	1.0431	1.0565	1.1126	0	0	0
Worst	−1.2374	0	0	0	0	0	−1.5322	−1.2548	−0.7527
	Design	Cost	Warranty	Cost of repair	Lifetime	Ecosystem PlugIn	Viscosity	Use of collision detection system	Financing system
A	−1.1058	1.1441	−0.3858	0.9822	−0.7872	0	0	−0.6911	1.4735
B	0	0.2019	0	0.5457	0	−0.3665	−0.1276	−0.4319	0.3767
C	−0.6635	0	−0.5401	0.7421	−0.5248	−1.8325	−0.1915	−0.1728	0.9634
D	−1.5481	0.7511	−0.4629	0.5457	−0.8396	−0.733	−0.2553	−0.0864	0.03
E	−1.327	0.6407	−0.5401	0	−0.9708	−0.3665	−0.6382	−0.1728	0.2809
F	−0.2212	0.5222	−0.2315	0.7421	−0.8921	−2.199	−1.1488	0	0
G	−1.7693	1.1845	0	1.1568	−0.8134	0	−1.4679	−0.5183	0.824
Ideal	0	1.1845	0	1.1568	0	0	0	0	1.4735
Worst	−1.7693	0	−0.5401	0	−0.9708	−2.199	−1.4679	−0.6911	0

Table 5 Ideal

	Capacity of load	Energy consumed	Water consume	Detergent	Washing cycle	Drying cycle	Electrical downloading	Engine	Rotary	Fuzzy Logic
A	0.8839	0.2229	0.2823	0.8928	1.0565	0.6954	0	0	0	0.45
B	0.7071	0.379	0.1375	0	0.9391	0.0695	0.4714	1.2026	0.3538	1.1251
C	0.3536	0.4682	0	0.1525	0.5869	0	0.6188	0.732	0.2831	0.225
D	0	1.0367	0.5284	0.3628	0.8804	0.6954	1.2375	1.2548	0.1737	0.6751
E	0.5303	0	0.6225	0.6415	0.8217	0.4868	0.2063	0.732	0.2895	1.3501
F	1.2374	0.379	0.8252	0.8629	0	0.6954	0.5598	1.2026	0.7527	0.45
G	0.8839	0	0.941	1.0431	0.9978	1.1126	1.5322	0.4706	0.3378	0
	Design	Cost	Warranty	Cost of repair	Lifetime	Ecosystem PlugIn	Viscosity	Use of collision detection system	Rinse cycle	Financing system
A	1.1058	0.0404	0.3858	0.1746	0.7872	0	0	0.6911	0.1172	0
B	0	0.9826	0	0.6112	0	0.3665	0.1276	0.4319	0.547	1.0968
C	0.6635	1.1845	0.5401	0.4147	0.5248	1.8325	0.1915	0.1728	1.0941	0.5101
D	1.5481	0.4334	0.4629	0.6112	0.8396	0.733	0.2553	0.0864	1.2894	1.4435
E	1.327	0.5438	0.5401	1.1568	0.9708	0.3665	0.6382	0.1728	1.0159	1.1926
F	0.2212	0.6622	0.2315	0.4147	0.8921	2.199	1.1488	0	0.7424	1.4735
G	1.7693	0	0	0	0.8134	0	1.4679	0.5183	0	0.6495

Table 6 Worst

	Capacity of load	Energy consumed	Water consume	Detergent	Washing cycle	Drying cycle	Electrical downloading	Engine	Rotary	Fuzzy Logic
A	0.3536	0.8138	0.6587	0.1503	0	0.4172	1.5322	1.2548	0.7527	0.9001
B	0	0.6577	0.1158	0.1802	1.0565	0.4172	0.9723	0.0523	0	0.9001
C	0.3536	1.0367	0	0	0.0587	0	0	0.7843	0.415	1.3501
D	1.2374	0	0.4126	0.6802	0.1761	0.4172	0.2946	0	0.579	0.6751
E	0	0	0	0	0	0	0	0	0	0
F	1.2374	0	0	0	0	0	1.5322	1.2548	0.7527	1.3501
G	0.3536	1.0367	0	0	0.0587	0	0	0.7843	0.415	1.3501
	Design	Cost	Warranty	Cost of repair	Lifetime	Ecosystem PlugIn	Viscosity	Use of collision detection system	Rinse cycle	Financing system
A	0.6635	1.1441	0.1543	0.9822	0.1837	2.199	1.4679	0	1.1722	1.4735
B	1.5481	0.5222	0.3086	0.7421	0.0787	0	0.3191	0.6911	0.547	0
C	0	1.1845	0.5401	1.1568	0.1574	2.199	0	0.1728	1.2894	0.824
D	0.2212	0.7511	0.0772	0.5457	0.1312	1.466	1.2126	0.6047	0	0.03
E	0	0	0	0	0	0	0	0	0	0
F	1.7693	0	0.5401	0	0.9708	2.199	1.4679	0.6911	0	0
G	0	1.1845	0.5401	1.1568	0.1574	2.199	0	0.1728	1.2894	0.824

Table 7 Ranking

	di +	di-	ci	Result rank
A	**2.4545**	**4.4748**	**0.6458**	**1**
B	2.7905	2.7798	0.4990	4
C	3.0373	3.7521	0.5526	2
D	3.8171	2.8578	0.4281	6
E	3.4831	0.0000	0.0000	7
F	4.0619	4.4357	0.5220	3
G	3.7827	3.7521	0.4980	5

6 Future Works

As future work, research and the incorporation of models that provide the essential tools for timely and reliable decision making will continue. In addition to evaluating the incorporation of optimal tracking devices and maximizing the specifications and performance of the creation of different decentralized virtual networks. And the incorporation of QR to archive the generated Hash and send it via text message. This QR, will be stored in the cloud, to be able to authenticate in time and form its use (Fig. 7).

QR barcode program code:

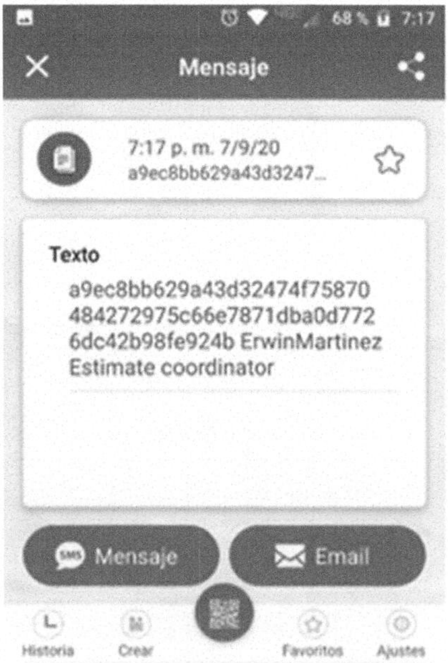

Fig. 7 Creation of QR in text message

Fig. 8 QR code

```
try:
        sql_command = """INSERT INTO USER VALUES (?, ?, ?, ?, ?, ?, ?);"""
        crsr.execute(sql_command,
(CKeyDB,namevalueMod,passwordvalue,mailvalueMod,departmentvalue,Hashvalue,timevalue))
        connection.commit()
        self.progress.setValue(60)
        NewNumDB=str(1+CKeyDB)
        CountKeyDB= open("Service/Users/Data/CountDB/CountUserDB.txt", "w")
        CountKeyDB.write(NewNumDB)
        CountKeyDB.close()
        self.progress.setValue(70)
        QrsValue= Hashvalue+" "+namevalueMod+" "+ departmentvalue
        QrKey= pyqrcode.create(QrsValue)
        QrKey.png('Service/Users/Data/SetQrs/%s.png' % Hashvalue, scale = 4)
        self.progress.setValue(80)
        self.name.setText('')
        self.password.setText('')
        self.confirm.setText('')
        self.mail.setText('')
        self.department.setCurrentIndex(0)
        self.progress.setValue(90)
        with open("Service/Users/DB/DB_U/DataBaseUser.db", "rb") as f:
        dbxDBUser.files_upload(f.read(), '/Aplicaciones/DB_USERS/DataBaseUser.db',
mode=WriteMode('overwrite'))
        print("DB_User")
```

(See Fig. 8).

References

1. Goodwin, R., Akkiraju, R. ., Wu, F.: A decision-support system for quote generation (2002)
2. Gonczol, P., Katsikouli, P., Herskind, L., Dragoni, N.: Blockchain implementations and use cases for supply chains–a survey (2020)
3. Wang, Y., Han, J.H., Beynon-Davies, P.: Understanding blockchain technology for future supply chains: a systematic literature review and research agenda. Supply Chain Manag. (2019)
4. Al-Jaroodi, J., Mohamed, N.: Blockchain in industries: a survey. IEEE Access (2019)

5. Pournader, M., Shi, Y., Seuring, S., Koh, S.L.: Blockchain applications in supply chains, transport and logistics: a systematic review of the literature. Int. J. Prod. Res. 1–19 (2019)
6. Meijer, C.R., Carlo R.W.: The UK and blockchain technology: a balanced approach. J. Paym. Strat. Syst. 9 (2016)
7. Yang, D., Pan, Z.D.: The blockchain bring finance and law optimization. China Finance (2016)
8. Wang, J., et al.: Block Chain-Based Data Security Sharing Network Architecture Research. J. Comput., Develop (2016)
9. Lambert, D.M., Cooper, M.C.: Issues in Supply Chain Management. Ind, Marketing Manage (2000)
10. Azzia, R., Chamouna, R.K., Sokhn, M.: The power of a Blockchain based supply chain (2019)
11. Wust, K., Gervais, A.: Do you Need a Blockchain? In: Proceedings of the Crypto Valley Conference Blockchain Technology (CVCBT) (2018)
12. Wang, Y., Singgih, M., Wang, J., Rit, M.: Making sense of blockchain technology: how will it transform supply chains? Int. J. Prod. Econ. **211**, 221–236 (May 2019)
13. Mathew, M., Thomas, J.: Interval valued multi criteria decision making methods for the selection of flexible manu-facturing system. Int. J. Data. Netw. Sci. 3(4):349–358 (2019)
14. Mendez Hinojoza, L.M.: Manual Práctico Para El Diseño De La Escala Likert. Editorial Trillas (2006)

Sociodemographic Analysis of the Location of MSW Collection Centers in Mexico City

Javier Gómez-Maturano and Benito Sánchez-Lara

Abstract Environmental pollution mitigation resulting from inadequate management of urban solid waste [MSW] is one of the challenges large cities are facing, mainly in emerging countries. Recovering the valuable materials of the MSW, from the flow that leads them to its final disposal, is the disruptive activity in the linear economy model that occurs in Mexico. MSW collection centers are key elements of reverse supply chains of MSW valorization. This article presents the results of a longitudinal analysis of social and economic variables that determine the location of MSW collection centers in Mexico City. The analysis is carried out with the support of geographic information systems with data from the National Institute of Statistics, Geography and Informatics. It identifies MSW recovery logistics networks and describes the intensity of MSW recovery operations on those networks. Networks energize the economy of groups in the bottom of the pyramid.

Keywords Reverse supply chain · Geographic information systems · Municipal solid waste · Recycling · Geospatial analysis

1 Introduction

The generation of Urban Solid Waste [MSW] is due to the pattern: production—consumption—waste. It depends on how goods are produced from virgin raw materials, how they are sold, used and discarded [1]. This pattern is associated with what has been called linear economics. Today, the rate of consumption of natural resources is as great as waste generation. Globally, 130 million tons of waste were generated in 2012, 2,010 million tons were generated in 2016 and an estimated 3.4 billion tons [2] will be generated in 30 years.

J. Gómez-Maturano · B. Sánchez-Lara (✉)
Departamento de Ingeniería de Sistemas, Facultad de Ingeniería, Universidad Nacional Autónoma de México, Mexico City, Mexico
e-mail: blara@unam.mx

J. Gómez-Maturano
e-mail: jgomezma@uaemex.mx; ingmaturano2010@comunidad.unam.mx

A. Ochoa-Zezzatti et al. (eds.), *Technological and Industrial Applications Associated With Industry 4.0*, Studies in Systems, Decision and Control 347,
https://doi.org/10.1007/978-3-030-68663-5_20

In Mexico, MSW generation was 53.1 million tons in 2015, an increase of 61.2% compared to 2003 [3]. In the Metropolitan Zone of Valle de México, about 27,067 tons of MSW were collected in 2014 [4], an increase of 16% compared to what was reported in 2010 [5].

The amount of waste generated per capita and its composition changes in relation to the following factors: income level, consumption patterns, promotion towards reduction, population changes, consumption behaviors, weight reduction policies, re-use of products and diverse economic conditions [6].

Globally, MSW management systems, focused on treatment and disposal, involved the emission of 1.6 billion tons of carbon dioxide equivalent, representing about 5% of total global emissions [2]. On the other hand, in Latin America the average municipal expenditure for the management of MSW is 2–8% [7]. The acceleration of urbanization has led to unplanned and high-cost management, which is accentuated in emerging economies [8].

Mexico City requires a 2,552-truck vehicle park that runs through the streets of 1,753 neighborhoods across 1,751 routes [9]. Collecting MSWs required thirteen transfer stations over twelve delegations and a staff of 4,084 people to drive 70% of MSWs to final disposal sites [9]: four in the State of Mexico and two in the State of Morelos. Thus, the collection system generates 1,253 tons of carbon dioxide annually [9].

The movement of waste in cities, not included in urban logistics studies, has negative, economic, environmental and social impacts. In environmental terms, the impact [10] is triple: soil and air water pollution, waste of resources and need for spaces for the final disposal of waste.

MSW's public management systems have focused on increasing collection and final disposal and have set aside their use and valorization. The percentage of MSW recycled in Mexico is less than 10% [3] in contrast to the global average percentage of 13% [2] and the European Union, by 2017, of 46%. The valorization of MSW materials is an alternative to reduce the amount of waste before final disposal and to reduce the derived impacts [11]. For the World Bank, MSW management systems must be transformed into a circular economy, targeting reuse and recycling [2].

In emerging countries, waste collection and recycling rates are small in contrast to the percentage sent to landfills estimated between 82 and 98% [12]. For example, in China the separation of recyclable materials is carried out by incipient reverse supply chain [RSC] incorporated formal enterprises while informal companies are the ones that collect, process and market them [13]. The acquisition of MSW is carried out by marginalized social groups, this often being their only source of income and survival strategy. These groups are called scavengers, recyclers and itinerant buyers. They are characterized by their intensive work, low technology and low income [14].

This work reports the results of a longitudinal analysis of the location of collection centers in Mexico City in 2009, 2014 and 2019. It describes the relationships between the location of collection centers, population income and economic activity in the area. The description of relationships is a geospatial analysis supported by geographic information systems. Quantitative elements and relationships are provided from the MSW acquisition network.

The work consists of five sections. The first section is an introduction on the MSWs, their acquisition and the collection centers. The second section raises the conceptual framework of sociodemographic analysis carried out, in particular, the role of RSCs on MSW valorization and its dynamics in urban areas. The third section describes the methodological elements for analysis, specifically the application of geographic information systems for the study of MSW logistics recovery networks. The fourth section integrates the results of geospatial analysis and its discussion. Finally, the conclusions are stated.

2 Theoretical Framework

Emerging cities are exposed to a variety of challenges arising from future market development, increased environmental requirements, new technologies, and the evolution of complex supply chains that meet the needs of their population [15]. To these we must add the problem of managing and leveraging the MSW generated in them.

The objective of sustainable urban logistics is to respond, as far as possible, to how society intends to provide the means of opportunity to meet economic, environmental and social needs in an efficient and equitable manner, and to minimize adverse, avoidable and unnecessary effects and their associated costs in the corresponding space and timescales [16]. A robust and cost-effective MSW treatment and recycling system is a requirement for a sustainable society [6].

A "complete" city logistics system would deal with reverse movements, from origins within the city to destinations outside, as well as movement between origins and destinations within the city [17]. Currently most Urban Logistics contributions are directed only to in-bound distribution activities, following the imbalance between the inbound and outbound flows that characterize most cities [17]. Comprehensive urban logistics should also include reverse flows, reverse business logistics and waste management [18], operations of RSC would also be added [19].

An RSC is a series of processes and activities that are necessary to recover post-consumer, used products, their components or associated materials, either to value them (reuse, recycle, restore, re-manufacture, repurpose, etc.) or to send it to final disposal [11, 20].

The valuable materials of these MSWs are recovered from the public management system through MSW Logistics Acquisition Operations [19], which will be abbreviated as LOA-MSW. These operations are part of the process of acquiring materials for reverse supply chains aimed at the valorization of MSW.

ALO-MSW is concentrated in two types of establishments or Economic Units [EU]: collection centers [CC], retail collectors and transfer centers [TC], wholesale copiers. These actors are private and can adopt formal and informal schemes. A CC is a commercial establishment where MSW eligible materials are received and conditioned, it is also where waste is received, quantified, collected, transferred and temporarily accumulated for treatment, recycling, reuse, reprocessing or sent to final disposal [9]. TCs are temporary storage facilities for waste to later transport it to an industrial valorization site, to another transfer center or to its final disposal. Eventually, some other processes could be applied to the materials received, such as separation, compaction and crushing. These TCs are private in nature and focus on the commercialization of high-scale MSW materials.

CCs and TCs are included in the North American Industrial Classification System, economic sub-branch 43431, including UE dedicated mainly to specialized wholesale trade in recycling waste materials, such as metal, paper, cardboard, glass, plastic [21].

Geographic information systems (GIS) are a powerful tool for the analysis of the location of CC and CT in urban areas and their relationship with social and economic variables. GIS has been used to develop regional information analysis, which can be divided into two basic categories [22]: (a) spatial decision support applications and (b) spatial statistics support applications. A GIS is a software designed to enable users to collect, manage, analyze, and retrieve large volumes of spatially referenced data and associated attribute data collected from a variety of sources [23]. A GIS allows the user to visualize and interpret data for a better understanding of relationships, trends, and patterns [8].

Although operational improvement of MSW management systems is common when GIS is included in research, they are only applied in the early stages of research, as they are the preamble to other mathematical modeling. This is the case with [6, 24, 25].

GIS allows visual and analytical representation of MSW's comprehensive management systems, such representations can show the location of CCs and spatial distribution pattern in particular study areas [24]. Geospatial location in an urban model is one of the main GIS applications, georeferencing by means of GPS the location of the generation and collection sources of the MSW, as well as the collection trucks of MSW allows to model the network with a view to its improvement. While it is the most basic it is a powerful tool [25].

Few studies apply GIS to model social or environmental aspects associated with MSW management and valorization. A sustainable approach is developed in [26] by location of MSW separation plants that are optimal for reducing environmental pollution, lowering costs and improving the service system for society. The approach proposed in [26] is one of the few that takes up the social dimension in the analysis of MSW networks in cities, which is partly integrated into this work.

3 Materials and Methods

For the analysis of this study area GIS was addressed, which allows the realization of statistical data analysis and the incorporation of geospatial analyses in some specific places, which is a form of modeling [25]. Obtaining a model of MSW's material acquisition network in Mexico City is the use GIS has been given in this work.

The representation models built in GIS for this work were vector models[1] with georeferenced input information layers (shapefile). Layers are files with attributes of geographic features and their locations, to make map compositions through layering the information contained in them. In this work the input layers were:

National Urban System (SUN) with the limits of Mexico City, integrating the 16 mayors with population data from 1990 to 2018.
Daily collection of MSW in the 16 mayors, of the years: 2011, 2013, 2015 and 2017.
Basic Geostatistical Areas (BGA) of Mexico City with population attributes, percentage of poverty and extreme poverty, human development index.
CC and CT in Mexico City in 2009, 2014 and 2015 with attributes of size, valued material, social reason, year of registration.
Mexico City's road network with road name attributes, road type, width, sense, restrictions on circulation, among others.

The input layers were made up of the following INEGI databases:

National Geostatistical Framework. National system that allows to georeference the statistical information of censuses and surveys, at different levels of disaggregation [27]:

National Urban System (NUN).
State Geostatistical Area (SGA).
Municipal Geostatistical Area (MGA)
Basic Geostatistical Area (BGA)

National Statistical Directory of Economic Units (NSDEU). National register of millions of economic establishments, derived from the Economic Census 2009, 2014 and 2019, and which is regularly updated and enriched with other sources [28].

Main results of the population and housing census 2010 by BGA and urban block. It is the set of indicators on the population and housing of the BGAs and blocks that make up the urban towns of the country, as well as the totals by entity, municipality and urban town, coming from the Census of Population and Housing 2010 [29].

Intercensal Survey 2015. This survey updates information on the volume, composition and distribution of the population residing in the national territory, and renews various socio-economic and cultural indicators of their territory, as well as the prevailing conditions in their homes [30].

Urban Poverty in Mexico, 2015. The study classified in ranks, according to the percentage of poverty and extreme poverty, each non-confidential, located in the urban towns of the municipalities with 15 thousand habitants or more, even those that were not visited by the 2015 Intercensal Survey. In these municipalities, 94% of the population and nine out of ten people in poverty live in these areas [31].

[1]The vector model is a data structure used to store geographic data from basic geometries: lines, points, and polygons.

4 Results and Discussion

In 2019, 1751 CC and TC were in operation, 130 more than those reported in 2014, an 8% increase in five years [28]. Growth is higher compared to 1161 CCs and TCs accounted for in 2009, with the number of CCs and TCs increasing by 50% in just ten years. This increase is not related to population growth, which for the same period had a reduction of −0.7%. While the increasing volume of MSW generation influences the establishment of more CCs in the City, the increase is much higher, as collection in the City only grew at a rate of 1.4% per year, while CC growth is 5% per year.

Economic growth in Mexico City also cannot explain the increase in the number of CCs and TCs: the EU in Mexico City went from 416,594 in 2009 to 465,097 in 2019, an increase of 11%. ALO-MSW in CC and CT is one of the most growing economic sub-branches in Mexico City.

The majority of EEs focused on the acquisition of MSW in MEXICO CITY are nano-collector employing between 0 and 5 people (90% in 2019), which has been almost constant since 2010 (87%). Micro collectors employ between 6 and 10 people, making up an average of 7.5% of CCs. These collectors constitute low-capacity neighborhood CAs that concentrate the materials collected mainly in the surrounding areas where they are located. Recovery networks start with retail warehouses that make up about 97% of collection points in Mexico City. Less than 3% of the EU in 2019 can be classified as small businesses (up to 50 workers) and constitute local TCs, wholesale warehouse that purchase MSWs from neighborhood CCs or organized collectors. Only three companies with more than 50 employees were identified in 2019, which constitute the regional transfer centers.

Only 150 CCs and TC (8.5%) had a Federal Taxpayer Registry in 2019, which shows the level of informality. This situation was no different in previous years, in 2010 the percentage of formality of CCs was 12%, falling to 5.7% in 2014. All medium and large companies that make up transfer centers are formalized.

Metal waste is the material that most CCs focus on over the past ten years. Plastics and paper and cardboard waste follow the metal waste. From a logistical perspective, the main difference of these materials is the so-called value density, which according to [19] is the value of the product in relation to its weight and volume.

The classified materials by INEGI are actually families of materials that are marketed indiscriminately in the CCs and TCs. Table 1 shows the Type A rated materials for each material family with their average value density. This table partially

Table 1 Type A wastes in the material families of MSW

Type A waste	Density (kg/m^3)	Sale price ($M.N./kg)	Value density ($/m^3)
Paperboard	130	2.00	$260/m^3
Polyethylene terephthalate	930	4.50	$4185/m^3
Glass	2,500	0.85	$ 2125/m^3
Old iron	7,850	3.50	$27475/m^3

explains why there is predominance in CCs dedicated to the marketing of metal waste, as these have higher density of value, which makes their movement through the city more profitable.

With regards to the spatial dispersion of CCs and the variables that have determined their location in the last ten years, the analysis shows that the dispersion in the City is neither homogeneous nor random, but influences its social and economic variables. Contrary to what could be the most important criterion in the logistics field, seek to minimize cost, whether in distance, time or money, CCs are not located in the areas of higher generation of MSW or in the vicinity of transfer centers (raw materials vs market), but their location seems to be explained by criteria and softer variables, such as the socioeconomic level of the area (see Fig. 1). Figure 1 is a photograph of the spatial distribution of CAs in 2018 in the eastern part of Mexico City. The estimated MSW generation volume and the amount of EU per BGA are also displayed.

Figure 1 shows the spatial dispersion of CS in the eastern part of Mexico City, showing at least six years. The Map shows cluster in which CCs are grouped more intensely, they are more dispersed in other areas, but what is clear is that the location of the CCs is not homogeneous, and as you will see later, it is neither arbitrary nor random. In addition to CC localization in a part of Mexico City, Fig. 1 shows BGA types established according to two criteria: estimated MSW generation volume and number of economic units, both set variables at the BGA level.

The red stripes on Fig. 1 are used to differentiate BGA according to the amount of UE all inside them, the areas of stripes with darker hue show great economic activity. On the map it is seen that CCs are rarely located in BGA with high intensity, rather

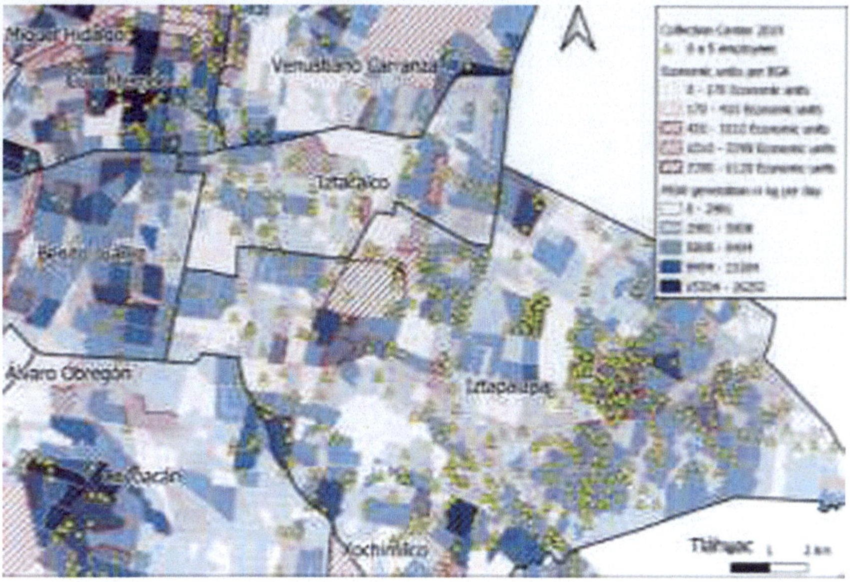

Fig. 1 Influence of the type of BGA in the location of CC in Mexico City

they are clumped to the shore of them, in adjoining BGA but of lower economic intensity. Numerical elements that support these spatial impressions are provided below.

In Fig. 1, BGAs are classified according to the estimated volume of MSW generation, determined by the per capita generation rate of 2018 and its estimated population in the same year. Categories are displayed in blue hues, the darker they are the more MSWs are generated inside. The clearest BGA is the one that generates the least MSW. The map easily identifies that they are not the most generation zones the one that have the most CAs established, statistical elements are provided later to support this assertion.

The results show that CCs have not been established in BGA with high MSW generation in the last ten years. Figure 2 shows the dispersion of the points corresponding to the ordered pair daily generation in kilograms per day (horizontal axis) and the amount of CC set in the BGA (vertical axis). Each point in Fig. 2 is one of the 2,421 BGA that make up Mexico City, for each economic census, 2009, 2014 and 2019, the dispersion of these 2,421 points is shown.

For the three moments in time that are analyzed, the density of the points towards the lower-generation MSW BGA is marked, suggesting that CCs, for the most part,

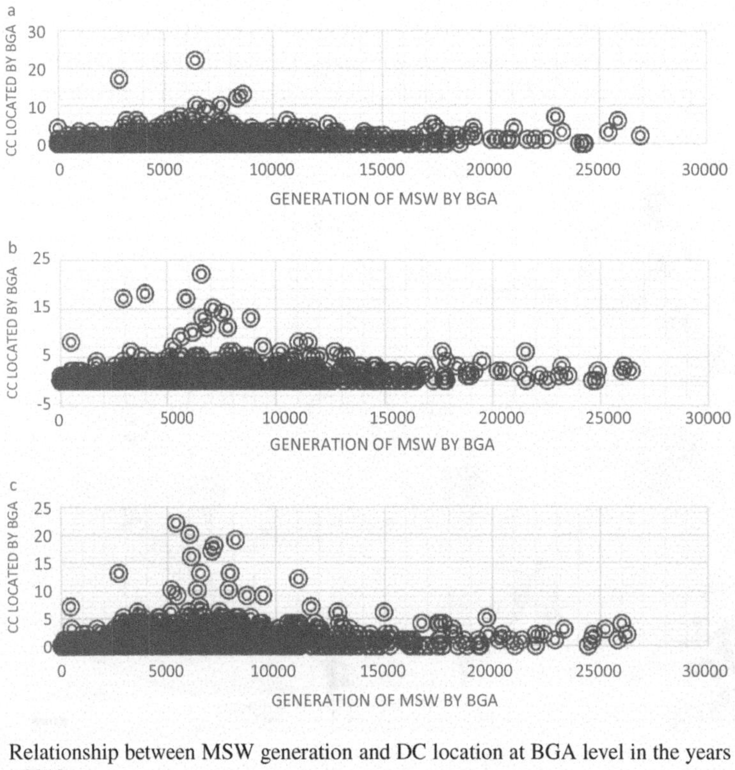

Fig. 2 Relationship between MSW generation and DC location at BGA level in the years a 2009 b 2014 c 2019

are not established in high-generation areas. BGAs with the highest AC quantity have yields less than 750 kg per day. The BGA-level correlation between the volume of MSW produced and the number of CCs established is very low, just 11% confirming that the amount of MSW generated in a zone does not strongly influence the location of the CCs.

A second criterion influencing the location of CCs at the BGA level is the intensity of the economic activities that are there, which can be measured with the number of UE established in the area. To make the analysis of the relationship between the quantity of UE y CCs at the BGA level, five intervals were established (see Fig. 3). The intervals were constructed from the natural breakout algorithm [32], this method calculates the differences in values between statistical individuals ordered increasingly and then places a boundary to separate the groups where the differences in values are high. Because three years are analyzed, 2009, 2014 and 2019 the limits set by the Jenks method [32] changed; however, this change is very small, which also tells us that the density of economic activities at the BGA level in Mexico City has changed very little in the last ten years.

In Fig. 3 the horizontal axis represents the intensity of economic activities in the BGA, grouped into five intervals in the intensity of economic activities: very low, low, medium, high and very high. As noted, the limits of each interval change slightly for the years 2009, 2014 and 2019. The vertical axis shows the total number of CAs set in each BGA type throughout Mexico City.

Figure 3 clearly shows that CAs are preferably located in areas with low economic activity, being very marked that their location rarely occurs in BGA with high or very high economic activity. This allows two inferences to be made, on one hand the value density of the materials contained in the MSW (see Table 1) is very low, suggesting

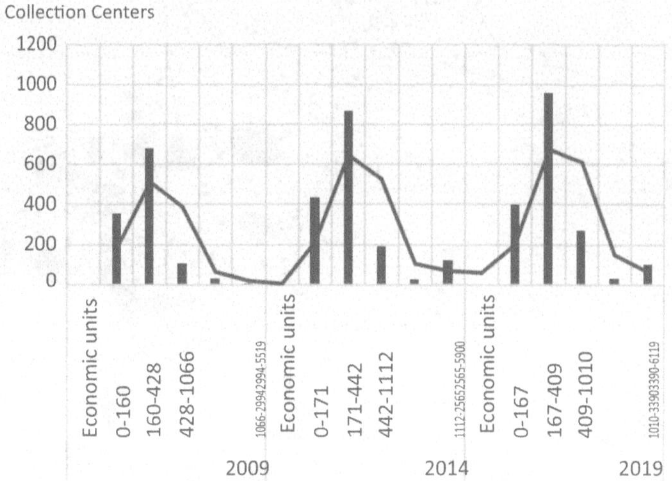

Fig. 3 Relationship between economic intensity and location of in Mexico City in the years 2009, 2014 and 2019

that the marketing of the materials does not have a large cash flow and the value of the income of the space occupied for storage must compensate for the low commercial value of the MSW. In areas with high or very high economic activity, the value of land use increases, putting pressure on those with less profitable economic activities. Those areas with high and very high economic activity have a strong demand for soil, the rent of it goes up making the location of CCs in these areas unworkable. At the same time, it is not true of areas of lower commercial activity, where ALOs can be established without the pressure that other, more profitable economic activities exert on land demand.

A third location criterion that is explored is the socio-economic level of the area in which CCs are established, because as the literature points out, ALO is usually developed by marginalized social groups. Unlike the previous two criteria, there is no consistent database to compare the location of CCs in the period 2009–2019 with the socioeconomic level of the population or some other indicator of the degree of marginalization or poverty of the population at the BGA level.

That is why two relationships for CC localization are explored based on indicators of marginality and poverty. The first is for 2010 and CC localization in 2009. The second is about the percentage of BGA poverty in [31] for 2015, which relates to CC location in 2014. These relationships should be taken as an exploration of the empirical relationship between CC location and marginality of the population living in the area in which they are established.

Figure 4 shows a BGA-level classification of Mexico City based on CONE-VAL's classification [31] of the poverty level in each BGA. Green areas show the areas with the lowest level of poverty, while the most red-ranked areas have the highest level of

Fig. 4 BGA classification by level of poverty and location of CC in Mexico City

poverty in the BGA. As it can be seen on the map, the Mayoralties of the center are the ones with the lowest percentage of poverty and as we move to the periphery the level of poverty increases. This is a dominant pattern throughout Mexico City.

Figure 4, which corresponds to the central area of Mexico City clearly shows a center-periphery pattern. Figure 4 shows the Mayoralties Benito Juárez, Cuauhtémoc, Miguel Hidalgo and Coyoacán with a predominant light color, white or dim pink, which means low levels of poverty. However, as we move to the west or east the tones change, increasing the level of poverty present in the BGAs. In addition to the fragmentation within Mexico City shown on the boundaries of mayoralties Cuauhtémoc and Venustiano Carranza, Map 5 also shows the CCs in which ALO develops, which tends to be located in BGA with higher levels of poverty. The black dots shown in Map 5 correspond to CCs and can be observed to be set more intensely in the most impoverished BGAs. Particularly in the eastern zone, it is observed how the shades of the BGA go from white to red as we move from center to periphery, but it also looks like CCs tend to settle in these poverty belts. Eligible MSWs move from the highest generation areas (in white and pink) to the most impoverished areas (in red) and have in the MSW ALO a way to earn income. Statistical support for the spatial pattern described is shown in Fig. 5.

Figure 5 shows the EU average focused on MSW ALO in each type of BGA, according to its poverty level, in Mexico City, it is highlighted in this figure that BGAs with higher levels of poverty tend to have more MSW EU. It is also noted that it is rare to have MSW UE in BGA with low levels of poverty.

Figure 6 is conclusive, in Mexico City there are more EU oriented to MSW ALO per capita in the most impoverished areas than in those with low levels of poverty. The social pyramid, in which the poorest population is located, is the context in which the MSW ALO develops.

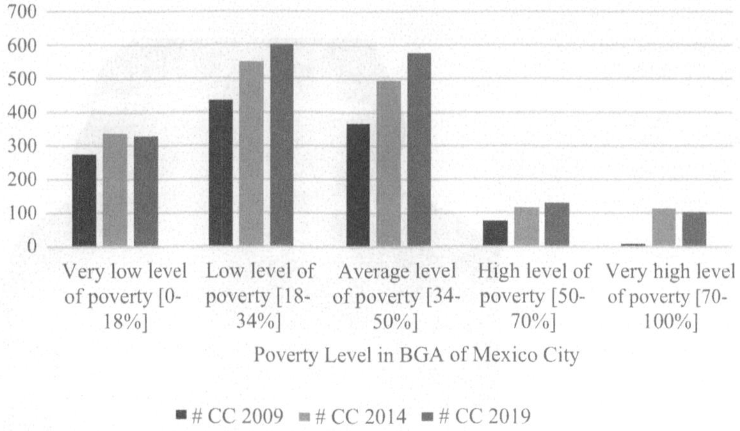

Fig. 5 Collection centers established in each type of BGA according to the poverty level

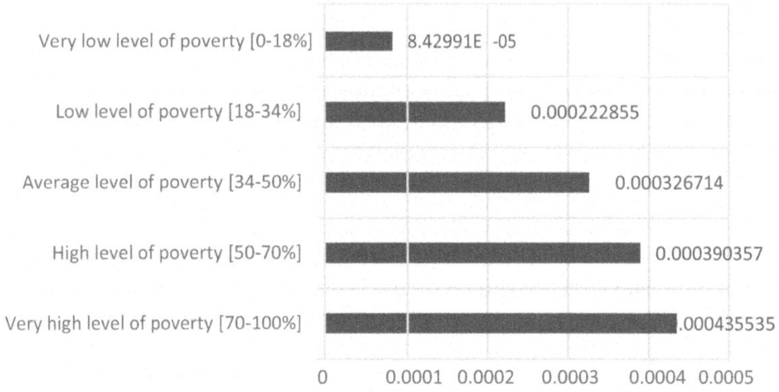

Fig. 6 CC per inhabitant according to poverty level in the BGA in 2019

Moreover, according to the trend shown in these ten years, the BGAs with the highest ALO growth rate are precisely the ones with the highest levels of poverty. Figure 7 shows the BGA with the highest number of CCs. The intensification of CA operations has taken effect in the most impoverished areas, such as the eastern zone. Madero 3.4%. These Mayors form a belt around the mayors with the lowest percentage of poverty, also establishing corridors through which the garbage leaves the city.

The EU is established throughout Mexico City, engaged in marginal locations, mostly in small commercial establishments employing fewer than five workers and are largely informal. It is through this network of collectors that MSW's recoverable materials begin to consolidate and form more robust procurement networks. The flow of materials ranges from micro retail warehouse to small and medium-sized EU that

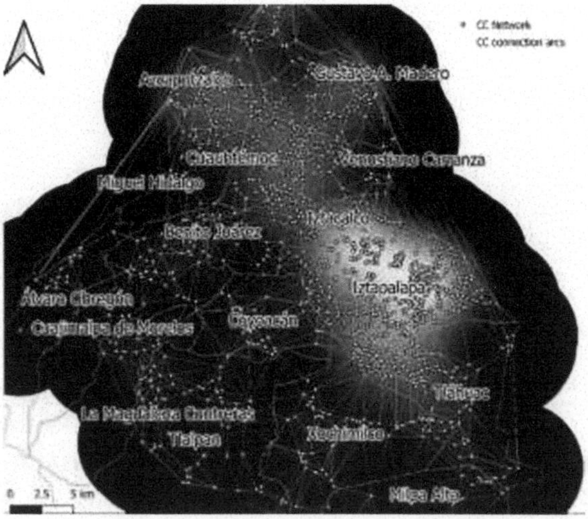

Fig. 7 CC network and ALO intensity in Mexico City

are responsible for consolidating and transferring. It is from these transfer centers that the recovery MSWs leave Mexico City. The red BGAs on Fig. 5 already show us the MSW departure zones in the City. In the network shown you can be moving more than three thousand tons per day of abled valorization materials, which is connected through 52 thousand kilometers of roads within Mexico City.

Figure 7 clearly shows these logistical clusters in which the MSW ALO is more intense and currently responsible for the only rupture links between a linear economy model and a circular economy model. The acquisition network shown in Map 6 is a high-impact network consisting of almost five thousand economic units, directly employing up to ten thousand workers and according to preliminary studies developed by the authors, could generate up to 40 thousand indirect jobs, considering only carriers and collectors. In the shown network, there can be moving more than three thousand tons of valuable materials per day, which is connected through 52 thousand kilometers of roads within the Mexico City.

In addition to the structural components of the network, functional elements are identified, mainly associated with the paths that follow the debris to exit the CDMX. The eastern corridors, whose flow includes the Iztacalco, Iztapalapa and part of Tláhuac demarcations. This corridor connects to the southern part with the Tláhuac and Xochimilco clusters which connect with the municipalities of Valle de Chalco, Chalco, Los Reyes La Paz and Chicoloapan. Another cluster is at the northeast of the CDMX that begins to take shape in the northern area of the Venustiano Carranza demarcation, it integrates Gustavo A. Madero, exiting in the Municipality of Ecatepec, and strengthening a flow through the municipalities of Coacalco, Tecámac, Tultepec and Cuautitlán. A third well-defined corridor is that of the northern area, which starts in the northern part of the Mayor Cuauhtémoc and Miguel Hidalgo, taking shape in Azcapotzalco and following the outward flow through the municipalities of Tlalnepantla, Tultitlán and Cuautitlán (show Fig. 7).

5 Conclusions

An MSW ALO system that is efficient and socially acceptable is essential for the sustainable development of Mexico City. What has been shown in this work is an accelerated growth in the intensity of MSW ALO, mainly in marginalized areas of the periphery. The growing rate of generation of MSW that characterizes the areas of central Mexico City and some BGAs with low levels of poverty, is bearable thanks to the acquisition networks that in a self-organized way have made the valorization of materials a form of subsistence.

Cost reduction, whether by distance, time or design, is one of the main localization criteria found in the literature; however, the location of CCs and TCs does not appear to fully follow this principle.

The contribution of these procurement networks is not quantifiable in this work, but the environmental services it provides must be recognized, because as far as registration is available, it may be the most important collection network operating in Mexico. A self-organized network that works despite the MSW's public management system. As an emphasis, these networks, break with the linear economy model and with the predominant MSW management model in Mexico City. Networks would not be able to function without large MSW-generating zones in small spaces, population density has helped make ALO profitable despite low value density.

A fair market for measurable materials is necessary to improve the working conditions of MSW ALO agents in Mexico. The social, economic and environmental importance of these networks must be recognized and included in zero-waste public policies. The concept of zero garbage and the implementation of the circular economy in Mexico, and other countries, must necessarily integrate these informal ALO networks.

The importance of using GIS for the representation and analysis of networks operating on a large scale has been shown in this work, including not only operational technical aspects, but also social aspects. The field of waste management and reverse supply chains can be enriched including the use of GIS tools.

References

1. Ellen Macarthur Foundation, M. & C.: Hacia una Economía Circular, 22 (2016). https://www.ellenmacarthurfoundation.org/assets/downloads/publications/Executive_summary_SP.pdf
2. Kaza, S., Yao, L., Bhada-Tata, P., Van Woerden, F.: What a Waste 2.0: A Global Snapshot of Solid Waste Management to 2050. The World Bank (2018)
3. Secretaría de Medio Ambiente y Recursos Naturales: Informe de la situación del medio ambiente en México: compendio de estadísticas ambientales. SEMARNAT, Sistema Nacional de Información Ambiental y de Recursos Naturales. México (2015). https://apps1.semar-nat.gob.mx:8443/dgeia/informe15/tema/pdf/Informe15_completo.pdf
4. INEGI: Censo Nacional de Gobiernos Municipales y Delegacionales 2015 (CNGMD), Módulo 6: Residuos Sólidos Urbanos. INEGI. México (2015). Recuperado el 11 de junio de 2019 de: https://www.inegi.org.mx/programas/cngmd/2015/default.html#Datos_abiertos
5. INEGI: Censo Nacional de Gobiernos Municipales y Delegacionales 2011 CNGMD), Módulo 6: Residuos Sólidos Urbanos. SNIEG Información de Interés Nacional, México (2011). Recuperado el 11 de junio de 2019 de: https://www.inegi.org.mx/programas/cngmd/2011/default.html#Datos_abiertos
6. Fujii, M., et al.: Regional and temporal simulation of a smart recycling system for municipal organic solid wastes. J. Clean. Prod. **78**, 208–215 (2014)
7. INEGI: Censo Nacional de Gobiernos Municipales y Delegacionales 2017 (CNGMD), Módulo 6: Residuos Sólidos Urbanos. México (2017). Recuperado el 11 de junio de 2019 de: https://www.inegi.org.mx/programas/cngmd/2017/default.html#Tabulados
8. Sanjeevi, V., Shahabudeen, P.: Optimal routing for efficient municipal solid waste transportation by using ArcGIS application in Chennai, India. Waste Manag. Res. **34**(1), 11–21 (2016)
9. SEDEMA: Inventario de Residuos Sólidos. Ciudad de México. 2016. Tomo I, Enciclopedia de México. Ciudad de México (2017). http://www.caepccm.df.gob.mx/doctos/transparencia2012/art15fracIX/IAT_TRIM_2_2014.pdf

10. Barradas-Rebolledo, A.: Gestión integral de residuos sólidos municipales. Instituto Tecnológico de Minatitlán **167** (2009). http://oa.upm.es/1922/1/Barradas_MONO_2009_01.pdf
11. Srivastava, V., Ismail, S.A., Singh, P., Singh, R.P.: Urban solid waste management in the developing world with emphasis on India: challenges and opportunities. Rev. Environ. Sci. Bio/Technol. **14**(2), 317–337 (2015)
12. Bing, X., Bloemhof, J. M., Ramos, T. R. P., Barbosa-Povoa, A. P., Wong, C. Y., van der Vorst, J. G. A. J.: Research challenges in municipal solid waste logistics man-agement. Waste Manag **48**, 584–592 (2016). https://doi.org/10.1016/j.wasman.2015.11.025
13. Fei, F., Qu, L., Wen, Z., Xue, Y., Zhang, H.: How to integrate the informal re-cycling system into municipal solid waste management in developing countries: Based on a China's case in Suzhou urban area. Resour Conserv Recycl (2016). https://doi.org/10.1016/j.resconrec.2016.03.019
14. Sembiring, E., Nitivattananon, V.: Sustainable solid waste management toward an inclusive society: Integration of the informal sector. Resour Conserv Recycl **54**(11), 802–809 (2010). https://doi.org/10.1016/j.resconrec.2009.12.010
15. Morana and Gonzalez-Feliu (2014)
16. Van Duin and Quak [21]
17. Benjelloun and Crainic (2009)
18. Callaba, J.P.A.: Logística de distribución urbana de mercancias: tendencias corporativas y bases para políticas públicas. Gaceta Instituto de Ingeniería, UNAM **1**(123), 10–12 (2017)
19. Gómez Maturano, J., Sánchez Lara, B.: Procedimiento para caracterizar cadenas de suministro inversas que valorizan residuos sólidos urbanos. Congreso Internacional de Logística y Cadena de Suministro. Asociación Mexicana de Logística y Cadena de Suministro, A.C., Ciudad de México (2018)
20. Larsen, S.B., Masi, D., Jacobsen, P., Godsell, J.: How the reverse supply chain contributes to a firm's competitive strategy: a strategic alignment perspective. Prod. Plan. Control **29**(6), 452–463 (2018). https://doi.org/10.1080/09537287.2017.1390178
21. INEGI: Sistema de Clasificación Industrial de América del Norte, México (SCIAN 2018) (2018). https://www.inegi.org.mx/contenidos/app/scian/estructura2018.pdf
22. Li, M.: A spatial-temporal analysis of hotels in urban tourism destination. Int. J. Hosp. Manag. **10** (2015)
23. Karadimas Nikolaos, V., Papatzelou, K., Loumos, V.G.: Optimal solid waste collection routes identified by the ant colony system algorithm. Waste Manag. Res. **25**, 139–147 (2007)
24. Hossein, A., Shahrooz, S., Prem, C.: Integrated municipal solid waste management under uncertainty: a tri-echelon city logistics and transportation context. Sustain. Cities Soc. **50**, 101–606 (2019). Recuperado el 19 de mayo de 2020
25. Hiramatsu, A., Honda, R., Sekiyama, M., Hara, Y.: Municipal solid waste flow and waste generation characteristics in an urban–rural fringe area in Thailand. WM&R **27**, 951–960 (2009)
26. Farahbakhsh, A., Forghani, M.A.: Sustainable location and route planning with GIS for waste sorting centers, case study: Kerman, Iran. Waste Manag. Res. **37**(3), 287–300 (2019)
27. Instituto Nacional de Estadística y Geografía. "Marco geoestadístico nacional." (2018)
28. INEGI: Directorio Estadístico Nacional de Unidades Económicas, DENUE Interactivo 11/2019. Documento metodológico (2019). Recuperado el 29 de mayo de 2020 de: http://internet.contenidos.inegi.org.mx/contenidos/Productos/prod_serv/contenidos/espanol/bvinegi/productos/nueva_estruc/702825109196.pdf
29. INEGI, XII. "Censo de población y vivienda 2010." Paracho, Michoacán, México 906 (2010)
30. INEGI: Encuesta intercensal 2015. Indicadores Demográficos y Socioeconómicos del Estado de México, Distrito Federal e Hidalgo (2015)
31. Consejo Nacional de Evaluación de la Politica de Desarrollo Social. "Pobreza Urbana en México, 2015". Disponible en https://www.coneval.org.mx/Medicion/Paginas/POBREZA-URBANA-EN-MEXICO-2015.aspx

32. Jenks, G.F.: The data model concept in statistical mapping. In: International Yearbook of Cartography, nº 7, pp. 186–190 (1967)
33. Kollikkathara, N., Feng, H., Stern, E.: A purview of waste management evolution: special emphasis on USA. Waste Manag. **29**(2), 974–85 (2009 Feb). https://doi.org/10.1016/j.wasman.2008.06.032. Epub 2008 Sep 14. PMID: 18796347
34. Williams, E., Kahhat, R., Allenby, B., Kavazanjian, E., Kim, J., Xu, M.: Envi-ronmental, social, and economic implications of global reuse and recycling of personal computers. Environ Sci Technol **42**(17), 6446–6454 (2008). https://doi.org/10.1021/es702255z

Classification System to Detect Diseases in Apples by Using a Convolutional Neural Network

Alejandro Acosta, Alberto Ochoa, Erick Rodriguez-Eparza, Diego Oliva, Angel A. Juan, and Gonzalo Pajares

Abstract Apples are a great fruit that is well known and is consumed around the world by all people. The importance of having a good quality in these fruits is huge because of to the nutrients for human consumptions and to the versatility that it offers. This work presents a full detection system created by a Convolutional Neural Network (CNN) based on a classifier using Keras that detects the diseases on apples. The CNN presented had an accuracy of the model as 0.7429 after 10 Epochs of training and having the number of data set used for testing of 319 images from 4 different classes. Also, it is described along with a diagram of the architecture where the system is allocated and working in the cloud. Finally, and for the purpose of using the system properly, a mobile app is proposed to be created in order to use the capabilities of the system to detect in a high-speed API by having fast responses when the disease detections are requested through the app.

A. Acosta (✉)
Universidad Cuauhtémoc, Aguascalientes, Mexico
e-mail: alejandro.acosta@uabc.edu.mx; alex.acosta.me@gmail.com

A. Ochoa
Doctorado en Tecnología, Universidad Autónoma de Ciudad Juárez, Ciudad Juárez, Mexico
e-mail: alberto.ochoa@uacj.mx

E. Rodriguez-Eparza · D. Oliva
División de Electrónica y Computación, Universidad de Guadalajara, CUCEI, Guadalajara, Jalisco, Mexico
e-mail: erick.rodrigueze@alumno.udg.mx

D. Oliva
e-mail: diego.oliva@cucei.udg.mx

D. Oliva · A. A. Juan
IN3 - Computer Science Department, Universitat Oberta de Catalunya, Castelldefels, Spain
e-mail: ajuanp@uoc.edu

G. Pajares
Departamento de Ingeniería del Software e Inteligencia Artificial, Facultad de Informática, Universidad Complutense de Madrid, Madrid, Spain
e-mail: pajares@ucm.es

A. Ochoa-Zezzatti et al. (eds.), *Technological and Industrial Applications Associated With Industry 4.0*, Studies in Systems, Decision and Control 347,
https://doi.org/10.1007/978-3-030-68663-5_21

Keywords Apple · Convolutional Neural Networks · Image classification · Disease · Cloud computing · Mobile apps

1 Introduction

Convolutional Neural Networks (CNN) are a powerful tool used around the world these days. Not only in the tech industry, but in any other with the purpose to enhance the changes to achieve more goals and take the industries to the next level. That is what Industry 4.0 is all about.

Now, consider a regular CNN to classify images. That is a common approach that is used very often to solve image classification problems. With that in mind, the work described here aims to show how this typical task can solve a real-world problem which is apple diseases.

All over the world we can find data about food waste and some of its most important causes. To cite some data, every year in the Madrid region, 30% of the products intended for human consumption are lost or wasted by improper handling in the food supply chain (CSA) mentioned by Gustavsson et al. (2011) [1]. A study conducted in the United States by the Natural Resources Defense Council (NRDC) found that up to 40% of food is lost from the producer's farm to the consumer's table (Gunders 2012) [2].

Apple trees are affected by several biotic and abiotic factors that limit their production, especially root diseases. In agroecosystems, the increased diversity of pathogenic microorganisms (Oomycetes, fungi, bacteria, nematodes, among others) can lead to root rot diseases. The root diseases caused by fungi and Oomycetes in apple trees, represent one important economic problem worldwide. These microorganisms destroy and collapse the root system of susceptible rootstocks, reducing the absorption of nutrients and water, blocking vascular bundles and, eventually, causing the death of apple trees, therefore the apples. This was a study performed by Rios-Velasco and colleagues [3] in Chihuahua, Mexico.

Based on the destruction caused by those agents (microorganisms), this is needed in this particular apple producing area to develop a better-integrated disease management in order to reduce the excessive use of broad-spectrum chemical pesticides to control these diseases. Undoubtedly, this improved integrated disease management include the use of microorganisms with antagonistic activity, which are efficient colonizers, produce metabolites that inhibit the growth of phytopathogens and act as plant growth promoters [4].

Thus, it is important to know what type of microorganisms are affecting apples in order to know better the cause of the disease. By detecting the diseases in early stages, it would be possible and helpful to reduce the amount of waste applying the correct preventive method.

2 The Problem

According to Natalie Golden from the University of New Mexico State [5], there are several factors that cause infections in apple trees, therefore in the fruit as well. Some of those disease agents are biotic pathogens such as fungi, bacteria, viruses, nematodes. Also, there are other agents that are non-infectious factors like temperature, moisture, nutrients, soil conditions and some chemicals.

This is a situation that needs to be controlled in a way that allows apple trees to produce good fruit that can be used for human consumption.

For this to be manage, a good solution can be to start monitoring the trees to find what kind of diseases exists in there at some specific point. Another way might be to revise the product after harvesting them and try to get a sample to be evaluated. There could be several more ways to do this.

Jedermann et al. [6] mentions that there are few apple diseases. Here are three of the most popular listed:

1. Black Rot. This disease is caused by a fungus that grows on many woody plants and trees and is spread by wind. Leaves may have many large brown spots that start round and become patches as they grow together. The spots are dry and like paper. Spots on fruit may appear any time during the season and infected fruit rot and hang on the tree. The fungus also causes cankers on branches.
2. Powdery Mildew or Blotch. Powdery mildew is a serious problem on susceptible apple varieties. The fungus may attack twigs, fruit, leaves and flowers. On leaves a white powder like covering appears. If the entire tip of a branch is affected, then growth is stunted, or new growth may be killed. Infected flowers do not produce fruit. Fruit with powdery mildew are of reduced quality and have a rough discolored skin.
3. Scab. Apple Scab is widespread in most apple growing regions. The fungus survives the winter on infected fallen apple leaves from the previous year.

Having defined 3 of the most common diseases on apples, next a solution will be described in a way that this can be controlled and helps to tackle these problems around apples.

3 The Solution

Technology is always a great option to solve problems. In the field of Artificial Intelligence, CNNs are becoming a great problem-solving tool for image classification situations where other tools are not the best option.

For the case of diseases on apples, with the definition of a CNN that allows to classify patterns inside an image where can the image be labeled and later used within a model to be used to test images in the real life, this will be an important start point.

Let's divide the full solution in steps. The first one is about the CNN model which was trained with a database of 319 different images between scab, rot, blotch diseases and normal apples (4 classes).

The second part is about the creation of a REST Web Service API which will be in charge of attend any kind of request to consume the CNN model. The web service is based on the Python's framework, CherryPy. It's a minimalistic web framework that allow to create rapid API applications in Python.

The last part, but not least, is a mobile app. This part of the classification system is the user interface that the users will need to use to upload images directly to the web service API to be processed by the CNN model to classify the image and return the final results.

(a) Apple Classifier Using a Keras CNN Model

The CNN that is going to classify images was created with the Keras framework. In order to create an accurate model, a few parameters were defined in order to get better results during the training and as result, have a great test results before putting this into a production instance.

For a first training (the one so far), only were defined 10 Epochs. Each Epoch throws some results that were good for the purposes of the first training model.

The first Epoch had a loss of 1.4840 and the final Epoch had a 0.6588. Having as a final accuracy of the model as 0.7429. Not so bad for the first training considering the number of images for testing, 319 from 4 different classes (Table 1).

The CNN is using 2 layers of convolutions to compute the images during the different epochs. Each convolution contains a Max Pooling of 2 in a 2D approach due to the images.

After the CNN model was trained, the resultant model was stored along with the weights. This way the results are going to be used for any other kind of application.

For the test part of the model, there were no better results as the final average percentage of accuracy was lower that the evaluated during the training. 68% during

Table 1 Training step of the model using during 10 epochs and 32 steps with 200 validation steps per epoch

Epoch	Elapsed time per step (ms)	Loss	Accuracy
1	346	1.4840	0.3950
2	345	1.0562	0.5016
3	344	0.9242	0.5799
4	344	0.8946	0.6176
5	353	0.8296	0.6458
6	345	0.7938	0.6771
7	345	0.8055	0.6740
8	347	0.7395	0.6708
9	343	0.7790	0.6740
10	346	0.6588	0.7429

Table 2 Test step of the model evaluating 47 test images across the 4 classes having different results each class

	Scab	Rot	Blotch	Normal	Total
Evaluated	15	15	15	2	47 (100%)
Predicted	12	7	12	1	32 (68%)
Failed	3	8	3	1	15 (32%)
Total %	80%	47%	80%	50%	100%

the test phase versus 74% during the training. However, if each class of image is evaluated individually, it shows that 2 of the 4 classes are great predicted that surpasses the original training accuracy percentage (Table 2).

(b) **REST API Web Service with CherryPy Framework**

CherryPy is an open-source web development framework based on the Python programming language. CherryPy, according to its website, allows developers to build web applications rapidly.

To understand better, a REST API as a Web Service is a web application that allows different clients to interact with it by allowing data exchange. Even better, according to AWS (Amazon Web Services) definition, an API enable communications between disparate software applications.

A normal REST API creation can be considered as the following diagram. See Fig. 1.

DigitalOcean is one of the best players in the market for rapid Cloud-Based server creation. This company offer services to create Virtual Machines (VM) that are called Droplets. A Droplet enables capabilities to install a full operating system based on Linux. Inside the Droplet, the CNN model needs to be installed in order to be used by running an internal process triggered by CherryPY (Fig. 2).

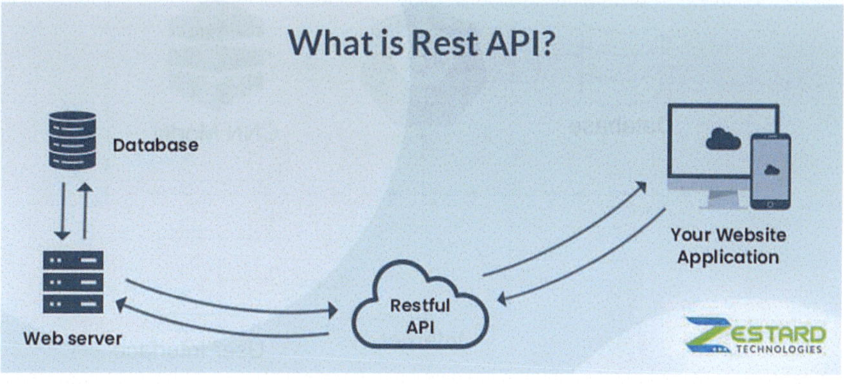

Fig. 1 API basic architecture. https://api.zestard.com/wp-content/uploads/2015/12/What-is-Rest-API-02-1.jpg

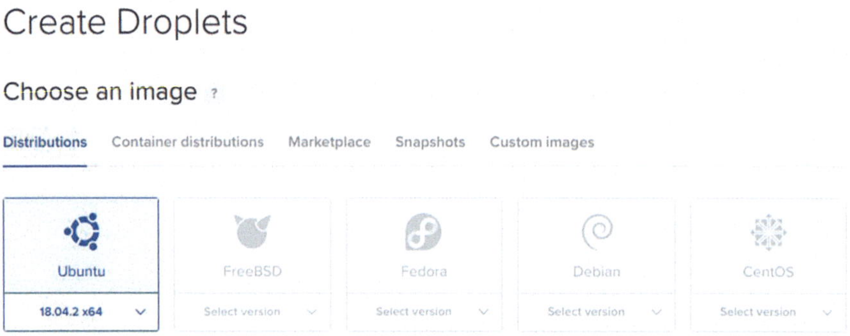

Fig. 2 DigitalOcean Droplet Creation. https://www.digitalocean.com/docs/images/droplets/create/choose-image.6a839626 11e90c5b3066c908cad05bac2a2a2d0a65a4accd4403a31a021b3519.png

The application created with CherryPy is basically the bridge between the images that a user will upload to the system and the CNN model that predicts the disease in the images considering the training and test process already performed and stored.

(c) **Mobile App**

The idea of a mobile app is that smartphone usage is so important for the present days. As the Fig. 3 shows, a User Interface is very important because is practically the client that is going to consume the full system mounted on the cloud of DigitalOcean and will use the CNN model trained to predict apple disease images.

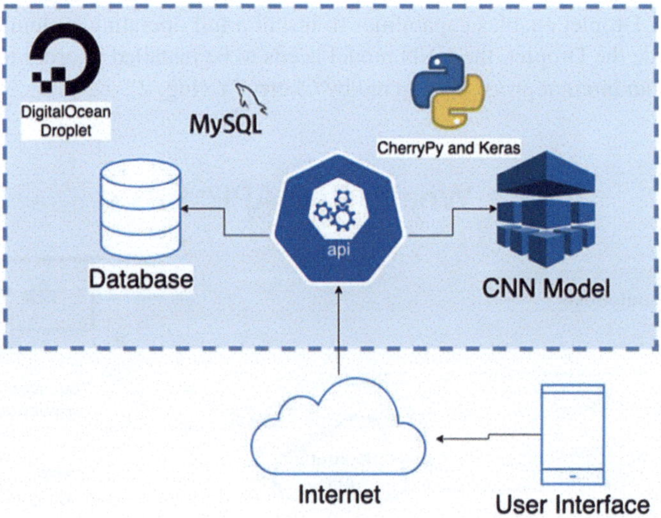

Fig. 3 REST API Architecture using DigitalOcean as provider and the rest of the components

For the app creation, it was considered the two best platforms that exists. Apple and Google. To tackle this out, the best way to achieve it to create an hybrid mobile app that is compatible with both platforms.

Flutter is the framework created by Google that supports Android and iOS development. Much more, and according to Flutter website, is a UI toolkit for building beautiful, natively compiled applications for mobile from a single codebase.

The mockups for this app are very simple and consistent to support a significantly simple design.

5 screens of the app make the whole experience from being able to enter the app, see a pretty dashboard, capture new images, process them and finally see the prediction results.

In Fig. 4 it is described the 2 initial screens where the user can login or register it self to be automatically logged in into the app and the user will see the main dashboard as firs screen once is inside the app.

The dashboard contains information about all the captures the user has been done and showing the accuracy of the prediction based on what the CNN model predicts.

The user can do a new capture by clicking in the Capture button located at the bottom of the screen and that will take the user to the camera app of its phone to take the picture of an apple.

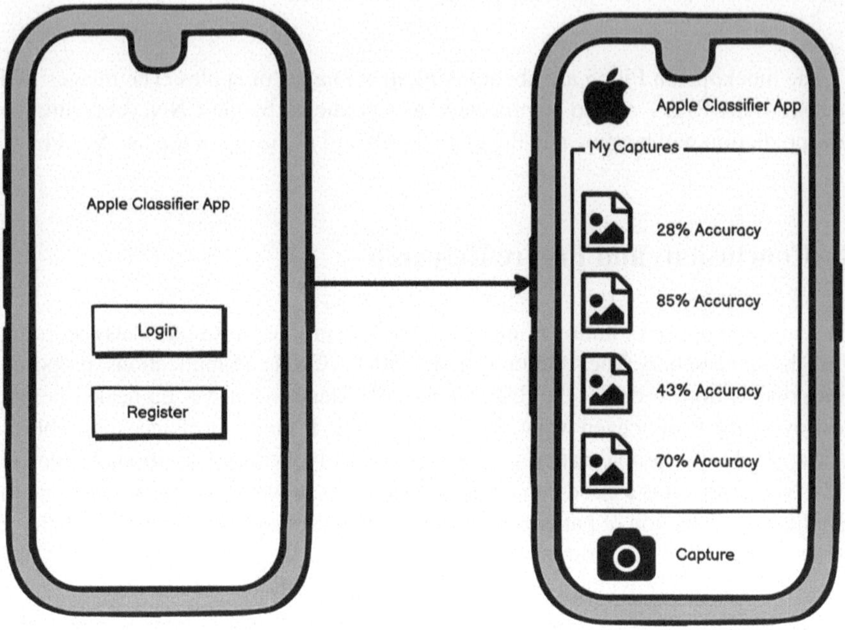

Fig. 4 Initial screen to enter the app and dashboard

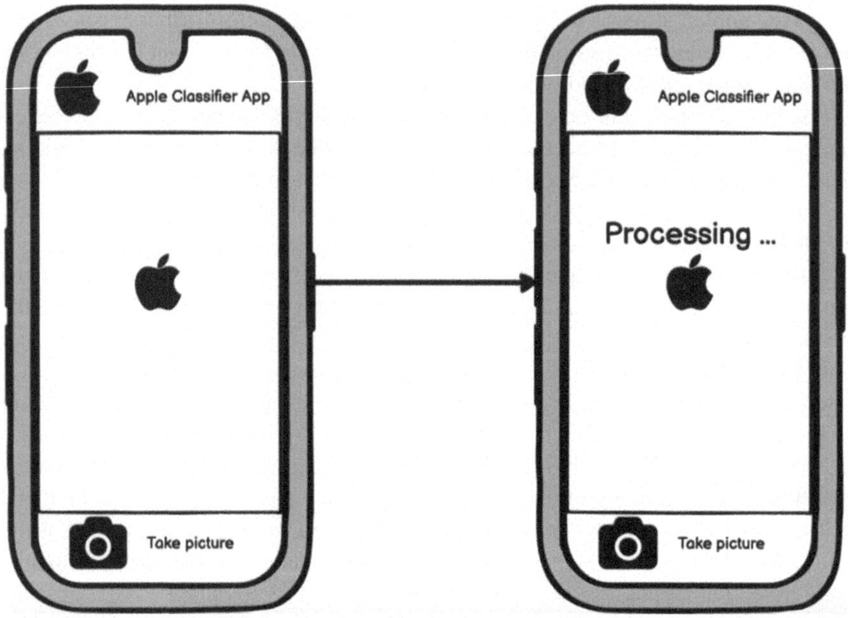

Fig. 5 Screens to take a new picture of an apple to be processed

The mockups on Fig. 5 are about taking new images of apples. The images will be sent to the REST API to be processed and predicted by the CNN. As result, the new prediction will be shown in the main dashboard at the top of the list. See Fig. 6.

4 Conclusions and Future Research

The current proposed solution is intended to solve a problematic that exists according to all the cited authors. The solution system (with the CNN and applications) proposed is a good option to detect the diseases in early stages so it can be healed before destroying the fruit or even worst, the complete tree. If there is a chance to be able to face apple disease problematics, technology is clearly the biggest option. As seen in the CNN results, it is a good percentage of accuracy, however, it can be improved to re-adjust the detection of patterns or features on images and be more precise to have better results. With the addition of the mobile app and the API to consume the CNN power, the system provides a powerful tool enhancing the whole experience.

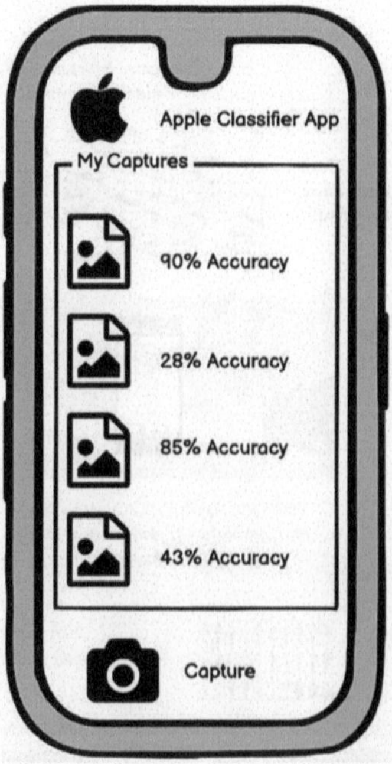

Fig. 6 Main dashboard showing the last predicted image

For future research, the main idea is to fully develop the mobile app and the REST API so the system complete can work together and as a united platform.

The main objective of this stage was to empathize with microentrepreneurs to analyze the problems they encounter when distributing, handling and storing the perishable products they market.

Also take a look after and work further with more fruits or other products to increase the area of knowledge. In order to analyze the freshness in diverse healthy products considering apple salad in sandwiches as is possible determine with a mobile device and deep learning, see Fig. 7.

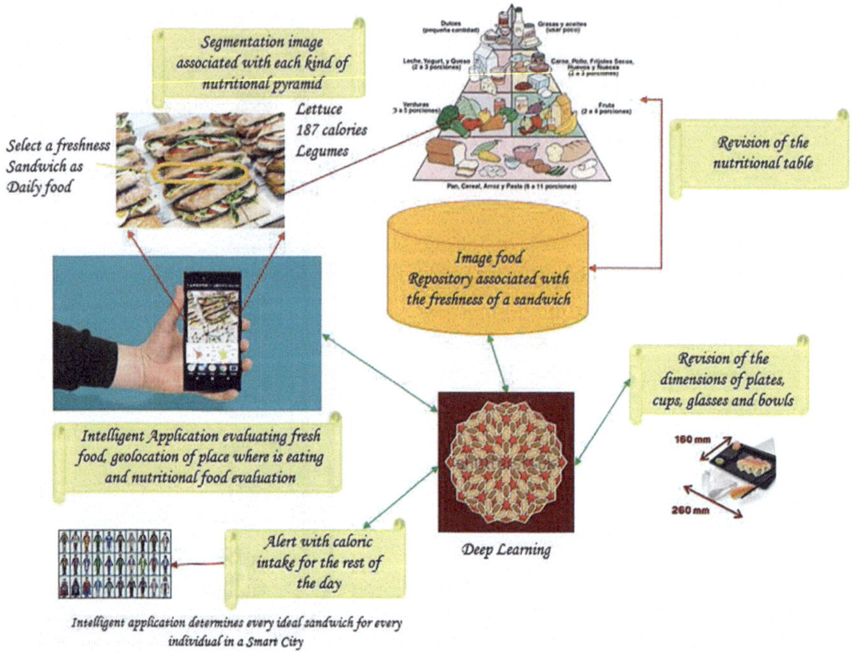

Fig. 7 Conceptual diagram of the implementation of an intelligent system that determinates the greatest freshness in the presentation threshold and color analysis of various sándwich issues—including apple salad—in a stock of a store selling healthy products

References

1. Gustavsson, J., Cederberg, C., Sonesson, U., van Otterdijk, R., Meybeck, A.: Global food losses and food waste: extent, causes and prevention. Study conducted for the International Congress SAVE FOOD! at Interpack 2011, Düsseldorf, Germany. Food and Agriculture Organization of the United Nations, Rome (2011)
2. Gunders, D.: Wasted: how America is losing up to 40 percent of its food from farm to fork to landfill. NRDC Issue Paper IP 12-06-B. New York City: Natural Resources Defense Council 2012
3. Ruiz-Cisneros, M.F., Rios-Velasco, C., Berlanga-Reyes, D.I., de Jesús Ornelas-Paz, J., Acosta-Muñiz, C.H., Romo-Chacón, A., Zamudio-Flores, P.B., Pérez-Corral, D.A., Salas-Marina, M.Á., Ibarra-Rendón, J.E., Fernández-Pavía, S.P.: Incidence and causal agents of root diseases and its antagonists in apple orchards of Chihuahua, México (2017). http://www.scielo.org.mx/pdf/rmfi/v35n3/2007-8080-rmfi-35-03-00437.pdf
4. Ezziyyani, M., Sánchez, C., Requena, M.E., Rubio, L., Candela, M.E.: Biocontrol por Streptomyces rochei -Ziyani-, de la podredumbre del pimiento (Capsicum annuum L.) causada por Phytophthora capsici. Anales de Biología **26**, 69–78 (2004). Disponible en línea: https://www.um.es/analesdebiologia/numeros/26/PDF/08-BIOCONTROL.pdf
5. Goldberg, N.P: Apple Disease Control (2000). New Mexico State University. https://aces.nmsu.edu/pubs/_h/H317.pdf
6. Jedermann, R., Nicometo, M., Uysal, I., Lang, W.: Reducing food losses by intelligent food logistics. Philos. Trans. R. Soc. A **1**, 1–37 (2014)

The manufacturer's authorised representative in the EU is Springer Nature Customer Service Centre GmbH, Europaplatz 3, 69115 Heidelberg, Germany. If you have any concerns regarding our products, please contact ProductSafety@springernature.com

Printed and bound by CPI Group (UK) Ltd, Croydon, CR0 4YY
30/09/2026
02225691-0008